T0293435

Advances in Material Design for Regenerative Medicine

Advances in Material Design for Regenerative Medicine

Editor: Cindy Barron

FOSTER
ACADEMICS

www.fosteracademics.com

www.fosteracademics.com

Cataloging-in-Publication Data

Advances in material design for regenerative medicine / edited by Cindy Barron.
   p. cm.
Includes bibliographical references and index.
ISBN 978-1-64646-631-3
1. Regenerative medicine--Materials. 2. Biomedical materials. 3. Regenerative medicine.
4. Medical instruments and apparatus. I. Barron, Cindy.
QH499 .A28 2023
610.28--dc23

Foster Academics,
118-35 Queens Blvd., Suite 400,
Forest Hills, NY 11375, USA

ISBN 978-1-64646-631-3 (Hardback)

Contents

Preface

Every book is a source of knowledge and this one is no exception. The idea that led to the conceptualization of this book was the fact that the world is advancing rapidly; which makes it crucial to document the progress in every field. I am aware that a lot of data is already available, yet, there is a lot more to learn. Hence, I accepted the responsibility of editing this book and contributing my knowledge to the community.

Regenerative medicine refers to a field of medicine that develops methods to recreate, replace or fix injured or diseased cells, tissues or organs, without having to cope with immune complications and donor supply shortage. It involves the development and application of tissue engineering, therapeutic stem cells, as well as the creation of artificial organs. The development of materials is a crucial component of regenerative medicine. Materials play a significant role in various tissue repair and therapeutic processes, such as tissue scaffolds, decellularized tissue substitutes, and delivery systems. Regeneration effectiveness depends on the design and manufacture of materials with biocompatible, optimal structural, as well as physicochemical qualities. Regenerative medicine strategies rely on tissue engineering (TE), induced autoregeneration, and somatic cell treatment. This book strives to provide a fair idea about material design and helps develop a better understanding of the latest advances in their use for regenerative medicine. Those with an interest in this field would find it helpful.

While editing this book, I had multiple visions for it. Then I finally narrowed down to make every chapter a sole standing text explaining a particular topic, so that they can be used independently. However, the umbrella subject sinews them into a common theme. This makes the book a unique platform of knowledge.

I would like to give the major credit of this book to the experts from every corner of the world, who took the time to share their expertise with us. Also, I owe the completion of this book to the never-ending support of my family, who supported me throughout the project.

Editor

Physicochemical and Biological Characterisation of Diclofenac Oligomeric Poly(3-hydroxyoctanoate) Hybrids as β-TCP Ceramics Modifiers for Bone Tissue Regeneration

Katarzyna Haraźna [1,*][iD], Ewelina Cichoń [2][iD], Szymon Skibiński [2], Tomasz Witko [1][iD], Daria Solarz [3], Iwona Kwiecień [4][iD], Elena Marcello [5], Małgorzata Zimowska [1], Robert Socha [1], Ewa Szefer [6], Aneta Zima [2][iD], Ipsita Roy [7][iD], Konstantinos N. Raftopoulos [6][iD], Krzysztof Pielichowski [6][iD], Małgorzata Witko [1] and Maciej Guzik [1,*][iD]

1   Jerzy Haber Institute of Catalysis and Surface Chemistry Polish Academy of Sciences, Niezapominajek 8, 30-239 Kraków, Poland; tomasz.witko@ikifp.edu.pl (T.W.); Malgorzata.zimowska@ikifp.edu.pl (M.Z.); Robert.socha@ikifp.edu.pl (R.S.); malgorzata.witko@ikifp.edu.pl (M.W.)
2   Faculty of Materials Science and Ceramics, AGH University of Science and Technology, 30 Mickiewicza Ave., 30-059 Kraków, Poland; ecichon@agh.edu.pl (E.C.); skibinski@agh.edu.pl (S.S.); azima@agh.edu.pl (A.Z.)
3   Faculty of Physics, Astronomy and Applied Computer Science, Jagiellonian University, Lojasiewicza 11, 30-348 Kraków, Poland; daria.solarz@doctoral.uj.edu.pl
4   Department of Physical Chemistry and Technology of Polymers, Silesian University of Technology, M. Strzody 9, 44-100 Gliwice, Poland; Iwona.Kwiecien@polsl.pl
5   School of Life Sciences, College of Liberal Arts and Sciences, University of Westminster, New Cavendish Street, London W1W 6UW, UK; w1614733@my.westminster.ac.uk
6   Department of Chemistry and Technology of Polymers, Cracow University of Technology, Warszawska 24, 31-155 Kraków, Poland; ewa.szefer@doktorant.pk.edu.pl (E.S.); konstantinos.raftopoulos@pk.edu.pl (K.N.R.); kpielich@pk.edu.pl (K.P.)
7   Department of Materials Science and Engineering, University of Sheffield, Broad Lane, Sheffield S3 7HQ, UK; i.roy@sheffield.ac.uk
*   Correspondence: Katarzyna.harazna@ikifp.edu.pl (K.H.); Maciej.guzik@ikifp.edu.pl (M.G.)

Abstract: Nowadays, regenerative medicine faces a major challenge in providing new, functional materials that will meet the characteristics desired to replenish and grow new tissue. Therefore, this study presents new ceramic-polymer composites in which the matrix consists of tricalcium phosphates covered with blends containing a chemically bounded diclofenac with the biocompatible polymer—poly(3-hydroxyoctanoate), P(3HO). Modification of P(3HO) oligomers was confirmed by NMR, IR and XPS. Moreover, obtained oligomers and their blends were subjected to an in-depth characterisation using GPC, TGA, DSC and AFM. Furthermore, we demonstrate that the hydrophobicity and surface free energy values of blends decreased with the amount of diclofenac modified oligomers. Subsequently, the designed composites were used as a substrate for growth of the pre-osteoblast cell line (MC3T3-E1). An in vitro biocompatibility study showed that the composite with the lowest concentration of the proposed drug is within the range assumed to be non-toxic (viability above 70%). Cell proliferation was visualised using the SEM method, whereas the observation of cell penetration into the scaffold was carried out by confocal microscopy. Thus, it can be an ideal new functional bone tissue substitute, allowing not only the regeneration and restoration of the defect but also inhibiting the development of chronic inflammation.

Keywords: tricalcium phosphate composites; polyhydroxyalkanoates; polyhydroxyoctanoate oligomers; diclofenac; bone regeneration

1. Introduction

Tens of millions of people suffering from bone dysfunction are estimated to require surgical intervention. As life expectancy increases, so do the number of environmental factors contributing to bone disease, and this figure is expected to increase from year to year [1]. There is still a search for new therapies and multifunctional materials designed for tissue regeneration that fulfil the requirements of contemporary medicine.

Special attention is paid to local drug delivery systems in tissue engineering as promising tools for the treatment of infections, tissue regeneration and the fight against cancer. They have many advantages, including high drug concentrations combined with release at the target site at desired rates and prolonged therapeutic effects [2,3]. One common approach to the treatment of patients, who have suffered bone damage or are at risk of postoperative pain and rejection of the implant, is the use of non-steroidal anti-inflammatory drugs (NSAIDs) [4,5]. The benefits of these substances result from their anti-inflammatory and analgesic effects [5]. Interestingly, the pharmacodynamics of NSAIDs indicates that they affect all stages of bone and surrounding tissue healing [6]. Chang et al. reported that such drugs, i.e., celecoxib, inhibited proliferation and induced cell death in human osteoblasts (hOBs) models in concentrations, in the range of 10^{-6} to 10^{-5} M. In the case of drugs, i.e., indomethacin, ketorolac, diclofenac (DIC) and piroxicam, no significant cytotoxic effects were demonstrated for therapeutic doses (10^{-5} M) [7]. Similar reports were presented by Diaz-Rodriguez et al. who proved that an appropriate therapeutic dose of ibuprofen ($<25 \times 10^{-3}$ M) in osteosarcoma cell line culture (MG-63) does not inhibit the growth of osteoblasts and does not induce cell death [8].

Tissue engineering introduces the possibility to design bone substitutes in a way that allows the dosing of bioactive substances facilitating and improving the healing process. Most often bone tissue regeneration materials should be osteoinductive, osteoconductive, bioresorbable, biocompatible, easy to use, cheap and structurally similar to bone [9]. These properties can be provided by hydroxyapatite bioceramics ($Ca_{10}(PO_4)_6(OH)2$—HAp), tricalcium phosphate materials ($Ca_3(PO_4)_2$—α,β-TCP), amorphous calcium phosphate (ACP) as well as their composites (BCP—biphasic calcium phosphates) with organic polymers [10,11]. The combination of a ceramic matrix with polymers, both synthetic and of natural origin, creates great opportunities to improve the mechanical, physicochemical and biological properties of the obtained composites. What is more, their macro and microporosity of bioceramics bone scaffold promotes the formation of features that correspond to the growth of new tissue [12]. Numerous reports on the use of polymeric systems containing NSAIDs in the regeneration of bone and cartilage tissue have appeared in the literature. An example is the biodegradable polyurethane system used in cartilaginous tissue regeneration, which provides prolonged release of DIC to 120 days [13]. Subramanian et al. produced thin flexible electrically driven membranes consisting of a copolymer of salicylic acid/succinic anhydride (SAPAE) and polycaprolactone (PCL) for use in bone tissue regeneration. These membranes were able to release salicylic acid during hydrolysis of one of the polymeric components [14]. Some studies described the effects of DIC release on reducing host response to the implanted foreign body. Sidney et al. reported the use of polymeric systems based on a copolymer of DL-lactic acid and glycolic acid (PLGA) and poly(ethylene glycol) (PEG) with DIC, which are good for the treatment of acute inflammation, thanks to the possibility of releasing 80% of the dosed active substance during the first four days [5]. Composites based on poly(anhydride ester) and ceramics (85% β-TCP and 15% HAp) were presented as targeted bone regeneration systems that are also able to reduce inflammation, thanks to the possibility of releasing salicylic acid in a hydrolysis process [15].

Polyhydroxyalkanoates (PHAs) are a group of bacterial biopolyesters, which can be synthesized by numerous bacteria under unfavourable conditions from a variety of carbon sources [16]. These macromolecules are characterised by an extended resorption time, making them an excellent organic material for the production of hybrid organic-inorganic composites for bone tissue [17–19]. This is possible due to the numerous advantages of these biopolymers, which include: biocompatibility, nontoxicity to human tissues and blood; bone-like piezoelectric properties; and controlled biodegradation, which is dependent on the composition of the polymer chain.

In recent years, these compounds were the subject of interest of scientists involved in the production of materials for medical applications [17–34]. Moreover, the attention of many researchers was focused on the chemical modification of PHAs and their copolymers in order to reach a stable chemical bond between the active substance and the matrix of the polymer or oligomer. These approaches have led to the development of controlled release systems for active substances [25,28,35–38]. To the best of our knowledge, there are no reports in the literature on the synthesis of composites based on tricalcium phosphates and polyhydroxyalkanoates with covalently attached molecules of (NSAIDs). Such materials could be panacea in the regeneration of bone defects after surgical interventions caused by both tumours and other diseases, i.e., rheumatoid arthritis and osteoarthritis. The polymer layer containing a NSAID—diclofenac, would allow the delivery of the active substance directly to the implanted site. In addition, bacterial biopolymer degradation products—3-(R)-hydroxyalkanoic acids—nourishes the wound, reducing the risk of implant rejection. The obtained modified oligomers and their blends containing different amounts of modified oligomers were analysed by several techniques, i.e., NMR, IR, XPS, GPC, AFM, TGA and DSC. In vitro studies were carried out with the use of the MC3T3-E1. The in vitro biocompatibility studies and indirect cytotoxicity tests based on the Alamar Blue reduction were performed. Furthermore, cell proliferation and penetration into porous composites were analysed by SEM and confocal microscopy.

2. Materials and Methods

2.1. Materials

The following chemicals were used: the octanoic acid, butyric acid, p-toluenosulfonic acid monohydrate (p-TSA), lithium hydroxide monohydrate, fetal bovine serum (FBS) calcium hydroxide and diiodomethane (Sigma Aldrich, Poznań, Poland); phosphoric acid (POCH, Gliwice, Poland); the α-MEM modification essential medium, trypsin and Alamar blue (Thermofisher, Cambridge, UK); the fetal bovine serum, pencillin, streptomycin, Dulbecco's phosphate-buffered saline (Dulbecco's PBS), trypan blue, dimethyl sulfoxide, ethanol and phosphate buffer saline (Sigma Aldrich, Dorset, UK), the Dulbecco's Modified Eagle Medium (DMEM) and 4',6-diamidino-2-phenylindole (DAPI) (Thermofisher, Warsaw, Poland); the hydrochloric acid, methanol, ethyl acetate and chloroform (Chempur, Poland); the sodium chloride and the tetrahydrofuran chromatographic grade (Avantor Chemicals, Gdańsk, Poland); the methanol and acetonitrile HPLC grade (Chemsolve, Łódź, Poland), and the diclofenac acid (Fluorochem, Hadfield, UK).

2.2. Preparation of the Investigated Materials

2.2.1. Production of Poly(3-hydroxyoctanoate) (P(3HO)) Polymer

P(3HO) polymer, which was composed of 91 mol.% (R)-3-hydroxyoctanoic acid, 7 mol.% (R)-3-hydroxyhexanoic acid and below 2 mol.% in total of other (R)-3-hydroxylated fatty acids, was obtained in the process of controlled feeding of the *Pseudomonas putida* KT2440 [39] strain in a 5 L fermenter. Octanoic acid was used as the sole energy and carbon source. A detailed process of its production and characterization were described previously [21].

2.2.2. Preparation of Highly Porous Tricalcium Phosphate (TCP) Ceramics

The initial TCP powder was synthesised via a wet chemical method according to a procedure described previously [17]. The calcium hydroxide (Ca(OH)$_2$, ≥99.5%) and phosphoric acid (H$_3$PO$_4$, 85.0%) were applied as reagents. The obtained precipitate was dried, calcined at 900 °C and grounded in attritor into a grain size below 0.063 mm. Ceramic slurry with a high amount of β-TCP powder was used to obtain highly porous bioceramics scaffolds via the polyurethane sponge replica method.

2.2.3. Preparation of Oligo-DIC Conjugates (Fin-Dic-oliP(3HO))

In order to prepare oligo-DIC polymer conjugates (Fin-Dic-oliP(3HO)), modified protocol described by Kwiecień et al. was used [35]. For this purpose, 5 g of P(3HO) polymer, 2 g (40% *w/w*) of DIC and 0.75 g (15% *w/w*) of *p*-TSA were introduced into a round bottom flask equipped with a magnetic stirring bar. The reaction was performed at 125 °C under an argon atmosphere, and started when the molten state of all reagents was observed. The reaction time was 2 min, after which the mixture was cooled in an ice bath and 50 mL of chloroform was added. The organic layer was washed five times with 1 M sodium chloride aqueous solution and two times with water to remove residual p-TSA. Then, the solvent was evaporated under vacuum. The reaction was performed in triplicate.

2.2.4. Preparation of Ceramic-Polymer Composites

To obtain polymeric coatings, three blends of P(3HO) with drug-modified oligomers were prepared: two (2Dic-oliP(3HO)), one (1Dic-oliP(3HO)) and a half (0.5Dic-oliP(3HO)) of a therapeutic DIC dose [40] (1 dose = 0.19 g DIC per 1 g of polymer) (Table 1). In this order, an appropriate amount of P(3HO) was dissolved in 50 mL of dichloromethane at 60 °C, which followed the addition of an appropriate amount of oligo-DIC [41]. The constituents were mixed for 3 h at an ambient temperature after the solvent was evaporated. The blends were dried in a vacuum dryer for 24 h, then dissolved in ethyl acetate in the concentration of 5% *w/v* and used to infiltrate the ceramic scaffolds [17].

Table 1. The reaction mixture compounds per 1 g of blends.

Name	Fin-Dic-oliP(3HO)	P(3HO)
2Dic-oliP(3HO)	0.97	0.03
1Dic-oliP(3HO)	0.48	0.52
0.5Dic-oliP(3HO)	0.24	0.76

2.3. Physicochemical Analysis of Oligo-DIC Conjugates and Their Blends

2.3.1. Nuclear Magnetic Resonance (1H NMR) Analysis

^{1}H NMR spectra were obtained in CDCl$_3$ with tetramethylsilane (TMS) as the internal standard using a 300 Hz spectrometer (Bruker BioSpin GMbH, Rheinstetten, Germany). Number of scans: 56; pulse width: 14.5 s; acquisition time: 3.5 s.

2.3.2. Attenuated Total Reflection Fourier Transform Infrared (ATR-FTIR) Analysis

FTIR spectra were recorded using a Nicolet iS5 spectrometer (Thermo Fisher Scientific, Waltham, MA, USA) equipped with a diamond crystal attenuated total reflectance unit (ATR) with a resolution of 8 cm^{-1} from 4000 cm^{-1} to 400 cm^{-1} as well as with an average of 16 scans.

2.3.3. Gel Permeation Chromatography (GPC) Analysis

Molecular weight distributions of oligo-DIC and oligo-DIC/P(3HO) blends were analyzed using modified protocol described by Sofińska et al. [21] Spectroscopic-grade tetrahydrofuran was used as an eluent at a flow rate of 1.0 mL min^{-1}. Sample concentration was 30 g L^{-1}, whereas injection volumes were 50 μL. All measurements were performed in triplicate.

2.3.4. Thermal Analysis

Thermogravimetric analysis (TGA) was carried out using a Netzsch STA 409 PC Luxx according to protocols described by Haraźna et al. [42] Differential scanning calorimetry (DSC) experiment analyses were performed with a Mettler Toledo 822e calorimeter purged with argon. The samples were first heated from room temperature to 200 °C at 10 K min^{-1} to erase the thermal history, then the samples were cooled at 10 K min^{-1} to −80 °C and second heating run at 10 K min^{-1} were accomplished.

2.3.5. X-ray Photoelectron Spectroscopy (XPS)

X-ray photoelectron spectroscopy (XPS) was applied to determine the surface composition and surface interaction of P(3HO) with DIC deposited on a silicon substrate. For the analysis, the XPS spectrometer with hemi-spherical analyzer SES R4000 (Gammadata Scienta, Uppsala, Sweden) and a Mg Kα radiation source (1253.6 eV) was used. The energy resolution of the spectrometer was equal to 0.9 eV (Ag 3d5/2 at pass energy of 100 eV). The spectrometer was calibrated according to ISO 15472:2010. A thin layer sample was placed on a dedicated holder, after the holder was pumped (8 h) to a high vacuum and then introduced into the UHV. The analysis area was 3 mm2, the spectra were deconvoluted with CasaXPS 2.3.15.

2.3.6. Measurements of the Contact Angle and Determination of Surface Free Energy

Thin films were prepared by a solvent casting technique. The Fin-Dic-oliP(3HO) and its blends were dissolved in chloroform at 10% w/v concentration and 100 µL of prepared solutions were deposited twice onto a glass plate surface and such prepared films were left at room temperature at normal pressure for 2 weeks. The wetting angles were obtained with Drop Shape Analyzer KRUSS DSA100M optical contact angle measuring instrument (Kruss Gmbh, Hamburg, Germany). The surface free energies were determined using an Owens-Wendt (OW) method. The polar and dispersive components were calculated.

2.3.7. Atomic Force Microscopy (AFM) Imaging

The samples for AFM study were prepared on glass support. Briefly, the blends were dissolved in ethyl acetate at 5% w/v concentration. On a glass surface of plates an aliquoted 50 µL of prepared solutions were dispensed and dried in a dryer at 50 °C for 15 min. After that the plates were kept at room temperature, normal pressure for 1 week. AFM images of the prepared films were recorded with a Veeco Innova atomic force microscope in a tapping mode, with tips of resonant frequency of ca. 300 kHz. Images used for further evaluation were prepared in canvases of 10 × 10 µm, with resolution of 512 × 512 pixels (for 0.5Dic-oliP(3HO) and 1 Dic-oliP(3HO)) or 256 × 256 pixels (for 2 Dic-oliP(3HO)). Images were levelled by subtraction of a fitted plane. Minor mutual shifts between line profiles along the fast-scanning axis, due to instrument noise, were smoothed by subtraction of the mean of each line. Root mean square (RMS) roughness (Sq) was calculated on the corrected images as:

$$S_q = \sqrt{\frac{1}{N}\sum_{n=1}^{N}(z_n - \bar{z})^2} \tag{1}$$

where z_n is the height at point n and $\bar{z}$ the mean height [43].

All mathematical operations on the AFM images were conducted with the Gwyddion software [44].

2.3.8. SEM Imaging

Scaffolds were sputtered with gold and analysed with the JEOL JSM—7500F Field Emission Scanning Electron Microscope equipped with Retractable Backscattered-Electron detector (RBEI) and EDS (energy dispersive spectra) detection system of characteristic X-ray radiation INCA PentaFetx3 EDS system (JEOL Ltd., Tokyo, Japan).

2.3.9. Confocal Imaging

Confocal images were obtained using Zeiss Axio Observer Z.1 microscope with LSM 710 confocal module. Image postprocessing was performed in dedicated software using Zeiss ZEN Black version 8,1,0,484, PALMRobo V 4.6.0.4 and FluoRender 2.21.0. Observations were made with an air 10× objective. For DAPI stain excitement a wave with length of 405 nm was applied [45].

2.4. Biological Characterisation of Prepared Ceramic-Polymer-Scaffold Composites

2.4.1. The Mouse Calvaria-Derived Pre-Osteoblastic (MC3T3-E1) Cell Culture

MC3T3-E1 cells (CRL-2593, ATCC, London, UK) were cultured in standard tissue culture flasks in α-MEM supplemented with 10% vol. FBS and 5% vol. penicillin/streptomycin antibiotics and maintained at 37 °C in a humidified atmosphere of 5% CO_2. The media were replaced every 2 days. When 80% confluency was reached, cells were washed twice with Dulbecco's PBS and detached using 0.25% trypsin containing 1 mM EDTA. After centrifuging the solution (5 min, 1500 rpm), the cells were suspended in α-MEM, and the number of cells was counted using Neubauer chamber. The cells were either seeded on the scaffolds or cultured again in a standard tissue flask.

2.4.2. The Cytotoxicity Test of Scaffold Extracts (Indirect Cytotoxicity)

The indirect cytotoxicity studies were carried out to evaluate the presence of toxic compounds in the extracts obtained from the scaffolds, following the ISO 10993-5, using a protocol described by Skibiński et al. [41].

2.4.3. In Vitro Biocompatibility Studies

Each scaffold was sterilised with ethylene oxide at 37 °C for 12 h and placed into a well of a 24-well tissue culture polystyrene plate. After this, the scaffolds were pre-wetted with 1.7 mL α-MEM for 1 day. The 24-well plates were incubated at 37 °C in a humidified atmosphere of 5% CO_2. Next, for attachment and proliferation studies, the cells were seeded directly on the scaffolds at a density of approximately 60,000 cells per well. The positive control consisted of cells cultured directly on tissue culture plastic with 1 mL of α-MEM medium. For the proliferation study, after 1, 3 and 7 days of culturing, the viability of cells for each scaffold was determined using the Alamar Blue assay (note: plates incubated for 7 days had medium replaced on the 3rd day).

2.4.4. Alamar Blue Assay

The metabolic activity of the cultured cells was measured using Alamar Blue. To determine the viability of cells, each cell culture medium was replaced with an appropriate amount of 10% *v/v* Alamar Blue in α-MEM medium (i.e., for direct tests—1.7 mL and indirect tests—150 μL) and incubated at 37 °C in a humidified atmosphere of 5% CO_2 for 3 h. After this, 100 μL of each solution was transferred to 96 well plates and absorbance at 570 nm and 600 nm was measured on a plate reader. The absorbance of the samples was normalised concerning the positive control.

2.4.5. Cell Morphology Observations

Specimens obtained from direct cytotoxicity assays were fixed twice by soaking in a Dulbecco's PBS, followed by a treatment with 4% *v/v* formaldehyde PBS solution and then stored at 4 °C. Next, the cells were gradually dehydrated by immersing in 2 × PBS solution and ethanol aqueous solutions of 35%, 50%, 70%, 90% and 2× in 98% *v/v*; each for 15 min. The analyses of cell attachment, colonization and microstructure of the scaffolds were performed using scanning electron and confocal microscopies.

2.5. Statistical Analysis

The results were expressed as mean values ± standard error (SE). The analysis of variance (ANOVA) was performed among groups for two population comparisons. The differences were deemed statistically significant at probabilities of * $p < 0.05$, ** $p < 0.01$, *** $p < 0.001$ (Origin Pro 2019 Software).

3. Results

3.1. Spectroscopic Insights into Structure of Obtained Drug-oligoP(3HO) Conjugates and Their Blends

The ^{1}H NMR spectra analyses (Figures S1 and S2) reveal the presence of signals that are assigned to protons originating from the 3-hydroxyoctanoic acid repetitive unit and 10-fold less intense signals originating from the 3-hydroxyhexanoic acid. One should stress that it is possible to distinguish between chemically bonded DIC to the oligomers via an ester bond (Figure 1A, signal a) and the remaining, unreacted and physically mixed compound (Figure 1A, signal b). This is confirmed by adding pure DIC to the post-synthesis sample and observing the change in the signal intensity assigned to physically mixed DIC at 3.85 ppm (Figure 1B,C, see the increase of peak b).

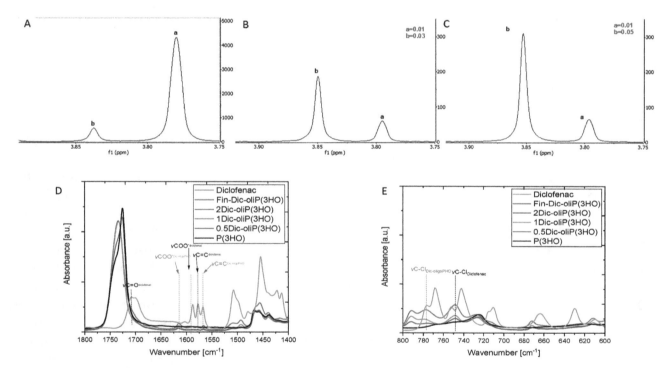

Figure 1. (**A**). ^{1}H NMR of Fin-Dic-oliP(3HO) analysis. (**B**). ^{1}H NMR analysis of the working sample before the addition of 4 drops of DIC solution (25 g L^{-1}). (**C**). ^{1}H NMR analysis of the working sample after addition of 4 drops of DIC acid solution (25 g L^{-1}), where (a) denotes the CH$_2$ groups from DIC attached to oligomers via an ester bond, (b) groups from free DIC. (**D**). Infrared spectra of the obtained compounds zoom in between 1800–1400 cm^{-1}. (**E**). Infrared spectra of the obtained compounds zoom in between 800–600 cm^{-1} ranges.

What is more, we have shown that linear oligomers without the attached modifier are also present in the mixture (Figures S1 and S2). Additionally, the high temperature of the reaction results in the formation of unmodified oligomers terminated with unsaturated end groups. Moreover, we have determined that 89% of DIC is attached to PHA oligomers and 11% remains free. The structure of the compounds included in the whole obtained mixture is presented in Figure 2. The formation of analogous structures as we present them here was also observed in other works [35,36].

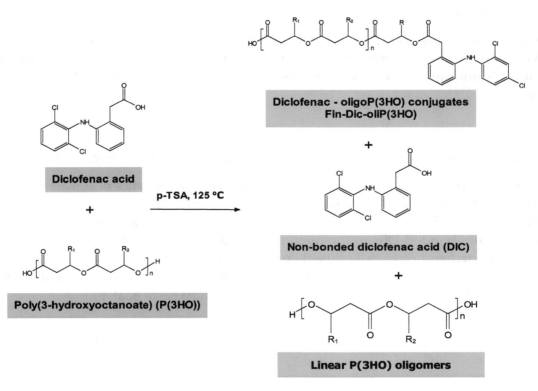

Figure 2. Species formed during the reaction of P(3HO) with DIC catalyzed by *p*-TSA.

From Figure S3 it is seen that the spectra of the final product (Fin-Dic-oliP(3HO)) and the resulting blends with the P(3HO) polymers (2Dic-oliP(3HO), 1Dic-oliP(3HO) and 0.5Dic-oliP(3HO)) contain vibrations characteristic mainly for the P(3HO) polymer as well as new vibrations and small changes compared with the drug molecules (Figure S3). The bands at 3000–2800 cm^{-1} are attributed to the vibrations of methylene (CH$_2$) and methyl (CH$_3$) groups, whereas the peak at 1732 cm^{-1} is attributed to the stretching vibrations of the carbonyl (C=O) group of the P(3HO) polymer [21,46]. The obtained DIC spectrum is similar to the one reported in the literature [47]. The visible absorption peaks at 1710 cm^{-1} and 1589 cm^{-1} indicated stretching vibrations of C=O and COO$^-$ groups. The stretching vibrations in the aromatic ring of C=C groups are observed at 1579 cm^{-1}. The vibrations of secondary amine groups are visible due to stretching of N-H and C-N groups at 3324 cm^{-1}, 1284 cm^{-1} and 1162 cm^{-1}. Additionally, the band at 752 cm^{-1} is attributed to the stretching vibrations of C-Cl groups. The visible change in the shape of the bands corresponding to the vibrations of the C=O groups at 1732 cm^{-1} is a premise confirming the formation of an ester bond between the drug molecule and the polymer (in Fin-Dic-oliP(3HO), Figure 1D). A similar phenomenon was observed by Gumel et al., who confirmed the successful modification of polymer from the mcl-PHA group by a biocatalytic approach using sucrose as a modifier [48]. Furthermore, the changes in the shape of the same bands are also observed in the prepared blends (2Dic-oliP(3HO), 1Dic-oliP(3HO)). Moreover, due to the low content of Fin-Dic-oliP(3HO) in the 0.5Dic-oliP(3HO) blend, the shape of the C=O bands did not differ much from the one of P(3HO). Modification of oligomers by DIC is also confirmed by the presence of the band characteristic for C-Cl stretching vibrations at 792 cm^{-1} (Figure 1E) and C=C and COO$^-$ stretching vibrations at 1565 cm^{-1} and 1615 cm^{-1} (Figure 1D). Additionally, bands visible at 1589^{-1} and 1579 cm^{-1}, which indicate vibrations of C=O and C=C groups of DIC, substantiate the presence of unbound DIC in the prepared samples.

Results of XPS used to analyze the elemental composition of investigated materials are summarized in Figure 3 and Figure S4. The modified polymeric oligomers (Fin-Dic-oliP(3HO)) exhibit two new peaks (N 1s and Cl 2p), indicating the presence of DIC molecules. Moreover, the O/C ratio decreases with the addition of modified oligomers. As shown in Table 2, the surface of the P(3HO) polymer is composed of 77.25% of carbon and 22.75% of oxygen, whereas the surface of modified oligomers

(Fin-Dic-oliP(3HO)) contain 79.27% of carbon, 20.11% of oxygen, 0.46% of nitrogen and 0.17% of chlorine. For P(3HO), the O 1s spectrum consists of two dominant peaks at 532.2 eV and 533.4 eV due to its presence in (C=O)-O* and (C=O*)-O-groups, respectively, and a smaller one at 534.4 eV assigned to oxygen in terminal OH groups in the P(3HO) chain (Figure S5A and Table S1). There is no significant difference observed for the modified oligomers with regards to O 1s (Figure S5B). However, when C 1s spectra for the P(3HO) and modified oligomers are compared, increased content of carbon atoms corresponding to C-O and C=O groups confirms the formation of an ester bond between oligomers and DIC (Figure 3A,B, Table S2). Liu with co-workers observed a similar phenomenon after cross-linking gentamicin with chitosan, where C-N, C-O and C=O peak intensity increased after modification [49].

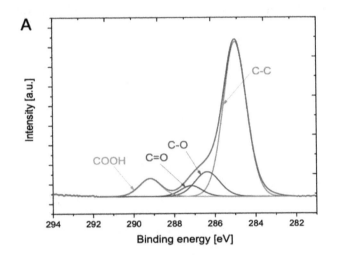 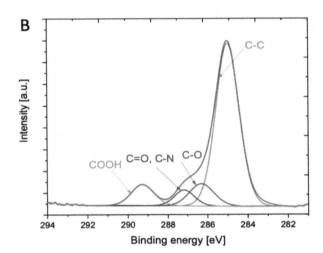

Figure 3. XPS analysis of deconvoluted C 1s spectra, (**A**) shows P(3HO) spectrum and (**B**) indicates Fin-Dic-oliP(3HO) spectrum.

Table 2. Surface elemental compositions of P(3HO) and Fin-Dic-oliP(3HO).

	C (%)	O (%)	O/C Ratio	N (%)	Cl (%)
P(3HO)	77.25	22.75	0.29	n.d.	n.d.
Fin-Dic-oliP(3HO)	79.27	20.11	0.25	0.46	0.17

Where: n.d.—not detected.

3.2. Physicochemical Characteristics of Fin-Dic-oligoP(3HO) Conjugates and Their Blends

It is known that the molecular weight of the polymer plays an important role in the context of carrier degradation and release of the active substance. We have found that the modification of P(3HO) using DIC yields compounds with an average molecular weight of 4.92 kDa, whereas the dispersity index (DI) for P(3HO) and Fin-Dic-oliP(3HO) are 2.19 and 1.66, respectively (Table 3).

Table 3. Physicochemical analysis of materials obtained and used in this study. The results are presented as mean ± SD ($n = 3$).

	M_n [kDa]	M_w [kDa]	Dispersity Index	T_{onset} (°C)	T_{DTG} (°C)	Residue (1%) (°C)	T_g (°C)	T_m (°C)
P(3HO) [21]	73	137	1.88	279	296	306	−41	53
Fin-Dic-oliP(3HO)	4.93 ± 0.0	8.21 ± 0.12	1.66 ± 0.02	285	293	298	−30	-
2Dic-oliP(3HO)	4.93 ± 0.0	8.21 ± 0.12	1.66 ± 0.02	285	293	367	−29	-
1Dic-oliP(3HO)	78.88 ± 4.55	133.74 ± 4.81	1.70 ± 0.04	280	294	305	−40	-
0.5Dic-oliP(3HO)	79.59 ± 0.12	139.26 ± 0.00	1.75 ± 0.00	281	288	308	−40	50
DIC	N/A	N/A	N/A	266	297	307	10	177

Where: N/A—not assayed.

Figure 4A,B shows changes in the processes related to the initiation of thermal degradation of the basic materials, i.e., DIC, P(3HO), modified oligomers (Fin-Dic-oliP(3HO)) as well as blends prepared using 2Dic-oliP(3HO), 1Dic-oliP(3HO) and 0.5Dic-oliP(3HO). In the case of inert thermal degradation, a one-stage process of material degradation is observed in all analyses (Figure S6). The reverse dependence was observed for the products of a high molecular weight medium chain length polyhydroxyalkanoates (mcl-PHAs) modification with sucrose as well as copolymer of 3-hydroxybutyric acid and 3-hydroxyvaleric acid (P(3HB)-co-(3HV)) with ascorbic acid—in both cases the degradation processes were two-staged [48,50]. Moreover, our analyses indicate a slight increase in temperature at which degradation started, in case of modified oligomers (Fin-Dic-oliP(3HO)) as well as blends containing two doses of active substance per 1 g of polymer (2Dic-oliP(3HO)). The onset of thermal decomposition processes (T_{onset}) of 1Dic-oliP(3HO) and 0.5Dic-oliP(3HO) blends show no significant differences (Figure 4A,B, Table 3).

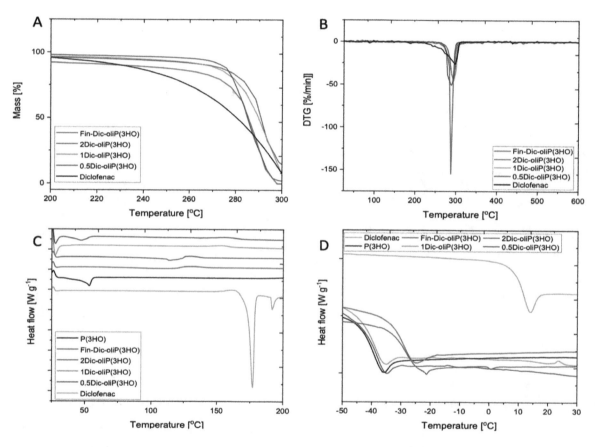

Figure 4. (**A,B**) Thermogravimetric (TGA); (**C,D**) Differential Scanning Calorimetry (DSC) results of the materials used in this study.

DIC is a strongly crystalline substance, showing two sharp melting endotherms at 177 °C and 192 °C (Figure 4C, Table 3). This crystallinity is not preserved when modified with the P(3HO) oligomer. P(3HO) curves show a weak and broad melting peak at 53 °C. Crystallinity is preserved when blended with a small amount of the modified DIC (sample 0.5Dic-oliP(3HO)), albeit the melting temperature (T_m) drops by a few degrees, indicating reduced quality of the crystals. However, more modified DIC completely disrupts the development of crystalline order as manifested by the absence of melting endotherms in the other blends. Glass transition temperature (T_g) of neat P(3HO) was observed at 41 °C (Figure 4D, Table 3). Pure DIC has a much higher T_g at 10 °C. Their compound (Fin-Dic-oliP(3HO)) has an intermediate T_g closer to that of the oligomeric decorations, as expected in the sense of mixing laws, such as that of Fox [51]. The 0.5Dic-oliP(3HO) and 1Dic-oliP(3HO) blends have T_g practically equal to pure P(3HO), while the mass fraction of Fin-Dic-oliP(3HO) in the blend is significant (24% *w/w* and 48% *w/w*; Table 3). This indicates a possible phase separation between the two components, despite the fact that no second glass transition is observed. Indeed, the AFM observations (Section 3.4) showed the development of Fin-Dic-oliP(3HO)-rich nano regions. On the contrary, the sample 2Dic-oliP(3HO) with a mass fraction of Fin-Dic-oliP(3HO) (96% *w/w*) has T_g practically equal to that of its majority component, as expected in the sense of mixing laws.

3.3. Wettability and Surface Free Energy Determination

In order to determine the wettability and surface energy of both the unmodified polymer and the obtained blends, we have measured the wettability angle for a polar liquid (water) and non-polar diiodomethane. One should stress that according to the literature data, concerning biomaterial analysis, hydrophobic properties of a given material are considered when wetting angle values are above 70° [32]. In our experiment, we observed a decrease in hydrophobicity of the material with modification of P(3HO) oligomers and wetting angle values for unmodified polymer and modified oligomers equal to 106 ± 2° and 69 ± 3°, respectively (Figure 5A). In other works, Lukasiewicz et al. demonstrated that the increase in the number of medium-chain oligomers (mcl-oliPHA) caused an increase in the hydrophilicity of the material [31]. In this case, for the initial material poly(3-hydroxybutyrate) P(3HB) and its blend (0.8 P(3HB)/0.2 mcl-oliPHA), wettability was 87.5 ± 7.4° and 118.0 ± 7.0°, respectively [31]. In our research, DIC modified oligomers and blends obtained from them have a wettability for water and diiodomethane of the values: for Fin-Dic-oliP(3HO) (69 ± 3°, 35 ± 5°), 2Dic-oliP(3HO) (74 ± 2°, 27 ± 3°), 1Dic-oliP(3HO) (91 ± 2°, 41 ± 4°) and 0.5Dic-oliP(3HO) (100 ± 1°, 62 ± 2°). Similar wettability for the unmodified polymer P(3HO) was determined by Cichoń et al., i.e., 100 ± 6° for water and 52 ± 3° for diiodomethane [17].

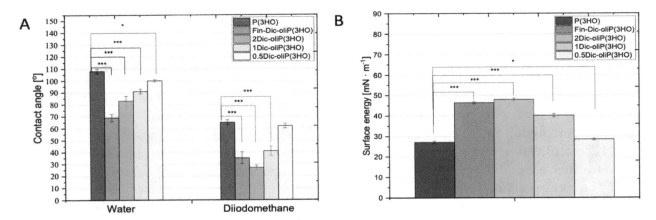

Figure 5. (A) Wettability characteristics of polymeric/oligomeric films in contact with water and diiodomethane ($n = 6$; error bars = ±SD). (B) Surface free energy of materials ($n = 36$; error bars = ±SD). The results are statistically significant, where: * $p < 0.05$, *** $p < 0.001$.

The analysis of surface free energy done by us revealed the similarity of the values obtained for both unmodified polymer and 0.5Dic-oliP(3HO)-coated composite. The values of the polar, dispersion and total free energy components are 0.38 ± 0.04, 27.71 ± 0.30, 27.06 ± 0.40 mN m^{-1} and 0.84 ± 0.09, 27.7 ± 0.44, 28.55 ± 0.42 mN m^{-1}, respectively, for the unmodified and modified materials (Figure 5b and Figure S7A,B). Strict dependence is also observed for the polar component and total surface energy. With increasing concentration of DIC modified oligomers in prepared blends both polar components and free surface energy increased. No biased relation is seen for the dispersion component. The values for P(3HO), 0.5Dic-oliP(3HO), 1Dic-oliP(3HO), 2Dic-oliP(3HO) and Fin-Dic-oliP(3HO) are 27.71 ± 0.30, 27.71 ± 044, 38.67 ± 0.79, 40.67 ± 0.45 and 35.14 ± 0.74 mN m^{-1}, respectively.

3.4. Surface Roughness Determination

Due to the fact that the surface of the materials may affect the activity of cells, allowing them to grow in a specific direction, it is necessary to determine the topographic parameters of the prepared blends [52]. Similar to the research described by Sofińska et al. of the surface of a non-modified P(3HO) polymer, all AFM images indicate the presence of micro- and nano-scale structures with different roughness [21]. Moreover, AFM reveals an impressive change in the morphology of the surface of the blends under investigation, as seen in the height images in Figure 6. The 0.5Dic-oliP(3HO) blend shows a rough topography, however, with characteristic microcavities. A similar surface topography for Poly(L-lactic acid) (PLLA)/Polystyrene (PS) composition was observed by Lim et al. [53]. The resulting microcavities can influence the cell activity in the regenerated tissue [52].

Figure 6. AFM images of Fin-dic-oliP(3HO)-based blend film (10×10 μm^2). (**A**): The height images in 3D representation Note the different scales on the z-axis. (**B**): The corresponding phase images.

In the other extremum, 2Dic-oliP(3HO) shows no features at all. In this blend, Fin-Dic-oliP(3HO) is the majority component (96% w/w) and presumably P(3HO) is fully miscible in it, creating a fully homogeneous material, as manifested in both the topographic and phase images. In the intermediate case of 1Dic-oliP(3HO), topography shows features of two types distributed in a flat background. The two features are globules of a radius in the order of 1 μm, and particles of a radius in the order of 100 nm. The corresponding phase image reveals that the two types of features have different mechanical properties. On a closer look, the smaller structures are visible to a much lesser extent in the phase image of 0.5Dic-oliP(3HO). We assume thus that they correspond to Fin-Dic-oliP(3HO)—rich areas where the phase separates from the polymer. This observation is in agreement with the observations about their T_g in Section 3.2.

Additionally, the root mean square roughness (S_q) of the samples surfaces was determined (Figure 6). A decreasing value of this parameter for the three prepared blends with an increasing content of Fin-dic-oliP(3HO) was observed. A similar phenomenon was described in the case of formulations based on polyvinyl alcohol and boric acid (PVAB) with DIC sodium salt [54]. Taking into account the fact that the 0.5Dic-oliP(3HO) blend consists of ca. 75% w/w of the P(3HO) polymer, the S_q is significantly higher (53.42 nm) compared to the research carried out by Sofińska et al. (13.2 ± 2.8 nm) on pure P(3HO) [21]. This may be due to the addition of Fin-dic-oliP(3HO), which acts as a co-crystallizer, reducing the polymer structure, thus allowing the formation of microcavities [55].

3.5. Cytotoxicity of Dic-oliP(3HO) Conjugates-Based Blends

Cytotoxicity of DIC-modified oligomers and prepared blends are evaluated in the indirect cytotoxicity test, in which the exudate from material incubation in the cellular medium is analyzed

(Figure 7A). Typical sources of cytotoxicity are residues of organic solvent from polymer synthesis, intermediate products used for synthesis and material degradation products as well as the additives themselves with which the polymer is modified [56]. Figure 7 summarizes indirect cytotoxicity studies and in vitro biocompatibility tests performed using an MC3T3-E1 model to assess the possibility of using the prepared bioceramic-polymer composites in bone tissue regeneration processes. According to ISO 10993-5, a 30% reduction in cell viability may indicate cytotoxicity of the material [57]. All studied material extracts show no cytotoxic effects. What is more, higher cell viability is achieved for the unmodified composites ($105 \pm 3\%$) compared to the negative control, which confirms that neither β-TCP nor P(3HO) contain or release harmful components for the MC3T3-E1 cell line. This substantiates the excellent biocompatibility of the composites components—according to the literature, both PHAs and TCPs show no cytotoxicity to human tissues [10,24]. Additionally, the release of polymer decomposition products—(R)-3-hydroxyacids—stimulates cell viability [58]. The lack of cytotoxicity of materials made of β-TCP is confirmed by the results obtained by Skibiński et al. Indirect cytotoxicity tests of β-TCP sponges performed in the MC3T3-E1 model showed high cell viability ($117.11 \pm 5.59\%$) [41]. In the case of composites infiltrated with blends containing different amounts of Fin-Dic-oliP(3HO), no cytotoxic effects with respect to the negative control are observed. For TCP/0.5Dic-oliP(3HO), TCP/1Dic-oliP(3HO) and TCP/2Dic-oliP(3HO) cell viability is equal to $78 \pm 4\%$, $99 \pm 1\%$ and $85 \pm 8\%$, respectively. From the analysis, it can be concluded that the amount of active substance released during the 24 h of incubation does not cause a cytotoxic effect when in contact with the MC3T3-E1 cell line. In the case of β-TCP/P(3HO) composites containing physically bounded DIC, analogous phenomenon was reported ($99.20 \pm 5.61\%$) [41]. Moreover, a similar effect was observed by Sidney et al. after a 14-day incubation of the mouse pre-osteoblast cells in the culture medium and further addition of 1 mg of sodium DIC per scaffold (8-fold lower than that used in our study) [5].

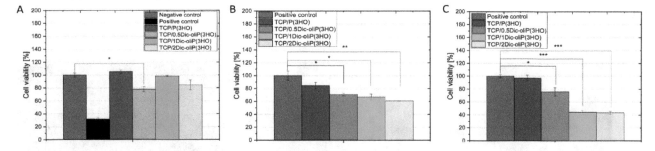

Figure 7. (**A**). Indirect cytotoxicity test showing no positive impact of the 1-day old leachate on MC3T3-E1 mouse pre-osteoblasts ($n = 3$; error bars = ±SD). In vitro biocompatibility study results of cells seeded on the materials on day 1 of the experiment (**B**) versus day 7 (**C**) ($n = 3$; error bars = ±SD). Besides the TCP/P(3HO) composite, only theTCP/0.5Dic-oliP(3HO) scaffold allowed for $76.0 \pm 6.4\%$ cell viability after 7 days of incubation, thus making it non-cytotoxic according to ISO 10993-5. The results are statistically significant, where: * $p < 0.05$, ** $p < 0.01$, *** $p < 0.001$.

3.6. In Vitro Biocompatibility and Cell Morphology on Hybrid Composites

The selection of an appropriate preparation method and the proper composition of the scaffold allows us to obtain the best possible biochemical and mechanical properties, which may lead to increased adhesion, differentiation and proliferation of the mammalian cells; it may simulate the environment around natural tissues. An important feature of the scaffold, apart from its porous structure, is its bioabsorption and biocompatibility [31–34]. For these reasons, we have carried out direct compatibility analysis, seeding the (MC3T3-E1) directly on our materials (Figure 7B,C). Cell proliferation ability on surfaces based on PHAs was evaluated already in various cell models [25–27,31–34,59]. Our study shows increased viability along with the time of cell incubation on the surface of TCP/P(3HO) composites. After 1 and 7 days of incubation, the cell viability for TCP/P(3HO) composites are $84.3 \pm 5.3\%$ and $97.1 \pm 4.4\%$, respectively, compared to the negative control. Taking into consideration the composites containing chemically bonded DIC, the cell viability at day 1 and 7 changes as follows:

for 0.5Dic-oliP(3HO) from 71.0 ± 1.9% to 76.0 ± 6.4%; for 1Dic-oliP(3HO) from 66.8 ± 4.5% to 44.2 ± 2.7%; and for 2Dic-oliP(3HO), from 60.9 ± 0.1% to 43.6 ± 2.4%. From the above data, we can conclude that the composite blends with the lowest dose of chemically bonded DIC (0.5Dic-oliP(3HO)) were able to facilitate the attachment and growth of MC3T3-E1 cells. For the composites with higher amount of chemically bonded DIC (1Dic-oliP(3HO) and 2Dic-oliP(3HO)), at day 1 the cell viability was comparable to 0.5Dic-oliP(3HO) samples. However, when the materials were incubated for a longer time, cell growth inhibition was detected.

Using SEM and confocal microscopy observations we have analyzed the pre-osteoblast mouse cell colonization over time on the studied materials. Microphotographs of TCP/P(3HO) scaffolds show the formation of a uniform cell layer covering the surface of the material after 7 days, indicating excellent cell viability (Figure 8B). Similar observations were shown by Lukasiewicz et al. in the (P(3HB))/mcl-(P(3HA)) blend study with the mouse myoblast (C2C12) cell line [31]. Our scanning electron and confocal microscopy analyses of 0.5Dic-oliP(3HO) material reveal single cells and fibers originating from their presence on the composite surface, however, with lower confluency compared to unmodified scaffolds (Figure 8E). Moreover, 3D reconstructions of confocal microscopy images suggest that cells seeded on TCP/0.5Dic-oliP(3HO) migrate slower and penetrate the material less invasively than TCP composites with pure P(3HO) (Figure 8C,F). In the case of the unmodified scaffold, the cells penetrate ~500 μm deep into the material, while in the case of TCP/0.5Dic-oliP(3HO), the migration range is reduced to a maximum of ~300 μm. Analyses of the other two composites (1Dic-oliP(3HO) and 2Dic-oliP(3HO)) do not show the presence of the cells in the middle of the scaffold's pores (Figure S9). However, their microphotographs reveal that chemical modification results in the crystallization of Fin-dic-oliP(3HO) in a different form than that for pure DIC. Fini et al. demonstrated that DIC sodium salt has the form of spindle-shaped crystals, while our research shows that modification of low molecular weight PHA oligomers with DIC results in the formation of hedgehog-like, crystalline structures [60] (Figure 8D).

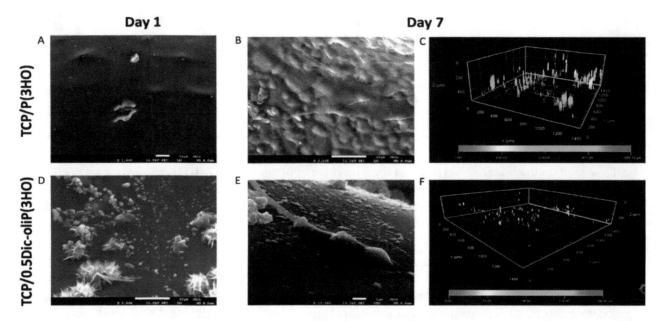

Figure 8. MC3T3-E1 cell colonization of TCP/P(3HO) and TCP/0.5Dic-oliP(3HO) composites. Materials on the 1st day (**A,D**) and on the 7th day (**B,E**) of incubation visualised using SEM. Bars in A,B and D correspond to 10 μm, whereas in E to 1 μm. (**C,F**): 3D reconstructions of cells migrating into the materials after 7-day incubation. Colour-coded depth penetration in C from 0 (red) to 556.12 μm (blue) and in F from 0 (red) to 292.0 μm (blue).

4. Discussion

The process of bone tissue regeneration requires the use of materials with osteoinductive and osteoconductive properties, which can simulate conditions occurring in natural bone. Moreover, after the implantation, inflammation is often observed, which can be eliminated by using appropriate bioactive agents [3]. In order to obtain a polymeric delivery system containing a non-steroidal, anti-inflammatory drug namely diclofenac, we presented a single-stage, solvent-free synthesis method. Its clue relies on a simultaneous reaction of polymer, hydrolyzing agent (aimed at transforming the polymer into oligomers) and a bioactive substance (DIC-modifier) and leads to new bioactive compounds—oligomers combined with DIC. From the literature it is known that the use of p-TSA is one of the methods enabling the simultaneous decomposition of the polymer as well as its modification [35,36]. Other methods supporting modification and partial decomposition of the polymer is transesterification reaction using a biocatalytic approach [48]. The methods resulting exclusively in the degradation of the polymer chain comprise acidic methanolysis, alkaline saponification or thermal degradation [31]. Kwiecień et al. demonstrated that a higher temperature and a higher amount of hydrolysing agent used led to oligomers with lower molecular weight [35,36]. Lukasiewicz et al. substantiated that the reactions in which smaller molecular weights are obtained have a higher homogeneity, characterised by lower DI, which is consistent with the results obtained in [31].

In the literature there is a correlation between the average molecular weight of the polymer and the release of the active substance. Furthermore, it was shown that with a decrease in the average molecular weight, the release rate of the active substance increased [61]. Other important parameters influencing the biodistribution and absorption of characterising the active substance are those of a given surface, i.e., hydrophilicity or hydrophobicity, surface charge [62]. As it is known, immediately after the implantation, an inflammation is observed which may even lead to the rejection of the newly inserted implant [63]. Therefore, it is important to use active substance carriers, which can release more drug molecules in the first days of application and then gradually release the residue, preventing possible side effects. In addition to the release of the active substance, degradation of the polymer carrier was shown, leading to a decrease in molecular weight and a change in dispersity [61]. The use of a low molecular weight matrix is also beneficial in terms of increasing the degradation rate of the polymeric material as shown in [64]. Rajaratanam et al. reported that the use of (P(3HB)-co-(3HHx)) oligoesters favours the process of the degradation of threads made of these materials [27].

Studies conducted by Witko et al. have shown that virgin P(3HO) is a non-toxic material in contact with a mouse embryonic fibroblast cell line (MEF3T3). Additionally, it was observed that this polymer affects cell morphology, tubular cytoskeleton and cytoskeletal components, i.e., actin. The authors reported that the surface properties of P(3HO), i.e., the coefficient of friction, combined with lower brittleness of the materials compared to other biodegradable and biocompatible polyesters such as poly(lactic acid) (PLA) and P(3HB), may indicate potential applications of P(3HO) in the construction of medical implants, joint endoprostheses. It should be underlined that the above materials are not only subjected to significant loads (hip joint, knee joint), but also high friction forces due to their movement in contact with bones or cartilage [30,45].

Naveen et al. reported that hydrophilic materials were characterised by a higher protein adsorption capacity and therefore improved cell adhesion and proliferation [65]. However, our studies on the scaffold colonization over time and the ability of the composites show an inverse relationship. The cell viability is found to be highest for the 0.5Dic-oliP(3HO) blend (on day 7 of in vitro biocompatibility study 76.0 ± 6.4%), characterized by the strongest hydrophobic properties among the prepared ceramic-polymer composites, containing modified oligomers. This is possibly due to the toxicity of DIC, not due to the hydrophobicity. Additionally, we can conclude that obtaining a similar value of surface free energy and wettability for all prepared composites does not mean a similar level of eukaryotic cell viability. More important parameters affecting viability seems to be surface roughness or production of specific mediators by the cell, but predominantly contact with toxic chemicals, as it is with DIC. For example, it was shown, that in hOB cell line studies, 10^{-5} M DIC concentration does not inhibit cell proliferation, stops the cell cycle in the G0/G1 phase and does not induce nor inhibit important mediators such as cyclin-dependent kinase inhibitor 1B (p27^{Kip1}), cyclin D2 and cell division protein kinase 2 (Cdk2). In the case of osteoblast growth inhibition, the expression of p27^{kip1} is increased, cyclin D2 expression decreased and cdk2 levels decreased [7].

The dynamic and complex environment that surrounds living cells enables the collection and efficient transport of signals necessary for their functioning. Complex interactions occurring at the cell–cell and cell-substrate levels affect the ability to adsorb proteins responsible for the regeneration of the appropriate tissue. However, an undoubted role in the abovementioned interactions is played by parameters relating to the surface characteristics—the as-mentioned wettability, surface energy, but also surface topography, its roughness as well as porosity and pore size distribution [66]. Taking into account only the topographic parameters of materials, i.e., roughness, it may have a different impact on the behaviour of living cells in the tested models [21]. For example, if the material is used as a bone regeneration scaffold, its surface structure should favour adhesion, proliferation and differentiation of the osteoblast, while the same parameters should be inhibitory for fibroblasts. This was confirmed in studies into changes of the topographic parameters of the TiO$_2$ surface consisting of micropites and nanodules described by Kubo et al. Moreover, a better adhesion and proliferation of osteoblasts was observed in materials characterized by a higher surface roughness [67]. A similar relationship was observed in our research. Among the composites containing various doses of DIC-modified oligomers, the 0.5Dic-oliP(3HO) composite with the highest roughness (53.42 nm) showed the highest cell viability in the seven-day direct biocompatibility experiment (76.0 ± 6.4%).

As shown by Sadowska et al., the sintered disc β-TCP surface roughness was 1.33 ± 0.23 μm [68]. This result is several times higher than that obtained in our study. However, it should be remembered that infiltration of the polymer into the ceramic sponges results in the formation of a polymer layer on the surface of the prepared composite. According to our previous studies, the thickness of the polymer layer on the surface of the composite infiltrated with a 5% w/v P(3HO) solution was about 30 μm [17]. Therefore, the roughness of the composite will depend on the surface layer, which is a polymer or polymeric blend. Remembering that the unmodified P(3HO) thin films had a roughness lower (13.2 ± 2.8 nm) than the highest roughness value obtained in this study for the 0.5Dic-oliP (3HO) blend (53.42 nm), the TCP/P(3HO) composite should be characterized by weaker adhesion and colonization of the MC3T3-E1 cell line compared to the TCP/0.5Dic-oliP(3HO) composite. However, our research showed an inverse relationship. Cell viability on the 7th day of the direct biocompatibility test was higher for TCP/P(3HO) (97.1 ± 4.4%) compared to TCP/0.5Dic-oliP(3HO) (76.0 ± 6.4%). In view of the above, the influence of characteristic markers and proteins involved in the inflammatory reaction of the cell lines studied should also be considered in future studies.

The analysis of literature data showed that DIC and other drugs from the NSAIDs group inhibits the viability of osteoblast cell lines—both MC3T3-E1 and MG63 as well as in the rabbit model—effectively preventing the processes associated with bone tissue regeneration [69–71]. In our study, the materials containing half of the therapeutic dose of chemically bonded DIC showed a cell viability over 70%, which, according to ISO 10993-5, confirms an absence of a cytotoxic effect. For the materials containing an amount of DIC equal or double the therapeutic dosage, a toxic effect could be detected. In other research, an inverse relationship in the case of cell scaffolds made of (P(3HB)-co-(3HV))/polyaniline composites with chemically bounded curcumin was shown [33]. Here, the in vitro biocompatibility analysis on the NIH 3T3 mouse fibroblast cell line resulted in better adhesion, adsorption and migration of cells on the surface of polymers with chemically bonded curcumin molecules. These results were associated with a change of roughness of the surface and the presence of permeating pores.

In the case of the in vitro biocompatibility studies, the results showed a decrease in cell viability, indicating a cytotoxic effect on the 3rd day of the experiment. However, it should be pointed out that the in vitro biocompatibility experiment was carried out under static conditions, and the culture medium was changed on the 3rd day. On the contrary, under natural conditions in the human body, a continuous circulation of body fluid occurs.

The flow of blood and other fluids would undoubtedly reduce the concentration of the eluted drug, thus reducing the cytotoxic effect. Therefore, we may postulate that analysis in a specially designed bioreactor, with a continuous exchange of cultured medium, could provide a realistic picture of the cytotoxicity of the investigated materials [72]. The incorporation of DIC in the scaffolds was investigated to overcome the negative side effects associated with the systematic delivery of such a drug. Nevertheless, further studies should be conducted to investigate the therapeutic effectiveness of the amount of the drug released, through the quantification of the drug released and the analysis of the in vitro effect on the cells in terms of inflammation markers (e.g., prostaglandin E2 and nitric oxide production) [5].

5. Conclusions

Chemical modification of polymers and molecules derived from their decomposition is a promising approach, which offers new functionalities that could not be provided by unmodified molecules. In this work, composites based on tricalcium phosphates and polyhydroxyalkanoates with covalently attached molecules of non-steroidal anti-inflammatory drugs are obtained. We have confirmed polymer modification using several physicochemical methods (i.e., NMR, IR and XPS). Moreover, the analysis of wettability of films using prepared blends has shown an increasing hydrophilicity of materials as a function of the number of modified oligomers. AFM analysis of the prepared Fin-Dic-oliP(3HO)-based blends showed a different surface topography, which was dependent on the concentration of the modified oligomers used. Additionally, the surface roughness analysis showed that the 0.5Dic-oli(P3HO) blend is characterized by the highest roughness, which affects the adhesion and colonization of the MC3T3-E1 cell line on the prepared composites. Moreover, the results obtained from a physicochemical analysis, in combination with biological tests, allowed us to perform an in-depth analysis of scaffolds with potential use in bone tissue engineering. Furthermore, we have performed studies related to indirect cytotoxicity and the determination of in vitro biocompatibility behaviour, as well as the rate of cell colonization on the scaffolds over time, in a mouse pre-osteoblast cell line model (MC3T3-E1).

The in vitro biocompatibility studies show no cytotoxicity effect for materials containing a half dose of the chemically bounded DIC (0.5Dic-oliP(3HO)). Thus, we can conclude that the oligomeric P(3HO) modified with DIC can provide a useful medical platform that opens up new therapeutic possibilities. Controlled release for long-term protection against chronic inflammation is essential and this is of particular importance in materials intended for the regeneration of bone defects, such as macroporous scaffolds, which are exposed to the occurrence of continual inflammation after implantation, associated with the human body immune system reaction. The use of covalently-linked NSAIDs such as DIC in this type of bone substitute is of crucial importance. It reduces the risk of rejection of the implanted foreign body. What is more, the use of biocompatible and bioactive materials such as TCP and P(3HO) provides a dual action: ceramic component due to its structural similarity to natural bone secures regeneration, while P(3HO) during natural enzymatic and physical hydrolysis processes, degrades to 3-(R)-hydroxycarboxylic acids, thus nourishing the implantation site.

Author Contributions: Conceptualization, M.G.; methodology, K.H., E.C., S.S., I.K. and E.M.; validation, K.H. and E.C.; formal analysis, K.H., T.W., D.S., R.S. and K.N.R.; investigation, K.H., E.C., S.S., T.W., D.S., E.M., I.K., M.Z., R.S., K.N.R. and E.S.; resources, A.Z., I.R. and M.G.; data curation, K.H. and E.C.; writing—original draft preparation, K.H. and M.G.; writing—review and editing, K.H., E.C., I.K., M.Z., R.S., A.Z., I.R., K.N.R., K.P., M.W. and M.G.; visualization, K.H., T.W., D.S. and K.N.R.; supervision, A.Z., I.R., K.P., M.W. and M.G; project administration, M.G.; funding acquisition, M.G. All authors have read and agreed to the published version of the manuscript.

Acknowledgments: K.H. acknowledges the support of InterDokMed project no. POWR.03.02.00-00-I013/16 as well as the PROM project—International scholarship exchange of doctoral students and academic staff, project no PPI/PRO/2018/1/00006/U/001. E.C. acknowledges financial support from the National Science Centre, Poland under Doctoral Scholarship No. 2019/32/T/ST5/00207. E.M. acknowledges support from the HyMedPoly project, Marie Skłodowska—Curie Grant agreement No 643050. We would like to thank Bogna Daria Napruszewska for help with TGA analysis.

Abbreviations

0.5Dic-oliP(3HO)	Blend containing half doses of active substance per 1 g of polymer
1Dic-oliP(3HO)	Blend containing one dose of active substance per 1 g of polymer
1H NMR	Proton nuclear magnetic resonance
2Dic-oliP(3HO)	Blend containing two doses of active substance per 1 g of polymer
ACP	Amorphous tricalcium phosphate
ATR-FTIR	Attenuated total reflection Fourier transform infrared
C2C12	mouse myoblast cell line
Cdk2	Cell division protein kinase 2
DI	Dispersity index
DIC	Diclofenac acid
DSC	Differential scanning calorimetry
Fin-Dic-oliP(3HO))	Oligo-diclofenac conjugates
GPC	Gel permeation chromatography
HAp	Hydroxyapatite
hOBs	Human osteoblasts
MC3T3-E1	The mouse calvaria-derived pre-osteoblastic cell line
mcl-PHAs	Medium chain length polyhydroxyalkanoates
MEF3T3	Mouse embryonic fibroblast cell line
MG-63	osteosarcoma cell line
NIH 3T3	Mouse fibroblast cell line
NSAIDs	Non-steroidal anti-inflammatory drugs
P(3HB-co-4HB)	Copolymer of 3-hydroxybutyric acid and 4-hydroxybutyric acid
P(3HB)-co-(3HV)	Copolymer of 3-hydroxybutyric acid and 3-hydroxyvaleric acid
P(3HO)	Poly(3-hydroxyoctanoate)
p27Kip1	Cyclin-dependent kinase inhibitor 1B
PCL	Poly(caprolactone)
PEG	Poly(ethylene glycol)
PHAs	Polyhydroxyalkanoates
PLGA	Copolymer of DL-lactic acid and glycolic acid
PLLA	Poly(L-lactic acid)
PS	Polystyrene
PVAB	Copolymer of polyvinyl alcohol and boric acid
SAPAE	Copolymer of salicylic acid and succinic anhydride

SEM	Scanning electron microscopy
Sq	Root mean square roughness
TCP	Tricalcium phosphate
Tg	Glass transition temperature
TGA	Thermogravimetric analysis
Tm	Melting temperature
Tonset	Onset of thermal decomposition process
XPS	X-ray photoelectron spectroscopy
α-TCP	Alpha-tricalcium phosphate
β-TCP	Beta-tricalcium phosphate

References

1. Iaquinta, M.R.; Mazzoni, E.; Manfrini, M.; D'Agostino, A.; Trevisiol, L.; Nocini, R.; Trombelli, L.; Barbanti-Brodano, G.; Martini, F.; Tognon, M. Innovative biomaterials for bone regrowth. *Int. J. Mol. Sci.* **2019**, *20*, 618. [CrossRef] [PubMed]

2. Madhumathi, K.; Rubaiya, Y.; Doble, M.; Venkateswari, R.; Kumar, T.S.S. Antibacterial, anti-inflammatory, and bone-regenerative dual-drug-loaded calcium phosphate nanocarriers—In vitro and in vivo studies. *Drug Deliv. Transl. Res.* **2018**, *8*, 1066–1077. [CrossRef] [PubMed]

3. Sarigol-Calamak, E.; Hascicek, C. Tissue Scaffolds As a Local Drug Delivery System for Bone Regeneration. In *Cutting-Edge Enabling Technologies for Regenerative Medicine*; Advances in Experimental Medicine and Biology; Chun, H., Park, C., Kwon, I., Khang, G., Eds.; Springer: Singapore, 2018; Volume 1078. [CrossRef]

4. Beck, A.; Krischak, G.; Sorg, T.; Augat, P.; Farker, K.; Merkel, U.; Kinzl, L.; Claes, L. Influence of diclofenac (group of nonsteroidal anti-inflammatory drugs) on fracture healing. *Arch. Orthop. Trauma Surg.* **2003**, *123*, 327–332. [CrossRef] [PubMed]

5. Sidney, L.E.; Heathman, T.R.J.; Britchford, E.R.; Abed, A.; Rahman, C.V.; Buttery, L.D.K. Investigation of localized delivery of diclofenac sodium from poly(D,L-lactic acid-co-glycolic acid)/poly(ethylene glycol) scaffolds using an in vitro osteoblast inflammation model. *Tissue Eng. Part A* **2015**, *21*, 362–373. [CrossRef] [PubMed]

6. Lisowska, B.; Kosson, D.; Domaracka, K. Positives and negatives of nonsteroidal anti-inflammatory drugs in bone healing: The effects of these drugs on bone repair. *Drug Des. Devel. Ther.* **2018**, *12*, 1809–1814. [CrossRef] [PubMed]

7. Chang, J.K.; Li, C.J.; Liao, H.J.; Wang, C.K.; Wang, G.J.; Ho, M.L. Anti-inflammatory drugs suppress proliferation and induce apoptosis through altering expressions of cell cycle regulators and pro-apoptotic factors in cultured human osteoblasts. *Toxicology* **2009**, *258*, 148–156. [CrossRef] [PubMed]

8. Díaz-Rodríguez, L.; García-Martínez, O.; De Luna-Bertos, E.; Ramos-Torrecillas, J.; Ruiz, C. Effect of ibuprofen on proliferation, differentiation, antigenic expression, and phagocytic capacity of osteoblasts. *J. Bone Miner. Metab.* **2012**, *30*, 554–560. [CrossRef] [PubMed]

9. Kwong, F.N.K.; Harris, M.B. Recent developments in the biology of fracture repair. *J. Am. Acad. Orthop. Surg.* **2008**, *16*, 619–625. [CrossRef]

10. Jeong, J.; Kim, J.H.; Shim, J.H.; Hwang, N.S.; Heo, C.Y. Bioactive calcium phosphate materials and applications in bone regeneration. *Biomater. Res.* **2019**, *23*, 1–11. [CrossRef]

11. Turnbull, G.; Clarke, J.; Picard, F.; Riches, P.; Jia, L.; Han, F.; Li, B.; Shu, W. 3D bioactive composite scaffolds for bone tissue engineering. *Bioact. Mater.* **2018**, *3*, 278–314. [CrossRef]

12. Iannazzo, D.; Pistone, A.; Salamò, M.; Galvagno, S. *Hybrid Ceramic/Polymer Composites for Bone Tissue Regeneration In Hybrid Polymer Composite Materials Processing*; Woodhead Publishing: Sawston, Cambridge, UK, 2017; pp. 125–155, ISBN 9780081007891. [CrossRef]

13. Sulistio, A.; Reyes-Ortega, F.; D'Souza, A.M.; Ng, S.M.Y.; Valade, D.; Quinn, J.F.; Donohue, A.C.; Mansfeld, F.; Blencowe, A.; Qiao, G.; et al. Precise control of drug loading and release of an NSAID-polymer conjugate for long term osteoarthritis intra-articular drug delivery. *J. Mater. Chem. B* **2017**, *5*, 6221–6226. [CrossRef] [PubMed]

14. Subramanian, S.; Mitchell, A.; Yu, W.; Snyder, S.; Uhrich, K.; Connor, J.P.O. Salicylic Acid-Based Polymers for Guided Bone Regeneration Using Bone Morphogenetic Protein-2. *Tissue Eng. Part A* **2015**, *21*, 2013–2024. [CrossRef] [PubMed]

15. Mitchell, A.; Kim, B.; Cottrell, J.; Snyder, S.; Witek, L.; Ricci, J.; Uhrich, K.E.; Connor, J.P.O. Development of a guided bone regeneration device using salicylic acid-poly (anhydride-ester) polymers and osteoconductive scaffolds. *J. Biomed. Mater. Res. Part A* **2013**, *102*, 655–664. [CrossRef] [PubMed]

16. Cerrone, F.; Duane, G.; Casey, E.; Davis, R.; Belton, I.; Kenny, S.T.; Guzik, M.W.; Woods, T.; Babu, R.P.; O'Connor, K. Fed-batch strategies using butyrate for high cell density cultivation of Pseudomonas putida and its use as a biocatalyst. *Appl. Microbiol. Biotechnol.* **2014**, *98*, 9217–9228. [CrossRef] [PubMed]

17. Cichoń, E.; Haraźna, K.; Skibiński, S.; Witko, T.; Zima, A.; Ślósarczyk, A.; Zimowska, M.; Witko, M.; Leszczyński, B.; Wróbel, A.; et al. Novel bioresorbable tricalcium phosphate/polyhydroxyoctanoate (TCP/PHO) composites as scaffolds for bone tissue engineering applications. *J. Mech. Behav. Biomed. Mater.* **2019**, *98*, 235–245. [CrossRef] [PubMed]

18. Misra, S.K.; Valappil, S.P.; Roy, I.; Boccaccini, A.R. Polyhydroxyalkanoate (PHA)/inorganic phase composites for tissue engineering applications. *Biomacromolecules* **2006**, *7*, 2249–2258. [CrossRef]

19. Francis, L.; Meng, D.; Knowles, J.C.; Roy, I.; Boccaccini, A.R. Multi-functional P (3HB) microsphere/45S5 Bioglass-based composite scaffolds for bone tissue engineering. *Acta Biomater.* **2010**, *6*, 2773–2786. [CrossRef]

20. Misra, S.K.; Watts, P.C.P.; Valappil, S.P.; Silva, S.R.P.; Boccaccini, A.R.; Roy, I. Poly (3-hydroxybutyrate)/Bioglass composite films containing carbon nanotubes. *Nanotechnology* **2007**, *18*, 1–7. [CrossRef]

21. Sofińska, K.; Barbasz, J.; Witko, T.; Dryzek, J.; Haraźna, K.; Witko, M.; Kryściak-Czerwenka, J.; Guzik, M. Structural, topographical, and mechanical characteristics of purified polyhydroxyoctanoate polymer. *J. Appl. Polym. Sci.* **2018**, 47192. [CrossRef]

22. Haraźna, K.; Szyk-Warszyńska, L.; Witko, M.; Guzik, M. Preparation, characterisation and modification of bacterial polyhydroxyoctanoate for construction of wound patches. In *7th International Seminar including Special Session "Recent Advances in Polymer Nanocomposites and Hybrid Materials"*; Pielichowski, K., Ed.; Tomasz Mariusz Majka Publisher: Tarnów, Poland, 2019; pp. 139–152, ISBN 978-83-937270-6-3.

23. van der Walle, G.A.M.; de Koning, G.J.M.; Weusthuis, R.A.; Eggink, G. Properties, Modifications and Applications of Biopolyesters. In *Biopolyesters. Advances in Biochemical Engineering/Biotechnology*; Springer: Berlin/Heidelberg, Germany, 2001; pp. 263–291, ISBN 978-3-540-40021-9. [CrossRef]

24. Szwej, E.; Devocelle, M.; Kenny, S.; Guzik, M.; O'Connor, S.; Nikodinovic-Runic, J.; Radivojevic, J.; Maslak, V.; Byrne, A.T.; Gallagher, W.M.; et al. The chain length of biologically produced (R)-3-hydroxyalkanoic acid affects biological activity and structure of anti-cancer peptides. *J. Biotechnol.* **2015**, *204*, 7–12. [CrossRef]

25. Zawidlak-Węgrzyńska, B.; Kawalec, M.; Bosek, I.; Łuczyk-Juzwa, M.; Adamus, G.; Rusin, A.; Filipczak, P.; Głowala-Kosińska, M.; Wolańska, K.; Krawczyk, Z.; et al. Synthesis and antiproliferative properties of ibuprofen—oligo (3-hydroxybutyrate) conjugates. *Eur. J. Med. Chem.* **2010**, *45*, 1833–1842. [CrossRef] [PubMed]

26. Chen, G.-Q.; Wu, Q. The application of polyhydroxyalkanoates as tissue engineering materials. *Biomaterials* **2005**, *26*, 6565–6578. [CrossRef] [PubMed]

27. Rajaratanam, D.D.; Ariffin, H.; Hassan, M.A.; Abd Rahman, N.M.A.N.; Nishida, H. In vitro cytotoxicity of superheated steam hydrolyzed oligo((R)-3-hydroxybutyrate-co-(R)-3-hydroxyhexanoate) and characteristics of its blend with poly(L-lactic acid) for biomaterial applications. *PLoS ONE* **2018**, *13*, e0199742. [CrossRef] [PubMed]

28. Nigmatullin, R.; Thomas, P.; Lukasiewicz, B.; Puthussery, H.; Roy, I.; Biotechnology, A. Polyhydroxyalkanoates, a family of natural polymers, and their applications in drug delivery. *J. Chem. Technol. Biotechnol.* **2015**, *90*, 1209–1221. [CrossRef]

29. Witko, T.; Guzik, M.; Sofińska, K.; Stepien, K.; Podobinska, K. Novel Biocompatible Polymers for Biomedical Applications. *Biophys. J.* **2018**, *114*, 363a. [CrossRef]

30. Witko, T.; Solarz, D.; Feliksiak, K.; Rajfur, Z.; Guzik, M. Cellular architecture and migration behavior of fibroblast cells on polyhydroxyoctanoate (PHO): A natural polymer of bacterial origin. *Biopolymers* **2019**, *110*, e23324. [CrossRef]

31. Lukasiewicz, B.; Basnett, P.; Nigmatullin, R.; Matharu, R.; Knowles, J.C.; Roy, I. Binary polyhydroxyalkanoate systems for soft tissue engineering. *Acta Biomater.* **2018**, *71*, 225–234. [CrossRef]

32. Basnett, P.; Ching, K.Y.; Stolz, M.; Knowles, J.C.; Boccaccini, A.R.; Smith, C.; Locke, I.C.; Keshavarz, T.; Roy, I. Novel Poly(3-hydroxyoctanoate)/Poly(3-hydroxybutyrate) blends for medical applications. *React. Funct. Polym.* **2013**, *73*, 1340–1348. [CrossRef]
33. Pramanik, N.; Dutta, K.; Basu, R.K.; Kundu, P.P. Aromatic π-Conjugated Curcumin on Surface Modified Polyaniline/Polyhydroxyalkanoate Based 3D Porous Scaffolds for Tissue Engineering Applications. *ACS Biomater. Sci. Eng.* **2016**, *2*, 2365–2377. [CrossRef]
34. Ang, S.L.; Shaharuddin, B.; Chuah, J.A.; Sudesh, K. Electrospun poly(3-hydroxybutyrate-co-3-hydroxyhexanoate)/silk fibroin film is a promising scaffold for bone tissue engineering. *Int. J. Biol. Macromol.* **2020**, *145*, 173–188. [CrossRef]
35. Kwiecień, I.; Radecka, I.; Kowalczuk, M.; Adamus, G. Transesterification of PHA to Oligomers Covalently Bonded with (Bio) Active Compounds Containing Either Carboxyl or Hydroxyl Functionalities. *PLoS ONE* **2015**, *10*, e0120149. [CrossRef] [PubMed]
36. Kwiecień, I.; Radecka, I.; Kwiecień, M.; Adamus, G. Synthesis and Structural Characterization of Bioactive PHA and γ-PGA Oligomers for Potential Applications as a Delivery System. *Materials* **2016**, *9*, 307. [CrossRef] [PubMed]
37. Francis, B.L.; Meng, D.; Locke, I.C.; Mordan, N.; Salih, V.; Knowles, J.C.; Boccaccini, A.R.; Roy, I. The Influence of Tetracycline Loading on the Surface Morphology and Biocompatibility of Films Made from P (3HB) Microspheres. *Adv. Biomater.* **2010**, *7*, 260–268. [CrossRef]
38. Francis, L.; Meng, D.; Knowles, J.; Keshavarz, T.; Boccaccini, A.R.; Roy, I. Controlled Delivery of Gentamicin Using Poly (3-hydroxybutyrate) Microspheres. *Int. J. Mol. Sci.* **2011**, *12*, 4294–4314. [CrossRef] [PubMed]
39. Guzik, M.; Narancic, T.; Ilic-Tomic, T.; Vojnovic, S.; Kenny, S.T.; Casey, W.T.; Duane, G.F.; Casey, E.; Woods, T.; Babu, R.P.; et al. Identification and characterization of an acyl-CoA dehydrogenase from Pseudomonas putida KT2440 that shows preference towards medium to long chain length fatty acids. *Microbiology* **2014**, *160*, 1760–1771. [CrossRef] [PubMed]
40. Small, R.E. Diclofenac sodium. *Clin. Pharm.* **1989**, *8*, 545–558.
41. Skibiński, S.; Cichoń, E.; Haraźna, K.; Marcello, E.; Roy, I.; Witko, M.; Ślósarczyk, A.; Czechowska, J.; Guzik, M.; Zima, A. Functionalized tricalcium phosphate and poly(3-hydroxyoctanoate) derived composite scaffolds as platforms for the controlled release of diclofenac. *Ceram. Int.* **2020**. [CrossRef]
42. Haraźna, K.; Walas, K.; Urbańska, P.; Witko, T.; Snoch, W.; Siemek, A.; Jachimska, B.; Krzan, M.; Napruszewska, B.D.; Witko, M.; et al. Polyhydroxyalkanoate-derived hydrogen-bond donors for the synthesis of new deep eutectic solvents. *Green Chem.* **2019**, *21*, 3116–3126. [CrossRef]
43. Klapetek, P.; David, N.; Anderson, C. Gwyddion User Guide. Available online: http://gwyddion.net/download/user-guide/gwyddion-user-guide-en.pdf (accessed on 27 November 2020).
44. David, N.; Klapetek, P. Gwyddion: An open-source software for SPM data analysis. *Cent. Eur. J. Phys.* **2012**, *10*, 181–188. [CrossRef]
45. Witko, T.; Solarz, D.; Feliksiak, K.; Haraźna, K.; Rajfur, Z.; Guzik, M. Insights into in vitro wound closure on two biopolyesters—Polylactide and polyhydroxyoctanoate. *Materials* **2020**, *17*, 2793. [CrossRef]
46. Rai, R.; Yunos, D.M.; Boccaccini, A.R.; Knowles, J.C.; Barker, I.A.; Howdle, S.M.; Tredwell, G.D.; Keshavarz, O.T.; Roy, I. Poly-3-hydroxyoctanoate P(3HO), a Medium Chain Length Polyhydroxyalkanoate Homopolymer from *Pseudomonas mendocina*. *Biomacromolecules* **2011**, *12*, 2126–2136. [CrossRef] [PubMed]
47. Ramachandran, E.; Ramukutty, S. Growth, morphology, spectral and thermal studies of gel grown diclofenac acid crystals Growth, morphology, spectral and thermal studies of gel grown diclofenac acid crystals. *J. Cryst. Growth* **2017**, *389*, 78–82. [CrossRef]
48. Gumel, A.M.; Annuar, S.M.; Heidelberg, T. Single-step lipase-catalyzed functionalization of medium-chain-length polyhydroxyalkanoates. *J. Chem. Technol. Biotechnol.* **2013**, *88*, 1328–1335. [CrossRef]
49. Liu, Y.; Ji, P.; Lv, H.; Qin, Y.; Deng, L. Gentamicin modified chitosan film with improved antibacterial property and cell biocompatibility. *Int. J. Biol. Macromol.* **2017**, *98*, 550–556. [CrossRef] [PubMed]
50. Kant, S.; Wadhwa, P.; Won, J.; Gi, Y.; Jeon, J. Lipase mediated functionalization of poly (3-hydroxybutyrate-co-3-hydroxyvalerate) with ascorbic acid into an antioxidant active biomaterial. *Int. J. Biol. Macromol.* **2019**, *123*, 117–123. [CrossRef]
51. Fox, T.G. Influence of diluent and of copolymer composition on the glass temperature of a polymer system. *Bull. Am. Phys. Soc.* **1956**, *1*, 123–132.

52. Chen, L.; Yan, C.; Zheng, Z. Functional polymer surfaces for controlling cell behaviors. *Mater. Today* **2018**, *21*, 38–59. [CrossRef]

53. Yul, J.; Dreiss, A.D.; Zhou, Z.; Hansen, J.C.; Siedlecki, C.A.; Hengstebeck, R.W.; Cheng, J.; Winograd, N.; Donahue, H.J. The regulation of integrin-mediated osteoblast focal adhesion and focal adhesion kinase expression by nanoscale topography. *Biomaterials* **2007**, *28*, 1787–1797. [CrossRef]

54. Ailincai, D.; Gavril, G.; Marin, L. Polyvinyl alcohol boric acid—A promising tool for the development of sustained release drug delivery systems. *Mater. Sci. Eng. C* **2020**, *107*, 110316. [CrossRef]

55. Sarkar, N.; Aakeröy, C.B. Evaluating hydrogen-bond propensity, hydrogen-bond coordination and hydrogen-bond energy as tools for predicting the outcome of attempted co-crystallisations. *Supramol. Chem.* **2020**, *32*, 81–90. [CrossRef]

56. Mondschein, R.J.; Kanitkar, A.; Williams, C.B.; Verbridge, S.S.; Long, T.E. Polymer structure-property requirements for stereolithographic 3D printing of soft tissue engineering scaffolds. *Biomaterials* **2017**, *140*, 170–188. [CrossRef] [PubMed]

57. Díaz, E.; Puerto, I.; Ribeiro, S.; Lanceros-Mendez, S.; Barandiarán, J.M. The influence of copolymer composition on PLGA/nHA scaffolds' cytotoxicity and in vitro degradation. *Nanomaterials* **2017**, *7*, 173. [CrossRef] [PubMed]

58. Goreva, A.V.; Shishatskaya, E.I.; Volova, T.G.; Sinskey, A.J. Characterization of Polymeric Microparticles Based on Resorbable Polyesters of Oxyalkanoic Acids as a Platform for Deposition and Delivery of Drugs. *Med. Polym.* **2012**, *54*, 224–236. [CrossRef]

59. Rai, R.; Boccaccini, A.R.; Knowles, J.C.; Mordon, N.; Salih, V.; Locke, I.C.; Keshavarz, T.; Roy, I. The Homopolymer Poly (3-hydroxyoctanoate) as a Matrix Material for Soft Tissue Engineering. *J. Appl. Polym. Sci.* **2011**, *122*, 3606–3617. [CrossRef]

60. Fini, A.; Garuti, M.; Fazio, G.; Holgado, M.A. Diclofenac Salts. I. Fractal and Thermal Analysis of Sodium and Potassium Diclofenac Salts. *J. Pharm. Sci.* **2001**, *90*, 2049–2057. [CrossRef] [PubMed]

61. Xu, Y.; Kim, C.; Saylor, D.M.; Koo, D. Polymer degradation and drug delivery in PLGA-based drug—Polymer applications: A review of experiments and theories. *J. Biomed. Mater. Res. Part B Appl. Biomater.* **2016**, *105*, 1–25. [CrossRef] [PubMed]

62. Palacio, J.; Orozco, V.H.; López, B.L. Effect of the Molecular Weight on the Physicochemical Properties of Poly(lactic acid) Nanoparticles and on the Amount of Ovalbumin Adsorption. *J. Braz. Chem. Soc.* **2011**, *22*, 2304–2311. [CrossRef]

63. Yanez, M.; Blanchette, J.; Jabbarzadeh, E. Modulation of Inflammatory Response to Implantes Biomaterials Using Natural Compounds. *Curr. Pharm. Des.* **2017**, *23*, 6347–6357. [CrossRef]

64. Braunecker, J.; Baba, M.; Milroy, G.E.; Cameron, R.E. The effects of molecular weight and porosity on the degradation and drug release from polyglycolide. *Int. J. Pharm.* **2004**, *282*, 19–34. [CrossRef]

65. Naveen, S.V.; Tan, I.K.P.; Goh, Y.S.; Balaji Raghavendran, H.R.; Murali, M.R.; Kamarul, T. Unmodified medium chain length polyhydroxyalkanoate (uMCL-PHA) as a thin film for tissue engineering application—Characterization and in vitro biocompatibility. *Mater. Lett.* **2015**, *141*, 55–58. [CrossRef]

66. Xu, J.L.; Lesniak, A.; Gowen, A.A. Predictive Modeling of the in Vitro Responses of Preosteoblastic MC3T3-E1 Cells on Polymeric Surfaces Using Fourier Transform Infrared Spectroscopy. *ACS Appl. Mater. Interfaces* **2020**, *12*, 24466–24478. [CrossRef] [PubMed]

67. Kubo, K.; Tsukimura, N.; Iwasa, F.; Ueno, T.; Saruwatari, L.; Aita, H.; Chiou, W.A.; Ogawa, T. Cellular behavior on TiO_2 nanonodular structures in a micro-to-nanoscale hierarchy model. *Biomaterials* **2009**, *30*, 5319–5329. [CrossRef] [PubMed]

68. Sadowska, J.M.; Wei, F.; Guo, J.; Guillem-Marti, J.; Lin, Z.; Ginebra, M.P.; Xiao, Y. The effect of biomimetic calcium deficient hydroxyapatite and sintered β-tricalcium phosphate on osteoimmune reaction and osteogenesis. *Acta Biomater.* **2019**, *96*, 605–618. [CrossRef] [PubMed]

69. De Luna-Bertos, E.; Ramos-Torrecillas, J.; Garcia-Martinez, O.; Guildford, A.; Santin, M.; Ruiz, C. Therapeutic doses of nonsteroidal anti-inflammatory drugs inhibit osteosarcoma MG-63 osteoblast-like cells maturation, viability, and biomineralization potential. *Sci. World J.* **2013**, *2013*, 809891. [CrossRef]

70. Costela-Ruiz, V.J.; Melguizo-Rodríguez, L.; Illescas-Montes, R.; Manzano-moreno, F.J.; Ruiz, C.; De Luna-Bertos, E. Effects of Therapeutic Doses of Celecoxib on Several Physiological Parameters of Cultured Human Osteoblasts. *Int. J. Med. Sci.* **2019**, *16*, 1466–1472. [CrossRef]

71. Krischak, G.D.; Augat, P.; Blakytny, R.; Claes, L.; Kinzl, L.; Beck, A. The non-steroidal anti-inflammatory drug diclofenac reduces appearance of osteoblasts in bone defect healing in rats. *Arch. Orthop. Trauma Surg.* **2007**, *127*, 453–458. [CrossRef]

72. Ahmed, S.; Chauhan, V.M.; Ghaemmaghami, A.M.; Aylott, J.W. New generation of bioreactors that advance extracellular matrix modelling and tissue engineering. *Biotechnol. Lett.* **2019**, *41*, 1–25. [CrossRef]

2

Cell-free Stem Cell-Derived Extract Formulation for Regenerative Medicine Applications

Ashim Gupta [1,2,3,4,5], Craig Cady [1,6], Anne-Marie Fauser [6], Hugo C. Rodriguez [2,4,7,8], R. Justin Mistovich [1,9], Anish G. R. Potty [1,4,7,10] and Nicola Maffulli [11,12,13,14,*]

[1]   General Therapeutics, Cleveland Heights, OH 44118, USA; ashim6786@gmail.com (A.G.); ccady@fsmail.bradley.edu (C.C.); justin@mistovich.net (R.J.M.); anishpotty@gmail.com (A.G.R.P.)

[2]   Future Biologics, Lawrenceville, GA 30043, USA; hcrodrig2112@gmail.com

[3]   BioIntegrate, Lawrenceville, GA 30043, USA

[4]   South Texas Orthopaedic Research Institute, Laredo, TX 78045, USA

[5]   Veterans in Pain, Valencia, CA 91354, USA

[6]   Bohlander Stem Cell Research Laboratory, Department of Biology, Bradley University, Peoria, IL 61625, USA; afauser@mail.bradley.edu

[7]   School of Osteopathic Medicine, University of the Incarnate Word, San Antonio, TX 78235, USA

[8]   Future Physicians of South Texas, San Antonio, TX 78235, USA

[9]   Department of Orthopaedics, School of Medicine, Case Western Reserve University, Cleveland, OH 44106, USA

[10]   Laredo Sports Medicine Clinic, Laredo, TX 78041, USA

[11]   Department of Musculoskeletal Disorders, School of Medicine and Surgery, University of Salerno, 84084 Fisciano, Italy

[12]   San Giovanni di Dio e Ruggi D'Aragona Hospital "Clinica Orthopedica" Department, Hospital of Salerno, 84124 Salerno, Italy

[13]   Barts and the London School of Medicine and Dentistry, Centre for Sports and Exercise Medicine, Queen Mary University of London, London E1 4DG, UK

[14]   School of Pharmacy and Bioengineering, Keele University School of Medicine, Stoke on Trent ST5 5BG, UK

*   Correspondence: n.maffulli@qmul.ac.uk

Abstract: Stem cells for regenerative medicine purposes offer therapeutic benefits, but disadvantages are still ill defined. The benefit of stem cells may be attributed to their secretion of growth factors (GFs), cytokines (CKs), and extracellular vesicles (EVs), including exosomes. We present a novel cell-free stem cell-derived extract (CCM), formulated from human progenitor endothelial stem cells (hPESCs), characterized for biologically active factors using ELISA, nanoparticle tracking analysis and single particle interferometric reflectance imaging sensing. The effect on fibroblast proliferation and ability to induce stem cell migration was analyzed using Alamar Blue proliferation and Transwell migration assays, respectively. GFs including IGFBP 1, 2, 3, and 6, insulin, growth hormone, PDGF-AA, TGF-α, TGF-β1, VEGF, and the anti-inflammatory cytokine, IL-1RA were detected. Membrane enclosed particles within exosome size range and expressing exosome tetraspanins CD81 and CD9 were identified. CCM significantly increased cell proliferation and induced stem cell migration. Analysis of CCM revealed presence of GFs, CKs, and EVs, including exosomes. The presence of multiple factors including exosomes within one formulation, the ability to promote cell proliferation and induce stem cell migration may reduce inflammation and pain, and augment tissue repair.

Keywords: regenerative medicine; musculoskeletal injuries; osteoarthritis; stem cells; progenitor cells; growth factors; cytokines; extracellular vesicles; exosomes

1. Introduction

During the past few decades, there has been a tremendous growth in the use of biologics for regenerative medicine applications [1,2]. Biologics currently available for clinical use include platelet rich plasma, bone marrow aspirate, lipoaspirate, amniotic allograft suspension, umbilical cord-derived Wharton's Jelly, cord blood, and exosomes [3,4]. The efficacy of these biologics is attributed to the presence of stem cells, growth factors (GFs), cytokines (CKs), and extracellular vesicles (EVs), including exosomes [5,6].

The use of stem cells, including mesenchymal stem cells (MSCs), for clinical application in regenerative medicine, has gained substantial interest. MSCs can be obtained from several sources, including bone marrow, adipose tissue, trabecular bone, and deciduous teeth [7–10]. MSCs likely exert their therapeutic effect by migration to the sites of injury, engrafting, and interacting with other cells after administration [11]. Despite their therapeutic benefits, MSCs present several disadvantages, including establishing a reliable source with stable phenotype, genetic instability and chromosomal aberrations, intravenous administration-related toxicity caused by physical trapping of the cells in the lung microvasculature, rejection by the host, formation of ectopic tissue, and tumorigenicity [11–13].

The beneficial effects of MSCs may well not result from their ability to differentiate, but from their secretion of bioactive molecules such as GFs, CKs, and exosomes [14–16]. GFs are a heterogenous group of peptides/proteins and lipid soluble factors secreted by various cells including MSCs. GF receptor activation induces signal transduction pathways which initiate cell migration, proliferation, growth, and differentiation [17]. CKs are low molecular weight proteins responsible for regulation of inflammation, immune response, cellular differentiation, and tissue remodeling [18]. GFs and CKs frequently have overlapping and synergistic actions with immense potential in regenerative medicine [19]. These factors can act in an autocrine or paracrine manner: a single cytokine can promote the synthesis and release of additional CKs, leading to a cascade of molecules, influencing cell division, differentiation, and regeneration of various tissues and organs [6].

Exosomes are secreted by MSCs and act as paracrine mediators between MSCs and target cells, providing a regenerative microenvironment for damaged tissues [16,20,21]. Exosomes are small EVs with a diameter from 30–150 nm. They are formed from a sequential process of multivesicular body membrane remodeling. Exosomes are present in body fluids, including blood plasma, amniotic fluid, and umbilical cord-derived Wharton's jelly [22]. MSCs-derived exosomes can recapitulate the MSC's biological activity and can act as a cell-free therapeutic alternative to whole cell therapy with great regenerative potential [23–25]. The use of exosomes can offer advantages over whole cell therapy given their higher safety profile, lower immunogenicity, and inability to directly form tumors [26]. In addition, given their smaller size, exosomes can potentially migrate to target organs efficiently after injection, without getting trapped in the lung microvasculature [26,27].

Considering the benefits of cell-free GFs, CKs, and exosomes, and the disadvantages of whole cell therapy, innovative and effective cell-free products should be considered for clinical applications. We have formulated a novel cell-free, stem cell-derived extract (CCM) and characterized it for the presence of GFs, CKs, and EVs, including exosomes. We hypothesized that numerous GFs, CKs, and EVs, including exosomes, would be present in this formulation. This preliminary study describes the preparation and characteristics of this novel cell-free stem cell-derived extract.

2. Results

2.1. Enzyme-Linked Immunosorbent Assay (ELISA)

GFs, including growth hormone (GH), insulin-like growth factor binding protein (IGFBP) 1, 2, 3, and 6, insulin, platelet derived growth factor-AA (PDGF-AA), transforming growth factor alpha and beta-1 (TGFα and TGFβ1), and vascular endothelial growth factor (VEGF) were detected in the formulated CCM (Table 1). In addition, IL-1RA, an anti-inflammatory cytokine was identified at significant levels in CCM relative to the control which had undetectable IL-1RA levels (Figure 1).

Table 1. Growth factors expressed in the formulated stem cell extract.

	Growth Factors	Average Amount (pg/mL)
IGFBP-1	Insulin-like growth factor-binding protein-1	70
IGFBP-2	Insulin-like growth factor-binding protein-2	18.1
IGFBP-3	Insulin-like growth factor-binding protein-3	552.6
IGFBP-6	Insulin-like growth factor-binding protein-6	20.3
Insulin	Insulin	518.5
GH	Growth hormone	10.5
FGF-7	Fibroblast growth factor-7	45.5
HB-EGF	Heparin-binding EGF-like growth factor	42.2
PDGF-AA	Platelet derived growth factor-AA	58.1
TGF-α	Transforming growth factor alpha	2.4
TGFβ1	Transforming growth factor beta 1	1036.90
EG-VEGF	Endocrine-gland-derived vascular endothelial growth factor	86.6
VEGF	Vascular endothelial growth factor	30.8
VEGF R2	Vascular endothelial growth factor-receptor 2	20.6
VEGF R3	Vascular endothelial growth factor-receptor 3	300
VEGF-D	Vascular endothelial growth factor D	16.6
BDNF	Brain-derived neurotrophic factor	68.6
b-NGF	Beta-nerve Growth factor	9.5
NGF-R	Nerve growth factor receptor	21
NT-4	Neurotrophin-4	110.3

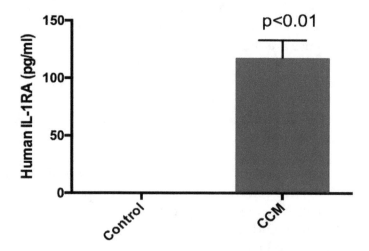

Figure 1. ELISA assay to determine the presence of human IL-IRA in control (media only) and cell-free stem cell-derived extract (CCM). Significantly high levels ($p < 0.01$) of IL-1RA identified in CCM compared to no detectable IL-1RA in control. Data represent mean + SEM. CCM was produced in two lots in duplicate from two vials and ELISA for IL-1RA was performed using CCM from each lot three separate times in triplicate.

2.2. Exosome Analysis

The nanoparticle tracking analysis demonstrated the presence of the average amount of 177 billion/mL of particles in the extracellular vesicle size range in the light scattering mode. Staining with the plasma membrane dye, CellMask Orange™, demonstrated the presence of the average amount of 3.2 billion/mL of particles in the fluorescent mode, indicative of true membrane-enclosed particles i.e., EVs. A representative image for concentration of particles/mL determined via nanoparticle tracking analysis in the light scattering mode and fluorescent mode is shown in Figure 2A.

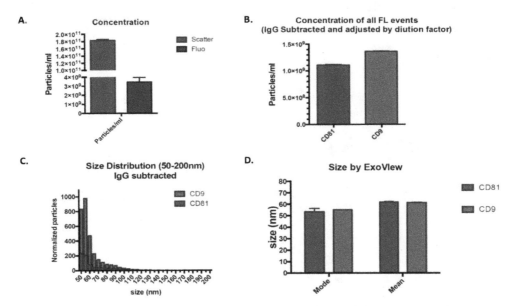

Figure 2. A representative nanoparticle tracking analysis showed the presence of 183 ± 3.45 billion particles/mL in the light scattering mode and 3.45 ± 0.51 billion particles in the fluorescent mode. Values are shown as mean ± standard error (**A**), and image for concentration of all fluorescent events showed presence of 1.1 billion/mL and 1.35 billion/mL CD81 and CD9 expressing particles, respectively (**B**). Representative histograms of the size distribution (**C**) and mean/mode values (**D**) of CD81- and CD9-expressing EVs measured by interferometry.

SP-IRIS (single particle interferometric reflectance imaging sensing) analysis with ExoView allowed high-resolution immunocapture of tetraspanin-positive (i.e., CD81/CD63/CD9) EVs and detection of the same EV markers by fluorescent counterstain, allowing for simultaneous size measurement, concentration, and phenotyping of EV subpopulations. Fluorescent marker detection by ExoView demonstrated an average amount of 1 billion/mL and 1.27 billion/mL CD81 and CD9 expressing particles, respectively. However, no CD63 positive EVs were detected. Representative image for concentration of all fluorescent events for CD81 and CD9 is shown in Figure 2B. EVs expressing CD81 and CD9 also exhibited typical size distribution with a mode of 53.34 and 55 nm as detected by interferometric measurement, which quantifies chip-bound particles in the 50–200 nm range. A representative image for size distribution and mean and mode size are shown in Figure 2C,D, respectively. Thus, expression of tetraspanins CD81 and CD9, and size of EVs <150 nm, are indicative of the presence of true exosomes.

2.3. Cell Proliferation

Qualitatively, cells treated with 20% CCM showed higher density over control (media only). Representative phase contrast images of human fibroblasts treated with media only and 20% CCM is shown in Figure 3A. Similarly, the Alamar Blue proliferation assay demonstrated that cells treated with 10% or 20% CCM exhibited significantly higher ($p < 0.0001$) rate of cell proliferation compared to control after 5 days (Figure 3B).

2.4. Cell Migration

Migration of BMSCs in response to CCM (the migration effector) was quantified as a percentage of the positive serum control using Transwell inserts. The control (media only) group induced the least number of BMSCs to migrate compared to all other groups tested. Qualitatively, images show a typical migration assay result after 24 h on the bottom of the insert with fewer migrated cells on the media only insert compared to the CCM (Figure 4A). Quantitatively, all treatment groups tested significantly increased BMSCs migration relative to the negative control, media only (Figure 4B). However, the heat inactivated CCM group had significantly lower capacity to induce BMSC migration compared to the 10% CCM and 20% CCM groups (Figure 4B).

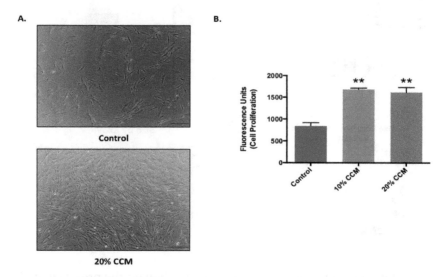

Figure 3. (**A**) Representative phase contrast images (size bar-100 μm) of human fibroblasts (CLL-171) treated with media only (control) and 20% CCM. (**B**) Alamar Blue assay for proliferation of CLL-171 treated with media only (control) and 10% or 20% CCM after 5 days. Data represents mean ± SEM, and ** represents significant difference ($p < 0.0001$) between CCM treated groups and control group, 5 days post-treatment with stem cell extract. CCM was produced in two lots in duplicate from two vials and cell proliferation assay was performed using CCM from each lot three separate times in triplicate.

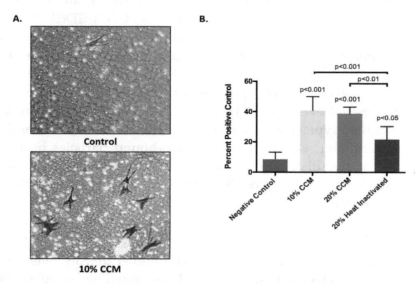

Figure 4. (**A**) Representative, typical images (20X) of migrated Bone marrow mesenchymal stem cells (BMSCs) on the bottom of the Transwell insert after 24 h suspension in negative control (media only) and 10% CCM. (**B**) Cell migration of BMSCs suspended in media only (negative control), 10% and 20% CCM, and heat inactivated (neutralized) 20% CCM for 24 h. Data represents mean ± SEM with significant difference between negative control versus CCM groups ($p < 0.001$) and heat inactivated CCM ($p < 0.05$); and significant reduction in inducing BMSCs migration of the heat inactivated CCM relative to the 10% CCM ($p < 0.01$) and 20% CCM ($p < 0.001$). CCM was produced in two lots in duplicate from two vials and cell migration assay was performed using CCM from each lot three separate times in triplicate.

3. Discussion

Over the last decade, the therapeutic use of biologics for regenerative medicine applications has led to a substantial increase in their marketing, patient demand, and clinical use [28]. While the use of biologics is promising, these treatments are still in their early stages of development [28]. Despite their widespread commercial use, there is still inadequate characterization of such biologically active formulations. In the present study, we describe the process of formulation of a novel cell-free stem cell-derived extract (CCM), and evaluated it for the presence of GFs, CKs, and EVs, including exosomes. The vital elements of regenerative medicine, namely GFs, CKs, and exosomes, are all present in this formulation. This characterization provides an initial step toward future in vivo preclinical and clinical studies to determine the safety and efficacy of CCM for regenerative medicine applications.

Numerous growth factors were identified in our CCM formulation including IGFBP 1, 2, 3, and 6, which acts as a carrier protein for IGF-1 (insulin-like growth factor-1). IGF-1 improves osteogenic differentiation of MSCs and stimulate production of extracellular matrix (ECM) [28]. Insulin was detected: this is an anabolic agent in bone, preserves and increases bone density and strength via direct and/or indirect effects on bone formation [29]. GH, which stimulates cell growth through an IGF-1 pathway and plays an essential role in cartilage regeneration, was identified [28]. PDGF-AA, which exhibits chemotactic effects towards human osteoblasts, was detected. The presence of PDGF-AA is important, as insufficient levels have been associated with cartilage degeneration [28]. TGF-α, a transforming growth factor ligand for epidermal growth factor receptor (EGFR), was also identified. EGFR promotes survival and proliferation of osteoprogenitors and plays an anabolic role in bone metabolism [28,30].

High levels of TGF-β1 were identified: TGF-β1 is a well characterized growth factor involved in the recruitment of stem/progenitor cell participation in tissue regeneration and remodeling. TGF-β stimulates cartilage repair through the stimulation of ECM production via collagen type-II and proteoglycans in chondrocytes and downregulates matrix-degrading enzymes [31]. Pre-clinical studies with TGF-β injections in the knee joint resulted in increased levels of proteoglycan of the articular cartilage. Additionally, the modulation of TGF-β signaling has been effective in several musculoskeletal pathologies including OA [32], and reported to counteract IL-1, known to induce cartilage degradation and therefore useful in cartilage repair [33].

We identified significant amounts of VEGF and its receptor VEGFR2, the main signaling receptor involved in the mediation of angiogenesis and vasculogenesis. The activation of VEGFR2 by VEGF promotes blood vessel permeability, influencing tissue repair [34]. VEGF is downregulated in patients with OA, contributing to degeneration, as VEGF is involved in new bone formation and bone tissue remodeling [28].

We also detected IL-1RA, which competitively binds IL-1 (both 1α and 1β) and blocks IL-1 mediated cellular changes. IL-1RA alleviates or prevents cytokine mediated hyperinflammatory hyperalgesia. Intra-articular injection of IL-1RA in patients with knee OA slowed its progression and improved pain and WOMAC (The Western Ontario and McMaster Universities Osteoarthritis Index) global scores [28,35]. The levels of IL-1RA in our cell-free formulation are higher compared to Wharton's Jelly, bone marrow-derived and adipose-derived stromal cell supernatant [28,36].

The analysis of this novel CCM extract indicated the presence of true exosomes. These particles were membrane-enclosed, within the EV size range with a mode size of <150 nm, and express exosome specific tetraspanins CD81 and CD9. Interestingly, we did not observe any expression of CD63, an exosome tetraspanin. However, this may be attributed to cell specificity, as phenotype and function of stem cell-derived exosomes may vary depending on the cell source [27]. Exosomes exhibit anti-inflammatory and pro-regenerative effects to stimulate healing in different tissue types [28]. Exosomes also improve cell viability and proliferation, angiogenesis, and immunomodulation in various physiological systems [28]. Exosome uptake by cells significantly decreases the expression of pro-inflammatory genes and M1 phenotypic markers, increases cell migration and osteogenic marker expression, and exerts an osteo-immunomodulatory role in the regulation of bone dynamics [37].

Exosomes increase the secretion of cellular factors needed to hasten the healing of tendon injuries, acting via paracrine and autocrine processes [38]. In addition, exosomes stimulate repair of cartilage and proliferation of chondrocytes in osteoarthritis and help alleviate pain in knee OA [39].

The stem cell-derived extract described in the present article significantly increases the rate of fibroblasts proliferation, suggesting a potential role for improving the proliferation of cells under stress including inflammation mediated stresses. In addition, CCM significantly induces BMSCs migration in comparison to heat inactivated CCM and media only control groups. Heat inactivation denatures hydrophilic factors, significantly reducing BMSCs migration below that of the non-heated CCM. Heat inactivation did not entirely arrest BMSCs migration, indicating that heat resistant hydrophobic factors also play a role in BMSCs migration. Induction of BMSCs migration suggests a projected benefit in treatment with CCM, facilitating endogenous stem cell migration to the site of application as occurs with tissue damage, inducing chemotaxic migration of stem cells to enhance repair and recovery [40].

These results confirmed our hypotheses that GFs, CKs, and EVs, including exosomes, are present in our formulated CCM and shown to increase cell proliferation, and induce stem cell migration for tissue repair. The presence of a high density of exosomes, with multiple factors known to enhance cellular growth and differentiation with anti-inflammatory function, suggests a possible opportunity for clinical application to reduce inflammation and pain, and perhaps augment healing of orthopedic injuries. Promotion of cellular proliferation and stem cell migration can be attributed to the presence of GFs, CKs, and EVs. In accordance with previously published preclinical and clinical studies, this would indicate that the presence of a combination of these factors may prove of benefit for regenerative medicine applications [28,41]. This novel extract has equal or improved potential efficacy compared with cell-based preparations in the ability to enhance cell proliferation and induce stem cell migration. When considering possible immunogenic reactions and challenges related to cell purification, the factors identified with this novel extract, including exosomes, provide a safer alternative to cell-based treatments.

Our study is not without limitations. The ELISA kit used in our assay was limited to 40 growth factors. Further studies will be required to determine whether other growth factors and cytokines are expressed in this formulation. Further in-depth investigations will be needed to determine the functional content of the exosomes identified to determine the possible molecular implications with this extract. In addition, culturing of cells up to eight passages may also induce some age-dependent modifications. Future studies to determine the biological characteristics of the cell population including cellular aging via aging markers will be performed, as cellular aging likely affects the cellular secretome. It would also be important to assess changes in gene and cell surface protein markers to rule out possible differentiation or lineage changes that may have occurred during the development of our hPESCs cell line. Further studies will also be necessary to explore more specifically the functional mechanisms for these biological effects and the efficacy of this extract in vitro, including comparing the ability of the CCM to induce cell proliferation and migration compared to whole stem cell before exploring its potential clinical advantages via in vivo preclinical and clinical studies.

4. Materials and Methods

4.1. Cells, Reagents, and Supplies

Human progenitor endothelial stem cells (hPESCs) were obtained from Celprogen at passage 2 (catalog # 36053-05; Torrance, CA, USA; characterized by the manufacturer expressing keratin 10, 14, 15, 16, and 19, ESA, CD29, MUCI, Alcian Blue, type IV collagen, E-cadherin, CD 18/19, CD 117/cKit, and VEGFR2/KDR/FLK-1), human fibroblasts (CLL-171) from ATCC (catalog # MRC-5; Manassas, VA, USA) and bone marrow mesenchymal stem cells (BMSCs) were kindly provided by Tulane Center for Gene Therapy, Tulane University (New Orleans, LA, USA; and sourced, expanded, and characterized according to the published studies [42,43]. BMSCs were sourced from a single, normal patient and characterized for positive expression of CD 44, CD 90, CD 105 and negative expression of CD 34 with multipotency for adipogenic and osteogenic differentiation). BMSCs vials were obtained at passage 4 with 10^6 cells/vial. These cells were further expanded for an additional 2–3 passages (P6–P7) and cryopreserved. These cells were thawed, expanded for 1–2 passages (P8–P9), and utilized for cell migration assay. Human progenitor endothelial stem cell complete media for the culture of progenitor stem cells was obtained from Celprogen (Torrance, CA, USA). Complete culture media for culturing BMSCs and fibroblast consisted of alpha MEM (Sigma, St. Louis, MO, USA), fetal bovine serum (Cytiva, Marlborough, MA, USA), Glutamax L-glutamine complex (Sigma, St. Louis, MO, USA), penicillin/streptomycin (Invitrogen, Waltham, MA, USA), and Versene (Lonza, Morristown, NJ, USA) for cell passaging. Alamar Blue Cell Proliferation Reagent was obtained from Life Technologies (Carlsbad, CA, USA), Transwell cell culture plate inserts for migration assay from Corning (Corning, NY, USA), and Transwell stain DIF Quick stain kit from MEB Inc. (San Marcos, CA, USA).

4.2. Cell Culture and Formulation of Novel Cell-Free Stem Cell-Derived Extract

hPESCs were cultured in serum or serum free media at 37 °C at 5% CO_2. The cell expansion, cryopreservation, and final subculture of hPESCs for generating cells for producing CCM was completed in accordance with the manufacturer instructions and recommendations. Cells provided at passage 2 by Celprogen were initially generated from multiple human sources, shipped on dry ice at 10^6 cells per vial. Cell expansion for banking and cryopreservation were passaged using enzyme free reagent for 3 times into multiple T-75 cell culture treated flasks with each passage initiated at 60–70% confluence. After the initial expansion process, cells were lifted in a similar manner described above, centrifuged at 750× g, resuspended into human endothelial progenitor cell freezing media (Celprogen #M36053-05M, Celprogen, Torrance, CA, USA), and cryopreserved under vapor phase liquid nitrogen. The resulting cells were considered a "progenitor/stem cell seed lot" and the source for all experiments and analyses performed. For the production of CCM, seed lot cells were recovered, passaged for an additional 2–3 times, and utilized to produce CCM as described below. The total number of passages including those provided by the manufacturer at passage 2 ranged from 7–8 passages. During the initial expansion and expansion prior to production of the CCM, the extract doubling time did not change from the manufacturer suggested range of 80–120 h. In addition, during cell passaging, we found no observable morphological changes to indicate cell stress, such as increases in intracellular vesicles, increases in the number of nonadherent cells or changes in general cell morphology relative to the initial cells of the seed lot. Although we did not assess the changes in gene or cell surface protein expression, we found no change in doubling time or morphology, indicating consistent estimates of confluence prior to passaging and seeding densities, increases in senescent cell density, or cell–cell contact inhibition during this process in the production of a stable hPESCs cell line. BMSCs and fibroblasts were cultured in FBS, L-glutamine in NaCl in αMEM with L-glutamine without deoxyribonucleosides media and expanded from an initial density of 50 cells/cm^2 and cultured to confluence (>90%) in serum-based media described above. At confluence, cells were cultured under serum free conditions for 24 h at 37 °C in 5% CO_2. After 24 h, the media was collected as conditioned media. Cells were then passaged enzyme free, pelleted at 750 g, resuspended, mechanically disrupted, and pelleted at 2000 g. The supernatant was then combined with the collected conditioned media and sterile filtered under conditions preserving growth factor and exosome functionality. All processing was performed on ice and under aseptic conditions.

4.3. Enzyme-Linked Immunosorbent Assay (ELISA)

Randomly selected samples from two different batches were sent to an independent laboratory, RayBiotech (Norcross, GA, USA), and were analyzed for the presence of GFs using Quantibody® Human Growth Factor Array 1 kit, according to the manufacturer's protocol. In addition, randomly selected samples from two different batches were analyzed for the presence of Interleukin 1 receptor antagonist (IL-1RA) using RayBio® Human IL-1Ra enzyme linked immunosorbent assay (ELISA) kit, according to the manufacturer's protocol.

4.4. Exosome Analysis

Randomly selected samples from two different batches were sent to an independent laboratory, Extracellular Vesicle Core at Children's Hospital Los Angeles (Los Angeles, CA, USA), and were analyzed by nanoparticle tracking analysis for the presence of particles in the extracellular vesicle size range using Malvern Panalytical Nanosight NS 300. These samples were also analyzed after staining with a general fluorescent membrane marker, CellMask Orange™ (Thermo Fisher Scientific, Waltham, MA, USA), as previously described [28]. The size, quantity, and surface protein characteristics of EVs were then further analyzed through single particle interferometric reflectance imaging sensing (SP-IRIS) using the ExoView platform (NanoView Biosciences, Boston, MA, USA), according to the manufacturer's instructions and as described in the literature [44]. Briefly, ExoView uses a multiplexed microarray chip for the immuno-capture of commonly expressed EVs tetraspanin proteins-CD9, CD63, and CD81. ExoView then analyzes EVs using visible light interference for size measurements and fluorescence for protein profiling [44].

4.5. Cell Proliferation Assay

Alamar Blue cell proliferation assay was completed according to the manufacturer's protocol. Briefly, human fibroblasts were plated at a density of 5000 cells/well into a 24 well plate and treated with either control (media only) or 10%, 20% CCM. Phase contrast images were taken using light microscopy (Olympus inverted fluorescent microscope with LCD DP71 Olympus digital camera, Olympus, Center Valley, PA, USA). After 5 days, cell viability assay was performed by replacing the media with the Alamar Blue reagent followed by incubation for 4 h at room temperature under light free conditions. The supernatant was then transferred to a 96 well plate and the fluorescence signal was measured at 540–570 nm excitation/580–610 nm emission using a microplate reader (Synergy LX, BioTek Instruments, Inc., Winooski, VT, USA).

4.6. Cell Migration Assay

Transwell migration assay was completed with passaged BMSCs across insert filters with a pore size at 8 microns against a migration effector consisting of either media only (negative control), 4% serum (positive control), 10% CCM, 20% CCM, or heat inactivated (neutralized) 20% CCM and transferred into the lower well. The insert filter membrane was then placed into the well in contact with the migration effector. Heat inactivation was accomplished by placing 20% CCM into a water bath at 100 °C for 30 min. A total of 5000 BMSCs were plated at the top of the insert and incubated at 37 °C for 24 h. After incubation, the remaining cells on the upper surface of the filter were removed with a cotton swab. Cells which migrated across the filter to the underside of the insert were fixed and stained with a Diff-Quik stain kit and counted in 5 fields at 20x for duplicate inserts. Migration was expressed as a percentage of the 4% serum positive control.

4.7. Statistical Analysis

CCM was produced in two lots in duplicate from two vials obtained from Celprogen. ELISA for IL-1RA was performed using CCM from each lot three separate times in triplicate and mean ± standard error of mean (SEM) values were calculated. Statistical significance was determined by unpaired *t*-test. Cell proliferation assay and migration assay were performed using CCM from each lot and experiments were repeated three separate times in triplicate and mean ± SEM values were calculated. Statistical analysis was determined using analysis of variance (ANOVA) with Tukey's post hoc test. The results were considered significant when $p < 0.05$.

5. Conclusions

Our novel cell-free, stem cell-derived extract formulation showed the presence of GFs, CKs, and EVs, including a high density of exosomes, and the ability to enhance the rate of cell proliferation and induce stem cell migration. The presence of multiple factors within one formulation and their ability to promote cell proliferation and induce migration of stem cells may eventually have a clinical role in reducing inflammation and pain and augment tissue repair and regeneration without the risk of using viable cells.

Author Contributions: Conceptualization, A.G. and C.C.; methodology, A.G., C.C. and A-M.F.; software, C.C.; formal analysis, A.G. and C.C.; data curation, A.G. and C.C.; writing—original draft preparation, A.G., C.C. and H.C.R.; writing—review and editing, A.G., C.C., A.-M.F., H.C.R., R.J.M., A.G.R.P. and N.M.; supervision, A.G. and C.C.; project administration, A.G. All authors have read and agreed to the published version of the manuscript.

Acknowledgments: The authors would like to thank Jarad Wilson (RayBiotech, Norcross, GA, USA) for his assistance with ELISA and Paolo Neviani (Extracellular Vesicle Core at Children's Hospital Los Angeles, CA, USA) for performing Exosome related assays. The authors would also like to thank Darwin Prockop at Tulane Center for Gene Therapy, Tulane University for providing BMSCs.

Abbreviations

ANOVA	Analysis of variance
BMSC	Bone marrow mesenchymal stem cells
CCM	Cell-free stem cell-derived extract
CKs	Cytokines
ECM	Extracellular matrix
ELISA	Enzyme linked immunosorbent assay
EGFR	Epidermal growth factor receptor
EVs	Extracellular vesicles
GFs	Growth factors
GH	Growth hormone
IGF-1	Insulin-like growth factor-1
IGFBP	Insulin-like growth factor binding protein
IL-1RA	Interleukin 1 receptor antagonist
MSCs	Mesenchymal stem cells
OS	Osteoarthritis
PDGF-AA	Platelet derived growth factor-AA
SP-IRIS	Single particle interferometric reflectance imaging sensing
SEM	Standard error of mean
TGFα	Transforming growth factor alpha
TGFβ1	Transforming growth factor beta-1
VEGF	Vascular endothelial growth factor

References

1. Navani, A.; Manchikanti, L.; Albers, S.L.; Latchaw, R.E.; Sanapati, J.; Kaye, A.D.; Atluri, S.; Jordan, S.; Gupta, A.; Cedeno, D.L.; et al. Responsible, Safe, and Effective Use of Biologics in the Management of Low Back Pain: American Society of Interventional Pain Physicians (ASIPP) Guidelines. *Pain Physician* **2019**, 22, S1–S74. [PubMed]
2. Lamplot, J.D.; Rodeo, S.A.; Brophy, R.H. A Practical Guide for the Current Use of Biologic Therapies in Sports Medicine. *Am. J. Sports Med.* **2020**, 48, 488–503. [CrossRef] [PubMed]
3. Le, A.D.K.; Enweze, L.; DeBaun, M.R.; Dragoo, J.L. Platelet-Rich Plasma. *Clin. Sports Med.* **2019**, 38, 17–44. [CrossRef]

4. Duerr, R.A.; Ackermann, J.; Gomoll, A.H. Amniotic-Derived Treatments and Formulations. *Clin. Sports Med.* **2019**, *38*, 45–59. [CrossRef] [PubMed]

5. Patel, J.M.; Saleh, K.S.; Burdick, J.A.; Mauck, R.L. Bioactive factors for cartilage repair and regeneration: Improving delivery, retention, and activity. *Acta Biomater.* **2019**, *93*, 222–238. [CrossRef]

6. Ioannidou, E. Therapeutic modulation of growth factors and cytokines in regenerative medicine. *Curr. Pharm. Des.* **2006**, *12*, 2397–2408. [CrossRef]

7. Caplan, A.I. Mesenchymal stem cells. *J. Orthop. Res.* **1991**, *9*, 641–650. [CrossRef]

8. Andia, I.; Maffulli, N. New biotechnologies for musculoskeletal injuries. *Surgeon* **2019**, *17*, 244–255. [CrossRef]

9. Minguell, J.J.; Erices, A.; Conget, P. Mesenchymal stem cells. *Exp. Biol. Med.* **2001**, *226*, 507–520. [CrossRef]

10. Potty, A.G.R.; Gupta, A.; Rodriguez, H.C.; Stone, I.W.; Maffulli, N. Intraosseous Bioplasty for a Subchondral Cyst in the Lateral Condyle of Femur. *J. Clin. Med.* **2020**, *9*, 1358. [CrossRef]

11. Wang, S.; Guo, L.; Ge, J.; Yu, L.; Cai, T.; Tian, R.; Jiang, Y.; Zhao, R.C.H.; Wu, Y. Excess Integrins Cause Lung Entrapment of Mesenchymal Stem Cells. *Stem. Cells* **2015**, *33*, 3315–3326. [CrossRef] [PubMed]

12. Jeong, J.O.; Han, J.W.; Kim, J.M.; Cho, H.J.; Park, C.; Lee, N.; Kim, D.W.; Yoon, Y.S. Malignant tumor formation after transplantation of short-term cultured bone marrow mesenchymal stem cells in experimental myocardial infarction and diabetic neuropathy. *Circ. Res.* **2011**, *108*, 1340–1347. [CrossRef] [PubMed]

13. Barkholt, L.; Flory, E.; Jekerle, V.; Lucas-Samuel, S.; Ahnert, P.; Bisset, L.; Buscher, D.; Fibbe, W.; Foussat, A.; Kwa, M.; et al. Risk of tumorigenicity in mesenchymal stromal cell-based therapies–bridging scientific observations and regulatory viewpoints. *Cytotherapy* **2013**, *15*, 753–759. [CrossRef] [PubMed]

14. Yao, Y.; Huang, J.; Geng, Y.; Qian, H.; Wang, F.; Liu, X.; Shang, M.; Nie, S.; Liu, N.; Du, X.; et al. Paracrine action of mesenchymal stem cells revealed by single cell gene profiling in infarcted murine hearts. *PLoS ONE* **2015**, *10*, e0129164. [CrossRef]

15. Deng, K.; Lin, D.L.; Hanzlicek, B.; Balog, B.; Penn, M.S.; Kiedrowski, M.J.; Hu, Z.; Ye, Z.; Zhu, H.; Damaser, M.S. Mesenchymal stem cells and their secretome partially restore nerve and urethral function in a dual muscle and nerve injury stress urinary incontinence model. *Am. J. Physiol. Ren. Physiol.* **2015**, *308*, F92–F100. [CrossRef]

16. Chang, Y.H.; Wu, K.C.; Harn, H.J.; Lin, S.Z.; Ding, D.C. Exosomes and Stem Cells in Degenerative Disease Diagnosis and Therapy. *Cell Transpl.* **2018**, *27*, 349–363. [CrossRef]

17. Mitchell, A.C.; Briquez, P.S.; Hubbell, J.A.; Cochran, J.R. Engineering growth factors for regenerative medicine applications. *Acta Biomater.* **2016**, *30*, 1–12. [CrossRef]

18. Onishi, M.; Nosaka, T.; Kitamura, T. Cytokine receptors: Structures and signal transduction. *Int. Rev. Immunol.* **1998**, *16*, 617–634. [CrossRef]

19. Barrientos, S.; Brem, H.; Stojadinovic, O.; Tomic-Canic, M. Clinical application of growth factors and cytokines in wound healing. *Wound. Repair. Regen.* **2014**, *22*, 569–578. [CrossRef]

20. Heldring, N.; Mäger, I.; Wood, M.J.; Le Blanc, K.; Andaloussi, S.E. Therapeutic Potential of Multipotent Mesenchymal Stromal Cells and Their Extracellular Vesicles. *Hum. Gene. Ther.* **2015**, *26*, 506–517. [CrossRef]

21. Gupta, A.; Kashte, S.; Gupta, M.; Rodriguez, H.C.; Gautam, S.S.; Kadam, S. Mesenchymal stem cells and exosome therapy for COVID-19: Current status and future perspective. *Hum. Cell* **2020**, *11*, 1–12. [CrossRef]

22. Raposo, G.; Stoorvogel, W. Extracellular vesicles: Exosomes, microvesicles, and friends. *J. Cell Biol.* **2013**, *200*, 373–383. [CrossRef] [PubMed]

23. Matei, A.C.; Antounians, L.; Zani, A. Extracellular Vesicles as a Potential Therapy for Neonatal Conditions: State of the Art and Challenges in Clinical Translation. *Pharmaceutics* **2019**, *11*, 404. [CrossRef] [PubMed]

24. Bagno, L.; Hatzistergos, K.E.; Balkan, W.; Hare, J.M. Mesenchymal Stem Cell-Based Therapy for Cardiovascular Disease: Progress and Challenges. *Mol. Ther.* **2018**, *26*, 1610–1623. [CrossRef]

25. Lou, G.; Chen, Z.; Zheng, M.; Liu, Y. Mesenchymal stem cell-derived exosomes as a new therapeutic strategy for liver diseases. *Exp. Mol. Med.* **2017**, *49*, e346. [CrossRef]

26. Liew, L.C.; Katsuda, T.; Gailhouste, L.; Nakagama, H.; Ochiya, T. Mesenchymal stem cell-derived extracellular vesicles: A glimmer of hope in treating Alzheimer's disease. *Int. Immunol.* **2017**, *29*, 11–19. [CrossRef]

27. Börger, V.; Bremer, M.; Ferrer-Tur, R.; Gockeln, L.; Stambouli, O.; Becic, A.; Giebel, B. Mesenchymal Stem/Stromal Cell-Derived Extracellular Vesicles and Their Potential as Novel Immunomodulatory Therapeutic Agents. *Int. J. Mol. Sci.* **2017**, *18*, 1450. [CrossRef]

28. Gupta, A.; El-Amin, S.F.; Levy, H.J.; Sze-Tu, R.; Ibim, S.E.; Maffulli, N. Umbilical cord-derived Wharton's jelly for regenerative medicine applications. *J. Orthop. Surg. Res.* **2020**, *15*, 49. [CrossRef]

29. Thrailkill, K.M.; Lumpkin, C.K.; Bunn, R.C.; Kemp, S.F.; Fowlkes, J.L. Is insulin an anabolic agent in bone? Dissecting the diabetic bone for clues. *Am. J. Physiol. Endocrinol. Metab.* **2005**, *289*, E735–E745. [CrossRef]

30. Zhang, X.; Tamasi, J.; Lu, X.; Zhu, J.; Chen, H.; Tian, X.; Lee, T.C.; Threadgill, D.W.; Kream, B.E.; Kang, Y.; et al. Epidermal growth factor receptor plays an anabolic role in bone metabolism in vivo. *J. Bone Miner. Res.* **2011**, *26*, 1022–1034. [CrossRef]

31. Edwards, D.R.; Murphy, G.; Reynolds, J.J.; Whitham, S.E.; Docherty, A.J.; Angel, P.; Heath, J.K. Transforming growth factor beta modulates the expression of collagenase and metalloproteinase inhibitor. *EMBO J.* **1987**, *6*, 1899–1904. [CrossRef] [PubMed]

32. van Beuningen, H.M.; van der Kraan, P.M.; Arntz, O.J.; van den Berg, W.B. Transforming growth factor-beta 1 stimulates articular chondrocyte proteoglycan synthesis and induces osteophyte formation in the murine knee joint. *Lab. Investig.* **1994**, *71*, 279–290. [PubMed]

33. Blaney Davidson, E.N.; Vitters, E.L.; van den Berg, W.B.; van der Kraan, P.M. TGF beta-induced cartilage repair is maintained but fibrosis is blocked in the presence of Smad7. *Arthr. Res. Ther.* **2006**, *8*, R65. [CrossRef] [PubMed]

34. Hu, K.; Olsen, B.R. The roles of vascular endothelial growth factor in bone repair and regeneration. *Bone* **2016**, *91*, 30–38. [CrossRef]

35. Frizziero, A.; Giannotti, E.; Oliva, F.; Masiero, S.; Maffulli, N. Autologous conditioned serum for the treatment of osteoarthritis and other possible applications in musculoskeletal disorders. *Br. Med. Bull.* **2013**, *105*, 169–184. [CrossRef]

36. Amable, P.R.; Teixeira, M.V.; Carias, R.B.; Granjeiro, J.M.; Borojevic, R. Protein synthesis and secretion in human mesenchymal cells derived from bone marrow, adipose tissue and Wharton's jelly. *Stem Cell Res. Ther.* **2014**, *5*, 53. [CrossRef]

37. Wei, F.; Li, Z.; Crawford, R.; Xiao, Y.; Zhou, Y. Immunoregulatory role of exosomes derived from differentiating mesenchymal stromal cells on inflammation and osteogenesis. *J. Tissue Eng. Regen. Med.* **2019**, *13*, 1978–1991. [CrossRef]

38. Connor, D.E.; Paulus, J.A.; Dabestani, P.J.; Thankam, F.K.; Dilisio, M.F.; Gross, R.M.; Agrawal, D.K. Therapeutic potential of exosomes in rotator cuff tendon healing. *J. Bone Miner. Metab.* **2019**, *37*, 759–767. [CrossRef]

39. Liu, Y.; Zou, R.; Wang, Z.; Wen, C.; Zhang, F.; Lin, F. Exosomal KLF3-AS1 from hMSCs promoted cartilage repair and chondrocyte proliferation in osteoarthritis. *Biochem. J.* **2018**, *475*, 3629–3638. [CrossRef]

40. Fu, X.; Liu, G.; Halim, A.; Ju, Y.; Luo, Q.; Song, A.G. Mesenchymal Stem Cell Migration and Tissue Repair. *Cells* **2019**, *8*, 784. [CrossRef]

41. Frizziero, A.; Vittadini, F.; Oliva, F.; Abatangelo, G.; Bacciu, S.; Berardi, A.; Maffulli, N.; Vertuccio, A.; Pintus, E.; Romiti, D.; et al. I.S. Mu. L.T. Hyaluronic acid injections in musculoskeletal disorders guidelines. *MLTJ* **2018**, *8*, 364–398. [CrossRef]

42. Sekiya, I.; Larson, B.; Smith, J.; Pochampally, R.; Cui, J.; Prockop, D. Expansion of human adult stem cells from bone marrow stroma: Conditions that maximize the yields of early progenitors and evaluate their quality. *Stem Cells* **2002**, *20*, 530–541. [CrossRef] [PubMed]

43. Colter, D.; Class, R.; DiGirolamo, C.; Prockop, D. Rapid expansion of recycling stem cells in cultures of plastic-adherent cells from bone marrow. *Proc. Natl. Acad. Sci. USA* **2000**, *97*, 3213–3218. [CrossRef] [PubMed]

44. Bachurski, D.; Schuldner, M.; Nguyen, P.H.; Malz, A.; Reiners, K.S.; Grenzi, P.C.; Babatz, F.; Schauss, A.C.; Hansen, H.P.; Hallek, M.; et al. Extracellular vesicle measurements with nanoparticle tracking analysis—An accuracy and repeatability comparison between NanoSight NS300 and ZetaView. *J. Extracell. Vesicles* **2019**, *8*, 1596016. [CrossRef]

3

Tooth Formation: Are the Hardest Tissues of Human Body Hard to Regenerate?

Juliana Baranova [1], Dominik Büchner [2], Werner Götz [3], Margit Schulze [2] and Edda Tobiasch [2,*]

[1] Department of Biochemistry, Institute of Chemistry, University of São Paulo,
     Avenida Professor Lineu Prestes 748, Vila Universitária, São Paulo 05508-000, Brazil; jbaranova@usp.br
[2] Department of Natural Sciences, Bonn-Rhein-Sieg University of Applied Sciences, von-Liebig-Straße 20,
     53359 Rheinbach, NRW, Germany; dominik.buechner@h-brs.de (D.B.); margit.schulze@h-brs.de (M.S.)
[3] Oral Biology Laboratory, Department of Orthodontics, Dental Hospital of the University of Bonn,
     Welschnonnenstraße 17, 53111 Bonn, NRW, Germany; wgoetz@uni-bonn.de
*   Correspondence: edda.tobiasch@h-brs.de

Abstract: With increasing life expectancy, demands for dental tissue and whole-tooth regeneration are becoming more significant. Despite great progress in medicine, including regenerative therapies, the complex structure of dental tissues introduces several challenges to the field of regenerative dentistry. Interdisciplinary efforts from cellular biologists, material scientists, and clinical odontologists are being made to establish strategies and find the solutions for dental tissue regeneration and/or whole-tooth regeneration. In recent years, many significant discoveries were done regarding signaling pathways and factors shaping calcified tissue genesis, including those of tooth. Novel biocompatible scaffolds and polymer-based drug release systems are under development and may soon result in clinically applicable biomaterials with the potential to modulate signaling cascades involved in dental tissue genesis and regeneration. Approaches for whole-tooth regeneration utilizing adult stem cells, induced pluripotent stem cells, or tooth germ cells transplantation are emerging as promising alternatives to overcome existing in vitro tissue generation hurdles. In this interdisciplinary review, most recent advances in cellular signaling guiding dental tissue genesis, novel functionalized scaffolds and drug release material, various odontogenic cell sources, and methods for tooth regeneration are discussed thus providing a multi-faceted, up-to-date, and illustrative overview on the tooth regeneration matter, alongside hints for future directions in the challenging field of regenerative dentistry.

Keywords: dentogenesis; amelogenesis; dentinogenesis; cementogenesis; drug release materials; scaffolds; odontogenic cells; stem cells; whole-tooth regeneration

1. Introduction

Dental injuries and diseases such as caries and periodontitis are affecting significant fractions of populations worldwide and are the main reason for dental tissue regeneration efforts [1,2]. Caries lesions cause local enamel resorption and dentin damage due to oral microbiota activities in the morbid tooth. Although relatively easily manageable at early stages, if left untreated caries causes excessive dentin damage and poses a need for reparative treatment [3]. Periodontitis is a complex inflammatory disease, where pathogenic oral microbiota and host immune response dysregulation lead to the gingiva, periodontal ligament, cementum, and alveolar bone damage [4]. Excessive periodontitis damage cannot be regenerated naturally, thus requires specialized soft and hard calcified tissues regeneration approaches. Next to infectious/inflammatory oral diseases, several heritable disorders of dental tissue formation exist (e.g., amelogenesis imperfecta, dentinogenesis imperfecta, and tooth agenesis), which affect tooth formation, eruption, calcification, or maturation [5–8]. In addition to disrupted teeth integrity, dental diseases often create an unaesthetically looking oral cavity, thus affecting patients emotionally, which makes dental tissues regeneration critical in both aspects: health and aesthetics.

Dental tissues have no or very limited capacity for self-regeneration [2,3,9,10]. Specifically, enamel becomes acellular after it is formed; dentin regeneration is limited and dependent on the dental pulp stem cell pool, which deteriorates in the case of an infection and inflammation; and cementum has no remodeling capacity and limited regrowth in the case of disease-induced resorption [10–13]. Each dental tissue contains a defined amount of inorganic matter (hydroxyapatite crystals), matrix proteins arranged in a scaffolding network, and microstructures such as lacunae in cellular cementum and microchannels, which accommodate cellular processes in dentin and cementum. The complex microarchitecture of the tooth poses a need for appropriate replacement materials, which have to be biocompatible and wear-resistant [14]. Additionally, the development of enamel and dentin relies heavily on mesenchymal–epithelial interactions, thus making it challenging to recapitulate the process in vitro even using existing odontogenic cell lines and adult stem cell culture methods [10,12,15–17]. Although a lot is already known about tooth formation and molecular cues shaping this process [5,6,18], signaling patterns involved in dental tissue differentiation in vitro, postnatal calcified tissue metabolism and regeneration are being actively studied and more research is expected in the future [18–47].

Efforts in whole tooth regeneration have been made for decades [48] and include biological, bioengineering, and genetic approaches. Revitalizing the odontogenic potency of the successional dental lamina (SDL) rudiment for lost tooth regeneration might be one possibility to induce tooth formation in vivo in the adult [49]. Whole-tooth restoration using autologous tooth germ cells and bioengineered tooth germ transplantation is another promising opportunity [50,51]. However, due to limited sources of tooth germ cells, the risk for immune rejection of allogeneic or xenogeneic cells, as well as ethical and legal constraints, adult stem cells of various sources or induced pluripotent stem cells (iPSCs) may be used instead [52–54]. Recently, combining cells of mesenchymal and epithelial origin of various plasticity is being actively explored for tooth regeneration using novel culture methods [55–57].

Although the implantation of recombined embryonic or adult cells may give rise to tooth-like organs in vivo, the combination with scaffold material may improve tooth formation. Scaffolds can influence the biological behavior of cells and can give mechanical support to tissue constructs. Their consecutive degradation should parallel the formation of the native extracellular matrix and promote the assimilation of constructs after implantation [58]. In contrast to periodontal bone and other bone grafts, where numerous scaffold compounds have been developed and tested within the last decade [59–64], studies on artificial scaffolds for tooth regeneration are still rare due to the rather complex nature of teeth [14,65]. Recent studies in biomaterial development involve hybrids and composites of inorganic/organic components to be used as scaffolds to mimic the complex composition of the natural tooth [55,58,66]. New investigations have shown also that the functionalization of scaffolds using cell-free methods is possible. Vesicles, small RNAs, or exosomes from cultured stem cells or embryonic cells can be used onto or within scaffold material to address regenerative functions [65]. Besides, scaffolds can be loaded with drugs, growth factors, and/or receptor ligands to guide the stem cell differentiation process during dentogenesis [19,64,67,68]. However, very few artificial materials have been tested thus far in clinical trials [9,69].

In this review, the most recent discoveries regarding cellular signals guiding dental tissue differentiation in vitro and in vivo are summarized. Current developments of biocompatible functionalized scaffolds, drug-release materials, and their applications are addressed as well. Finally, whole-tooth generation approaches using various cellular sources and dilemmas in tooth regeneration are elucidated. An interdisciplinary approach is taken to cover tooth regeneration issues from molecular, via structural to biological aspects.

2. Hard Dental Tissues and Their Genesis

2.1. The Complexity of Dental Tissues

The process of teeth formation starts within embryogenesis and proceeds in multiple phases throughout the prenatal period, childhood and adolescence resulting in an eruption of permanent teeth. Each dental tissue forms in a unique way and in a tightly regulated manner, where one tissue is guiding or supporting the formation of the other [11,30,40]. Early odontogenesis is characterized by an epithelial–mesenchymal interaction, which is also a blueprint for the formation of other organs such as hair follicles or exocrine glands [70]. The epithelium is derived from the embryonic endoderm, while the mesenchyme is derived from the cranial neural crest. Placodal thickenings of the oral epithelium along the dental lamina first induce a cellular condensation of the underlying mesenchyme. The tooth primordium then undergoes different morphological stages forming a bud, cap, and later bell stage. While the epithelium gives rise to enamel, the mesenchyme is the source of the later pulp, periodontal apparatus and hard substances such as dentine and cementum. Then, epithelial components lose their inductive odontogenic competence while a reciprocal induction starts from the mesenchyme. These reciprocal crosstalks are governed by a signaling program consisting of a large number of molecules interacting in signaling pathways. Major examples of these factors are families such as the Bone Morphogenic Proteins (BMPs), Fibroblast Growth Factors (FGFs), Wingless/Int1 (Wnt), Hedgehog (Hh), or Ectodysplasin (EDA) functioning as morphogenetic inducers [18,65,71]. The morphogenesis is driven by signaling centers, which orchestrate tissue interactions and are involved in the size and shaping of the single tooth. In addition to cellular signaling, tissue forces, e.g., through an epithelial contraction, mesenchymal condensation, or bone biomechanics, participate in the formation of tooth morphology [71,72].

During tooth development, several stem cell niches have been identified. Epithelial stem cells are located, e.g., in the cervical loop, which is the apical end of the advancing epithelium consisting of an outer and inner layer and active until the onset of tooth root formation. These stem cells play a role in continuously growing teeth, e.g., mouse incisors. The elongation of the cervical loop as a double-layered structure is named Hertwig's epithelial root sheath and is the signaling center for tooth root formation. It should also be mentioned that tooth formation depends on the interaction with the developing alveolar bone, which therefore should be considered in strategies for whole tooth regeneration [71,73].

The mature tooth is a complex organ consisting of non-vascularized hard tissues: enamel, dentin, and a soft vascularized innervated dental pulp. The dental pulp is closely associated with dentin and harbors odontoblasts, dental pulp stem cells (DPSCs), pericytes, and other cellular populations. Blood vessels penetrating the pulp nourish the resident cells, while nerves participate in the sensory information exchange between the pulp and oral environment (Figure 1B). In the case of excessive dental injury (e.g., deep caries), odontoblasts, their precursors, and DPCSs can be recruited from the dental pulp and participate in dentin repair [74]. The tooth is surrounded by the periodontal ligament, which is a complex attachment tissue harboring odontogenic stem cells [69,75], linking the tooth to the alveolar jawbone (Figure 1). The mature molar tooth macrostructure and microstructures of dental tissues containing cell niches are depicted in Figure 1.

2.2. Signaling Pathways Modulating Hard Dental Tissue Generation

Many signaling cascades such as FGF, sonic hedgehog (Shh), transforming growth factors beta (TGF-β), BMPs, and Wnt/β-catenin are involved in the regulation of dentogenesis during development and adulthood [11,45,76–78]. Specific functions elicited by activation of these pathways are noted during distinct phases of dental tissue differentiation, some of which are beneficial for cell stemness and proliferation (FGF, Shh) while others such as Wnt, TGF-β, and BMPs act in postnatal differentiation phases and promote polarization, migration, and calcification [23,25–28,30,31,37,77,79,80]. Next to this, purinergic signaling function is gaining research attention in dental tissues metabolism [32,81,82].

Most ligands activate transcription factors such as runt-related transcription factor 2 (Runx2), osterix (Osx or Sp7), and extracellular signal-regulated kinase 1/2 (ERK1/2 or MEK1/2), which are central regulators of gene sets crucial for calcified tissues [33–35,83]. Epithelial–mesenchymal interactions are also involved in odontogenic and cementogenic differentiation [23,24,30,38,39].

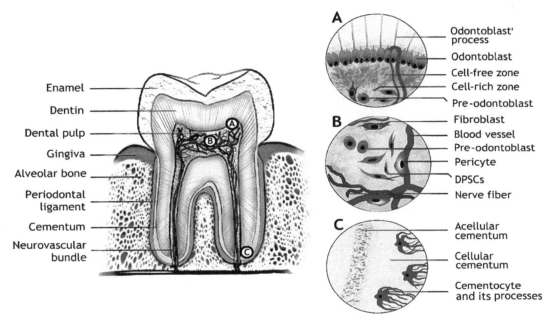

Figure 1. Tooth structure and dental tissues with the respective stem cell populations. (**A**) The odontoblast niche is bordering dental pulp beneath the dentin with odontoblast processes projecting towards enamel. (**B**) Diverse cell populations are found in dental pulp, DPSCs, which can give rise to odontoblasts. (**C**) Cementocytes are residing in the lacunae of cellular cementum at the root apex with their cellular processes projecting towards the periodontal ligament.

2.2.1. Amelogenesis

Tooth enamel formation or amelogenesis is the process of tooth enamel generation by ameloblasts, during which ameloblasts move towards the enamel surface and secrete proteins such as amelogenin, ameloblastin, and enameling. These proteins serve as scaffolds for calcium and phosphorus ions to be deposited on, thus guiding hydroxyapatite crystals aggregates—the enamel rods—generation. In this process, amelogenin and amelotin phosphorylation appears to be essential for correct enamel rod formation/organization [84–86]. The scaffolds are later degraded by matrix proteases and ameloblasts undergo apoptosis, which makes enamel the most mineralized acellular tissue in the human body, consisting of 95% hydroxyapatite crystals and 5% organic matter and water by weight [10,11]. Enamel is subjected to wear and tear throughout life. However, unlike other mineralized tissues of the human body, enamel cannot be regenerated due to its acellular nature. Although several cell sources were shown to have amelogenic capacity including keratinocyte stem cells, epithelial cell rests of Malassez (ERM) from periodontal ligament, odontogenic oral epithelial stem cells (OEpSCs), adipose tissue-derived mesenchymal stem cells (AT-MSCs), and iPSCs [87–92].

Since ameloblasts undergo apoptosis upon fulfilling their function of enamel production, studies of amelogenesis rely on in vitro models such as murine immortalized ameloblast-lineage cell (ALC) line [15], organotypic cultures, or rodent models. Many discoveries regarding ligands, their downstream transcriptional factors and responsive genes expressing core enamel proteins and matrix metalloproteinases were done using the mentioned ALC line. Shh, which is one of the major ligands expressed in the enamel knot during tooth morphogenesis, was shown to have a direct effect on the expression of the major enamel matrix proteins amelogenin and ameloblastin. The upregulation of these proteins is mediated by an activated glioma-associated transcription factor (Gli1) in the presence of Shh [42,77]. Notably, Gli1 was proposed as a marker for selecting stem cells with the odontogenic potential for tooth regeneration [93,94]. Runx2 together with odontogenic ameloblast-associated protein (ODAM) regulates matrix metalloproteinase 20 (MMP20) expression, the key enamel matrix-degrading enzyme [43], and has an affinity for the *Wdr72* (gene coding for maturation-stage ameloblast-specific protein) promoter [44]. WDR72 is an intracellular protein abundant in ameloblasts during enamel maturation with a proposed function in amelogenin endocytosis [20].

Studies in dental organotypic cultures and transgenic mice also point out the importance of the mentioned pathways in dentogenesis. For example, Shh in combination with FGF8 was recognized as a stemness promoting ligans for ameloblast precursors (human skin fibroblasts) in a human-mouse chimeric tooth [87], while Runx2 was shown to have an affinity for the amelotin promoter and regulates its expression during the enamel maturation stage [95]. Regarding amelogenin turnover, a novel role of cytoplasmatic B-cell CLL/lymphoma 9 protein (Bcl9), its paralog B-cell lymphoma 9-like protein (Blc9l) and interaction partners Pygopus 1/2 (Pygo1/2) is proposed to play a role in amelogenin secretion [96].

Timely expression of β-catenin in dental tissues shapes tooth development by modulating various developmental signaling pathways, leading to the proper tooth number and morphology [45]. It was demonstrated in vitro that the β-catenin pathway, which is regulated by Wnt ligands, is involved in ameloblast polarity and motility [97]. Overactivation of the β-catenin pathway in the dental epithelium during the earliest stages of tooth development results in hyperdontia, and ablation—in tooth agenesis [98]—while, if overactive in postnatal ameloblasts, it causes poorly structured, softened enamel and its delayed formation [46]. Additionally, β-catenin overactivation downregulates enamel matrix metalloproteinases MMP20 and kallikrein 4 (Klk4), which are important in the removal of scaffolding proteins from maturing enamel [46]. An important regulator of Wnt/β-catenin pathway activity in ameloblasts is glycogen synthase kinase 3 beta (GSK3β) [99].

TGF-β superfamily ligands such as BMPs and TGF-βs are regulating enamel structural genes and matrix metalloproteinases expression. MMP20 in turn regulates TGF-β isoforms activity [47,100]. All three TGF-β isoforms induce Klk4 expression, while TGF-β1 and β2 induce amelotin expression [47]. TFG-β1 regulates Runx2 and its downstream target Wrd72 gene [44]. Thus, it appears that TGF-βs are key ligands involved in the regulation of enamel scaffolding protein removal and endocytosis during enamel mineralization. BMP knock-outs result in downregulated matrix proteins and metalloproteinase expressions. In detail, BMP2 knock-out reduced amelogenin, enamelin, MMP20, and Klk4 expression, similarly to double-knockout of BMP2 and -4, which resulted in a significant reduction of MMP20 and Klk4 in ameloblasts [21,22]. Metalloproteinase insufficiency is detrimental for the enamel structure since excessive protein content in enamel does not allow properly organized crystalline structure formation, making the enamel softer and less shear-resistant.

From the above-reviewed studies, it is evident that timely regulation of ligands known to be important for cell stemness maintenance and calcified tissue metabolism are the keys to structurally and morphologically correct enamel formation. Enamel integrity depends on proper enamel scaffolding protein deposition, phosphorylation state, and timely cleavage, which allow ameloblast migration and crystals deposition in an organized oriented pattern. The summary of the major signaling pathways involved in amelogenesis is schematized in Figure 2A and pathways modulators listed in Table 1.

2.2.2. Dentinogenesis

Dentin is an acellular calcified tissue consisting of 70% hydroxyapatite, 20% organic phase, and 10% water by weight. Dentin formation is executed by odontoblasts (or dentinoblasts), which are cells of mesenchymal origin. During dentinogenesis, odontoblasts migrate towards dental pulp and deposit collagen types I, III, and V, proteoglycans, and other matrix proteins, which provide the nucleation base for hydroxyapatite crystals. Besides scaffold-mediated mineralization, minerals precipitation and cell-derived matrix vesicles-driven mineralization occur during various stages of dentinogenesis [12]. After dentin synthesis is complete, odontoblasts remain beneath it with tiny cellular projections called odontoblast processes protruding into the microscopic channels in the dentin (Figure 1A). These projections are involved in detecting environmental stimuli (pH, cytokines, inflammatory mediators, and other signaling molecules) by odontoblasts, which can be mobilized for dentin regeneration in a case of damage. Thus, dentin possesses a limited capacity for regeneration [5,12,71,76]. Therefore, finding the appropriate cell source and differentiation strategy for dentin regeneration is of crucial importance. Thus far, dental pulp stem cells (DPSCs), stem cells from human exfoliated deciduous teeth (SHEDs), AT-MSCs, bone marrow-derived MSCs (BM-MSCs), and iPSCs have been shown to have the dentinogenic potential [25,80,81,90,101–103].

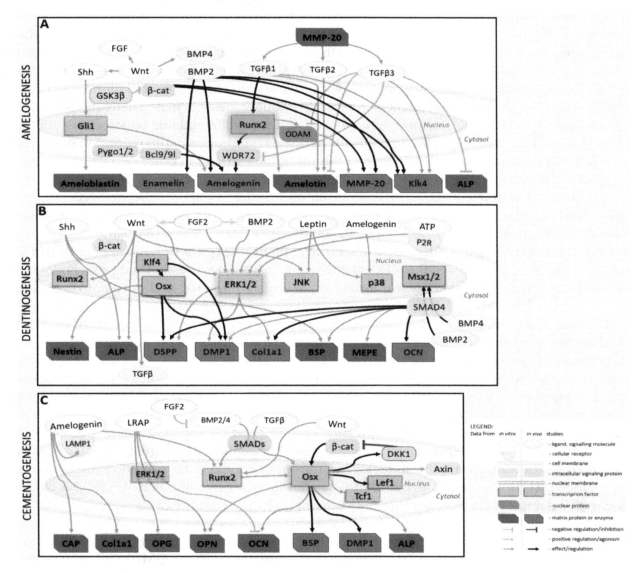

Figure 2. Major signaling cascades involved in amelogenesis, odontogenesis, and cementogenesis. (**A**) Signaling pathways modulating amelogenesis with TGF-β superfamily ligands (BMP2 and TGF-β1/2/3) playing the major role in matrix protein and metalloproteinases feedback-regulation and Runx2 being an important transcription factor. (**B**) Central signaling cascades of odontogenesis are depicted. The TGF-β superfamily ligands (BMP2/4 and TGFβs) regulate many odontogenic genes with ERK1/2 as convergence point and Klk4-Osx as important transcription factor tandem. (**C**) Major cementogenesis-related signaling cascades with Osx as the central transcription factor being regulated via Wnt/β-catenin in a feedback-loop. Ameloblast-derived products (LRAP and amelogenin) were shown modulate key cementogenic gene expression in vitro.

Shh is secreted by an epithelial cell layer, the zone of amelogenesis initiation, and serves as a paracrine differentiation signal for odontogenic cells [23,30]. It is later secreted by dentinoblasts during dentinogenesis and dental pulp stem cells (DPSCs), suggesting its autocrine function in odontogenic differentiation and dental pulp stem cell niche maintenance [30]. Amelogenin, secreted by ameloblasts, also participates in odontogenic differentiation of DPSCs by upregulating dentin sialophosphoprotein (DSPP) and dentin matrix acidic phosphoprotein 1 (DMP1) expression via the ERK1/2 and p38 pathways [104]. A similar effect could be achieved by the application of leptin: DSPP and DMP1 expression and ERK1/2, p38, and c-Jun N-terminal kinase (JNK) phosphorylation levels were markedly increased in leptin-treated DPCs [25,105]. Moreover, leptin application in the induced pulp cavity in rats leads to increased dentin formation during reparative dentinogenesis [106].

FGF exerts a time-dependent effect on dental-pulp derived odontoblast precursors. Transient exposure to FGF2 during the proliferation phase is beneficial for odontogenesis while no such effect is achieved upon constitutive FGF application until the maturation phase. FGF2 induces DSPP and DMP1 expression, which is also mediated via ERK1/2 pathway activation. Moreover, the agonistic effect on BMP2 and Wnt signaling during early odontogenesis were noted in cells treated with FGF2 [26,27].

BMP/TGF-β signaling is important during early odontogenesis, where it activates SMADs and regulates Msx-1/2 transcription factors expression, as well as in differentiated odontoblasts, for matrix gene expression [101,107–109]. BMP2 positively regulates odontogenic differentiation of stem cells from exfoliated deciduous teeth (SHEDs) by promoting the expression of DSPP, DMP1, and matrix extracellular phosphoglycoprotein (MEPE) [80]. BMP2 knock-out in dental mesenchyme results in dentin deposition and microstructure abnormalities indicating its pivotal non-redundant role in early dentinogenesis [28,107], while BMP2 together with BMP4 have redundant functions in mature odontoblasts where they regulate DSPP, DMP1, bone sialoprotein (BSP) and collagen type I alpha-1 (Col1a1) expression [108]. Smad4, the intracellular component downstream of BMP/TGF-β signaling, is also necessary for DSPP, Col1a1, and osteocalcin (OCN) expression and proper odontoblast maturation. If Smad4 is ablated, dentin formation is largely impaired and does not reach normal thickness in mice [29].

Wnt ligands are involved in odontoblast differentiation from mesenchymal precursor cells during the early stages of tooth development and later regulate dentin matrix deposition. It is proposed that at early stages of tooth development some Wnt ligands exert effects via the canonical Wnt/β-catenin signaling cascade and support odontoblast precursor cells stemness, while other Wnt ligands expressed at later developmental stages activate non-canonical pathways and promote the migration, proliferation, and mineralization of odontoblast precursors during dentinogenesis [31,83,110]. Experiments in vitro demonstrated that Wnt7b stimulates the expression of Runx2 and the key dentin matrix proteins DSPP, DMP1, and Col1a1 via ERK1/2-mediated activation during dentinogenesis [83]. Wnt7b can activate canonical Wnt/β-catenin, but also the JNK cascade, thus promoting cellular migration and odontogenic differentiation [31]. Notably, activation of Wnt/β-catenin signaling by inhibition of GSK3β is beneficial for reparative dentine formation during cavity repair [111].

Purinergic signaling mediated by adenosine receptors (P1 receptors, ARs) and purine receptors (P2X and P2Y) was also shown to play an important role in odontogenic differentiation of human DPSCs. P2 receptor activation by ATP promotes the expression of DSPP, DMP1 and mineralization of DPSCs via rapid phosphorylation of ERK1/2 [32]. Treatment of DPSCs with P1 receptor agonists in combination with ATP further improved odontogenesis by contributing to the upregulation of DSPP (mediated by A2BR and A3R) and DMP1 (via A1R and A2BR) and increased mineralization (via A1R and A2BR) [81]. Intracellular molecular events of P1 and P2 receptors agonistic action remain to be elucidated, but ERK1/2 is likely involved, at least partially, in the purinergic receptor-mediated odontogenic differentiation of DPCs as is the case with several other differentiations regulated by purinergic signaling.

Aside from the importance of activation of ERK1/2 and its downstream targets resulting in the expression of key dentin matrix genes, Tao and colleagues outlined Krüppel-like factor 4 (Klf4) as a major transcription factor regulating odontogenesis [33]. Klf4 induces TGF-β secretion, which together with BMPs positively regulates DMP1, the major dentin matrix protein expression. Moreover, Klf4 regulates odontogenesis-related gene expression temporally by interacting with histone deacetylase 3 (HDAC3) during early phases of odontoblastogenesis where it represses the expression of *osterix* and *DSPP*, while at later stages, when paired up with P300, it promotes their expression [33]. Osterix is a master-regulator of many structural genes of dentin and also of odontoblasts including DSPP, DMP1, nestin, and alkaline phosphatase (ALP) [34].

Studies regarding odontoblast differentiation outline the importance of signaling pathways and their interactions alike noted to be important for ameloblast differentiation with ERK1/2 being a convergence point for several signaling cascades involved in odontogenic differentiation of dental mesenchymal cells. Recently identified Klk-Osx transcriptional tandem, p38 and JNK are important in dentin structural genes regulation and odontoblast function (Figure 2B). Several dentinogenesis-promoting molecules (listed in Table 1) were already tested in vivo and shown promising results.

2.2.3. Cementogenesis

Cementum, a thin calcified avascular tissue between dentin and periodontal ligament, is produced by cementoblasts. Cementum contains collagen type I, bone sialoprotein, osteopontin, glycoproteins and proteoglycans arranged in a fibrous network with hydroxyapatite deposits. Various types of cementum are present in distinct regions of mature tooth roots: thin acellular cementum is deposited around the cervical tooth area and below, while thick cementum with entrapped cementocytes and their processes penetrating cementum locates at the root apexes (Figure 1C). Histological studies also indicate that a thin layer of dense acellular cementum lies beneath the cellular cementum at the root apex and plays an important role in cementum mineral metabolism. The cementum volume is enlarging over the lifespan and is not subjected to remodeling such as bones. Cementoblast precursors are present in the periodontal ligament and can be mobilized for cementum regeneration if needed [13,41,112]. Ex vivo, cementoblasts can be generated from periodontal ligament stem cells (PDLSCs), dental follicle stem cells (DFSCs), and iPSCs [75,113,114].

By analogy with dentinogenesis, TGF-β, and BMPs, Wnt and ameloblast-derived factors regulate cementum structural matrix protein expression. The central transcription factor of cementogenesis is Osx, which is activated by Wnt and TGF-β/BMP signaling. Osx is abundantly expressed in cementoblasts and cementocytes during cementum deposition, where it regulates DMP1, BSP, OCN, and ALP expression. It is proposed that Osx regulates cementogenic differentiation, while it inhibits cementoblast proliferation [35,115]. Stabilization of β-catenin leads to increased cementum formation via the upregulation of Osx, which is achieved by β-catenin binding to the Osx promoter, thus pointing to the direct regulation of Osx by β-catenin [36]. Additionally, Osx regulates the expression of dickkopf-related protein 1 (DKK1), an antagonist of β-catenin, and the transcription factors T-cell factor 1 (Tcf1) and lymphoid enhancer-binding factor 1 (Lef1), which together with β-catenin form a transcription initiation complex with β-catenin in the cell nucleus. It is therefore evident that cross-regulation of β-catenin and Osx plays a central role in cementogenesis [36,115].

In addition to Wnt/β-catenin regulation, Osx is regulated via the TGF-β/Smad axis, as Smad3 plays an important role in *Osx* gene expression during cementogenesis [37]. BMP2 and -4 likewise regulate Osx expression via a BMP-Smad-Runx2 cascade, but also Runx2-independently [34]. Despite the suggested beneficial role of Wnt/β-catenin in cementogenesis, another point of view has been expressed, according to which excess Wnt may inhibit cementogenesis under normoxic conditions, while hypoxia reverses this effect [24]. BMP2/4 signaling, which promotes cementogenesis in several ways, is negatively regulated by FGF2 in a concentration-dependent manner. This has been shown in periodontal ligament cells undergoing cementogenesis, thus implying that FGF2 is not beneficial for differentiation, but is important for cellular stemness [75]. This is in line with similar results in amelogenesis or very early stages of odontogenic differentiation [27,75,87]. Contrarily, in vivo, local FGF2 infusion was shown to promote cementum formation during periodontal injury regeneration by recruiting, enhancing and accelerating the proliferation of endogenous cemento/ostogenic cells [116].

Table 1. Cell Sources and signaling modulators useful for amelogenesis, dentinogenesis, and cementogenesis.

Tissue	Plausible Cell Sources	Signaling Pathway/Node	Interfering Molecule(s)	
			Stimulatory	Inhibitory
Enamel	Keratinocyte stem cells [87]; ERM from periodontal ligament [88]; OEpSCs [89]; AT-MSCs [90]; iPSCs [91,92,117]	Hh	Shh [42]; purmorphamine [118] [a]	cyclopamine [118] [a]
		FGF	FGF8 [87], *FGF10* [118] [a]	pan-FGF receptor inhibitor SU5402 [118] [a]
		Wnt/β-catenin	6-Bromoindirubin-3'-oxime (BIO) (GSK3βi) [45]	GSK3β [99], ICG-001 [97]
		BMP	BMP2/4 [21,22] [b]	Noggin (BMP4i) [117]
		TGFβ	TGF-β1,2,3 [47,100]	*SMAD7* [119] [a]
Dentin	DPSCs [25,81,101,102]; SHEDs [80]; AT-MSCS [102]; iPSCs [103]	Hh	Shh [23], purmorphamine [120]	–
		FGF	FGF2 [26,27,120]	PD173074 (FGFR1i) [120]
		Wnt/β-catenin	BIO, CHIR, Tideglusib (GSK3bi) [111,121] [b], Wnt7b [83];	XAV939 (tankyrasei) [31,101], rhDKK1 [101]
		BMP	BMP2 [28,108] [b], BMP4 [108] [b]	Noggin, LDN193189 [101]
		P2Rs	ATP, ARL 67156 (ATPasei) [32]	Suramin [32], iso-PPADS tetrasodium salt [82]
		ERK1/2	Leptin [105]	PD98059 (ERK1/2i) [105]
		ERK1/2	Amelogenin [104]	U0126 (ERK1/2i) [104]
Cementum	PDLSCs [75]; DFSCs [113]; iPSCs [114]	Wnt/β-catenin	LiCl, Wnt3a [35]	DKK1 [35]
		FGF	FGF2 [116] [b]	–
		BMP	BMP2/4 [75]	FGF2 [75]
		TGFβ	rhTGFβ-1 [78]	SIS3 (Smad3i) [37]
		ERK1/2	Amelogenin [39], LRAP [79]	U0126 (ERK1/2i) [79]

[a] studies of epithelial invagination/development; [b] studies in vivo; the rest are cell culture-based reports.

The enamel-derived signaling components, amelogenin and its alternatively-spliced isoforms, regulate cementogenesis by modulating the expression of various matrix proteins. Full-length amelogenin application induced the expression of osteopontin (OPN), cementum attachment protein (CAP), OCN, Cola1, BSP, DMP1, and ALP mRNA; upregulated OPN and Col1a1 proteins; and improved the mineralization of an immortalized mouse cementoblast cell line (OCCM-30). Moreover, amelogenin positively regulated its putative receptor lysosome-associated membrane glycoprotein 1 (LAMP1) in murine dental follicle cells and OCCM-30 cells, thus confirming its role as an important ligand regulating cementogenesis [38,39]. Amelogenin derivates, such as leucine-rich amelogenin peptide (LRAP), modulate gene expression in a slightly different manner: LRAP inhibited OCN expression, while promoted OPN and osteoprotegerin (OPG) expression in a dose-dependent manner and had a negative effect on cementoblast mineralization. The effects are probably mediated through the ERK1/2 pathway since ERK inhibition annuls the LRAP effects [79].

Similar to dentin and dentinogenesis, cementogenesis has a central transcription factor: Osx, which regulates cementogenesis-specific gene expression. Besides Osx, Runx2, and ERK1/2 are involved in cementogenic differentiation. In addition, ameloblast-derived proteins are important ligands positively regulating cementum matrix-associated gene expression (Figure 2C). Modulators of herein discussed cementogenic pathways are listed in Table 1.

Gained knowledge about molecular cues shaping dental tissue genesis may help to establish novel stem cell selection, culture, and differentiation methods and develop functionalized scaffolds and biomaterials, which will support and promote amelogenic, dentinogenic, and cementogenic differentiation in vitro. Thus, it will approximate the era of dental tissues regeneration using most suitable odontogenic cells with adequately functionalized biomaterials.

3. Scaffolds and Drug Release Materials for Tooth Regeneration

3.1. Scaffolds for Enamel, Dentin, and Cementum Regeneration

Scaffolds and biomaterials are essential components in dental tissue regeneration since they can be used as a template for tissue regeneration by serving as a site of attachment for the regenerative cells from the surrounding tissues or act as a delivery platform for implantable odontogenic cells with the ability to differentiate towards the desired cell type [122,123]. Additionally, the scaffold material may be used as a delivery platform for bioactive molecules such as drugs or proteins (especially growth factors) that further enhance the regenerative potential [60,61,63,124].

In general, scaffold materials used in tissue regeneration need to be readily available and meet criteria such as biocompatibility and biodegradability without any toxic metabolites. In the case of scaffolds for tooth regeneration, biomaterials are subjected to the challenging environment of the oral cavity—including mechanical forces due to mastication, the presence of microorganisms, and varying conditions regarding temperature and pH. The intended biomaterial has to face these challenges without limitations in its biocompatibility [125]. Since it is generally intended to mimic the native extracellular matrix by using biomaterials, properties besides biocompatibility are imposed by the tissue which should be regenerated. Thus, in the case of scaffold materials for dental tissue engineering, the used material systems differ greatly depending on whether enamel, endodontic, or periodontic tissue is intended to be regenerated. Categories for biomaterials used in tooth regeneration are natural organic, synthetic organic materials, or inorganic materials [126]. Natural organic materials involve peptides such as collagen or gelatin and polysaccharides such as chitosan, alginate, or agarose. Frequently used synthetic organic materials are poly(lactic acid) (PLA), poly(glycolic acid) (PGA), poly(lactic-co-glycolic acid) (PLGA), and poly(caprolactone) (PCL), while commonly used inorganic materials are bioactive glasses or calcium phosphates such as hydroxyapatite (HA), β-tricalcium phosphate (TCP), and cementitious systems of calcium phosphate (CPC) or calcium silicate (e.g., mineral trioxide aggregate, MTA). Polymeric materials often lack mechanical and biological properties but are able to establish three-dimensional porous structures, thereby providing a highly hydrated matrix in vivo that facilitates the transport of nutrients, anabolites, and catabolites. In turn, inorganic biomaterials used in tissue engineering often comprise preferable biological properties but have disadvantages such as brittleness and lacking in the supply of nutrients. Thus, composite materials comprising both organic and inorganic constituents gain increasing interest in recent years due to their inherent combination of the desirable properties of the single components [127]. In the following subsection, the challenges, approaches, and recent studies for the targeted and scaffold-assisted regeneration of enamel, dentin, and cementum are presented. Injectable biomaterials are a central and highly desirable class in the context of dental regeneration, but are not extensively reviewed here due to the very recent and detailed publication of a distinct review on this topic by Haugen and coauthors [128].

3.1.1. Enamel Formation

The main challenge in the regeneration of enamel is its acellular nature. Enamel forming ameloblasts go through apoptosis when amelogenesis is finalized and the in vitro culture of ameloblasts is yet unestablished in a scale needed for appropriate tissue regeneration [129]. Furthermore, although the synthesis of hydroxyapatites is widely investigated, attempts to model the unique assembling of HA-crystals in enamel were not yet successful [130]. Thus, many recently published studies follow a biomimetic approach by using amelogenin, peptide fragments of amelogenin, or various synthetic peptides as a template matrix to mimic the spatiotemporal environment for the deposition of enamel.

Recently, Zheng et al. used a peptide consisting of eight repetitive sequences of aspartate-serine-serine (8DSS) as a biomimetic template for enamel remineralization in an in vivo model. Their results indicate that 8DSS peptides serves as both inhibitor of further enamel demineralization and promoter of remineralization by entrapping calcium and phosphate from the surrounding medium. As a result, mineral density and enamel volume increased to a comparable extent as with a fluoride treatment [131]. Treating enamel surface with an elastin-like polypeptide (ELP) functionalized with glutamic acid residues to dissolve calcium and phosphate due to its acidic properties leads to a matrix consisting of ELP and amorphous calcium phosphate (ACP). After immersing the specimen in simulated oral fluid, a dense layer of highly orientated apatite nanorods is formed from the matrix with mechanical properties close to natural enamel and high chemical stability against acidic impacts [132]. The properties of poly(amidoamine) (PAMAM) dendrimers can be tailored by modification of their functional surface groups. Accordingly, the effect of amino-, carboxyl-, and alcohol-terminal groups has recently been studied in vitro. The results show that the electrostatic interactions between biomaterial and enamel surface affect the remineralization process. PAMAM-NH2, exhibiting interactions between pro-cationic amino groups and negatively charged enamel surface, shows the best results, followed by PAMAM-COOH due to interactions between carboxylate residues and calcium cations in hydroxyapatite, while neutral PAMAM-OH was not effective [133]. Additionally, Gao et al. evaluated the performance of amorphous calcium phosphate loaded PAMAM-dendrimers functionalized with an SN15 peptide sequence, which is known for its good adsorption on hydroxyapatite, for the use as adhesive in resin-based approaches of caries lesion treatments and achieved 90% higher remineralization compared to control [134].

3.1.2. Dentin Formation

Dentin regeneration is most often related to a treatment of the dentin-pulp complex. Since pulp vitality is essential for tooth homeostasis and stability, strategies to maintain this vitality are highly desirable. Presently, pulp capping is the main therapy maintaining the pulp vitality but is frequently accompanied by irreversible pulp inflammation and reinfections [16]. Thus, innovative approaches and biomaterials for the regeneration of the pulp–dentin complex are highly desirable.

In classical endodontic therapy via apexification, the pulp space is initially cleared and sealed with calcium hydroxide or MTA to induce a hard-tissue formation at the apical area that is used as a barrier for a permanent root filling material. Since this procedure does not promote further root development, root canal walls remain thin and fragile, leading to teeth that are prone to further issues [135]. To overcome these limitations, regenerative endodontic therapies including revascularization are being developed. Here, bleeding is induced to fill the endodontic canal and form an autologous blood clot that serves as a scaffold homing matrix proteins, (stem) cells, and growth factors, which consequently leads to the regeneration of the pulp–dentin complex due to root development, apical closure, and maintenance of the tooth vitality [17,136]. However, due to the presence of mesenchymal stem cells in the infiltrating blood, the generated tissue is more bone-like mixed with connective tissue instead of the desired pulp–dentin complex [137].

Recently, Mandakhbayar and colleagues used strontium-free and strontium-containing nanobioactive glass cement in a pulp capping approach to evaluate their potential to regenerate the pulp–dentin complex in vitro and in vivo [138]. The nanobiocement based on mesoporous calcium silicate nanobioactive glasses showed a fast release of Ca-, Sr-, and Si-ions, which are known for their bioactive properties in hard-tissue regeneration; promoted the odontogenesis of DPSCs in vitro; and showed promising results in vivo, especially for Sr-containing biomaterials [138]. Boron-modified bioactive glass nanoparticles were embedded in an organic matrix of cellulose acetate, oxidized pullulan, and gelatin by Moonesi-Rad and associates to build a dentin-like construct by freeze-drying and subsequent mold pressing [139]. The composite material induced the enhanced deposition of a calcium phosphate layer after immersion in simulated body fluid. Moreover, cell culture studies using DPSCs indicated the promotive effects of boron-modified bioactive glasses on attachment, migration, and odontogenic differentiation [139]. In a classical ternary system comprising an injectable collagen scaffold, DPSCs, and growth factors, Pankajakshan and coworkers evaluated the effect of mechanical properties of the collagen matrix [140]. Via concentric injection, the authors created a scaffold with an inner section of lower stiffness, which is covered with an outer section of higher stiffness to mimic the mechanical properties of the natural pulp–dentin complex. Additionally, they loaded the softer scaffold material with proangiogenetic vascular endothelial growth factor (VEGF) and the stiffer scaffold material with BMP2 to enhance the site-specific endothelial or odontogenetic differentiation of DPSCs, respectively. The results show that the stiffness of the materials regulates the direction of DPSCs differentiation. This effect is further enhanced by the loading of the collagen matrices with VEGF or BMP2, respectively [140].

3.1.3. Cementum Formation

Cementum regeneration is closely related to the treatment of the periodontal complex comprised of alveolar bone, periodontal ligament, gingiva, and cementum (Figure 1). Besides the structural support a scaffold material provides to the affected tissue, scaffolds used for regeneration of the periodontal complex are often used as a delivery vehicle for various bioactive compounds such as proteins, growth factors, or gene vectors to favor the regenerative process and induce the recruitment and homing of endogenous stem cells from surrounding tissues. The development of multicompartment scaffolds aims to meet the diverse challenges of the different tissues to be regenerated in periodontal defects in a single scaffold [141]. Additionally, besides synthetic scaffolds, cell-based scaffolds such as cell sheets are part of current research. In this approach, cell types that are relevant for the periodontal regeneration are cultivated in vitro extensively, until strong cell–cell interactions are established and an extracellular matrix has formed, thus allowing transplantation of the cell sheet as a scaffold-like material [142].

Recently, Fakheran and peers evaluated the regenerative potential of Retro MTA, a calcium silicate cement, in combination with tricalcium phosphate in vivo and showed that newly formed bone and cementum was significantly higher than in the untreated control group. Moreover, the poor biodegradation rate of MTA is improved due to the combination with biodegradable TCP [143]. In a preclinical study to treat periodontal defects in dogs, Wei et al. used an inorganic calcium phosphate-based scaffold material loaded with BMP2 [144]. The CaP-based biomaterial alone leads to a significantly increased regeneration of mineralized tissue as well as to an improved attachment of the teeth to the surrounding tissue compared to untreated control and a deproteinized bovine bone mineral that serves as commercial control. When loaded with BMP2, these positive results could even be improved two- and three-fold regarding height and area of the remineralized tissues, respectively. Noteworthy, the encapsulated BMP2 had a greater impact on osteogenesis than on cementogenesis [144]. Following the multicompartment-scaffold approach, Wang and collaborators applied a bilayered material containing growth factors. The hybrid material containing an FGF2-loaded propylene-glycol alginate gel coating the root surface for ligament regeneration and a BMP2-loaded (PLGA)/calcium phosphate cement for periodontal regeneration was tested in vivo with non-human primates. Following a promising study in rodents, the authors reported significantly enhanced regeneration of cementum and periodontal ligament and a high vascularization of the newly formed periodontal ligament (PDL), thereby confirming the positive results of the previous study [145,146].

Vaquette el al. developed bilayered scaffold materials based on polycaprolactone and combined them with cell sheets: while a fibrous three-dimensional compartment with macropores should favor alveolar bone regeneration, a flexible porous membrane aims at delivering the cell sheet and regenerates the periodontal ligament [147]. In their study, the authors evaluated the in vivo regenerative potential of the hybrid materials with different cell types forming the cell sheet, namely gingival cells, periodontal ligament cells (PDLCs), and bone marrow-derived mesenchymal stem cells (BM-MSCs). Results from histomorphometry and micro-computed tomography (μ-CT) show that scaffolds containing BM-MSCs and PDLCs had greater regenerative potential due to superior new bone and cementum formation compared to the scaffolds containing gingival cell sheets. However, the regenerative potential of scaffolds containing BM-MSCs and PDLCs did not differ significantly compared to the performance of the non-cellularized control scaffold. Thus, the biphasic scaffold alone is also a promising candidate for further studies [147]. Table 2 summarizes recently published studies emphasizing regenerative approaches of enamel, dentin, and cementum.

Table 2. Compilation of recently published studies emphasizing regenerative approaches of enamel, dentin, and cementum.

Tissue	Scaffold Material	Study Model	Results	Ref.
Enamel	8DSS: Oligopeptide of eight repetitive sequences of aspartate-serine-serine	In vivo model using Sprague-Dawley rats with induced caries.	Increased remineralization by 8DSS due to inhibited enamel demineralization and promoted remineralization.	[131]
	Elastin-like polypeptide functionalized with glutamic acid residues	In vitro remineralization of bovine enamel specimens by pH cycling after immersion in biomaterial solution.	Formation of a dense layer of highly oriented apatite nanorods with mechanical properties close to natural enamel and high chemical stability against acidic impacts.	[132]
	PAMAM-dendrimers with varying terminal groups: -NH2, -COOH, -OH	In vitro remineralization of bovine enamel specimens by pH cycling.	Remineralization is affected by electrostatic interactions between scaffold and enamel surface. PAMAM-NH$_2$ shows the best results, followed by PAMAM-COOH.	[133]
	ACP-loaded PAMAM dendrimers functionalized with SN15 peptide sequence.	In vitro enamel remineralization by cycling immersion in artificial saliva and demineralization solution.	Evaluated biomaterial achieves 90% higher remineralization compared to control.	[134]
Dentin	Nanobioactive glass cements with or without Sr	In vitro evaluation of biocompatibility and differentiation of DPSCs. In vivo evaluation using an ectopic odontogenesis model and a tooth defect model in rats.	Fast release of bioactive Ca-, Sr- and Si-ions. Promotion of the odontogenic differentiation of DPSCs in vitro. More new dentin formation by Sr-containing biomaterial in vivo.	[138]
	The organic matrix of cellulose acetate, oxidized pullulan and gelatin loaded with boron-modified bioactive glass nanoparticles.	In vitro evaluation of biomineralization, biocompatibility, proliferation, and differentiation with hDPSCs.	Boron-modified bioactive glass nanoparticles exhibit promotive effects on the deposition of a CaP as well as on adhesion, migration, and differentiation of hDPSCs.	[139]
	Biphasic collagen matrix: Inner section of lower stiffness loaded with VEGF covered by an outer section of higher stiffness loaded with BMP2.	In vitro evaluation using hDPSCs regarding biocompatibility, proliferation, and differentiation.	The direction of DPSCs differentiation is regulated by material stiffness and amplified by the respective growth factor.	[140]
Cementum	retroMTA + tricalcium phosphate	In vivo test using dehiscence periodontal defects in dogs.	Significantly increased the new bone and cementum formation. The biodegradability of retroMTA is enhanced by adding TCP.	[143]
	Calcium phosphate loaded with BMP2	In vivo periodontitis model using critical-sized supra-alveolar defects in dogs.	Significant increase in regeneration of mineralized tissues. Loading with BMP2 leads to a further 2–3-fold increase.	[144]
	Bilayered material: FGF2-propyleneglycol alginate gel covered by BMP2-PLGA/CaP cement.	In vivo test using three wall periodontal defects in non-human primates.	Significantly enhanced regeneration of cementum and periodontal ligament. Newly formed PDL is highly vascularized.	[145]
	PCL-based bilayered material: a flexible porous membrane delivers cell sheets and is covered by a fibrous and porous 3D compartment.	In vivo test using dehiscence periodontal defects in sheep to evaluate the potential of different cell types forming the cell sheets: Gingival cells (GCs), PDLCs, and hBM-MSCs.	Scaffolds containing BM-MSCs and PDLCs show superior new bone and cementum formation compared to scaffolds containing gingival cells.	[147]

3.2. Drug Release Systems Useful in Tissue Engineering—To be Adapted to Tooth Engineering

As discussed in the previous section, whole tooth regeneration is one of the most challenging fields in regenerative medicine—also regarding drug release aspects. In stem cell-based approaches, a cocktail of different drugs would be required to tightly tailor the differentiation of the corresponding cells involved in amelogenesis, dentinogenesis, and cementogenesis, respectively. This means that, besides appropriate scaffolds, compounds have to be developed for drug encapsulation and controlled release of those substances involved during tooth formation (such as growth factors and receptor ligands, as listed in Figure 2). Thus far, drug release approaches in tooth regeneration are mainly restricted to the delivery of antibiotics to avoid inflammation [66].

In analogy to other tissues and organs engineered using stem cell-based approaches, the drug delivery systems (DDS) are mainly classified into the following release mechanisms: diffusion through water-filled pores; diffusion through the polymer; osmotic pumping; and erosion [148]. In the past two decades, novel release materials have been designed and prepared that could be classified into the following three groups: (a) polymer-based systems; (b) ceramics-based systems; and (c) hybrid systems (e.g., organic/inorganic and polymer/ceramic) [62,149]. Many of them are prepared as nanomaterials (e.g., spheres, capsules, and rods) [64].

To develop a DDS that allows kinetically controlled release of drugs supporting the required stem cell differentiation processes, a variety of material characteristics would have to be considered. Parameters that influence the release behavior of polymer-based release materials include the following: molecular weight (number and weight average, respectively, M_n/M_w) and corresponding polydispersity index (PI), number and nature of end-groups, and the polymer morphology mainly determined by the monomer 3D structure (amorphous and crystalline/semi-crystalline with the degree of crystallinity). All of them are able to influence the size and shape, as well as density and porosity of the entire DDS that includes the encapsulated drugs. In addition, the active substance (drug) itself influences the release kinetics via interaction with the encapsulation material. Thus, the drug hydrophilicity/hydrophobicity (resulting from chemical composition, functional groups, hydrogen bonds, etc.) is one of the most limiting aspects, as well as its ability to act as surfactants or plasticizer which would interfere with the release mechanism. Huang et al. comprehensively reviewed the release mechanisms discovered within the last five years, including drugs for tooth regeneration [65]. Most recent developments include tunable conductive polymers to be used for controlled delivery [150]. As stated in Section 3.1, in tooth regeneration, drugs (such as growth factors and FGF-2) are usually simply added to the scaffold material—not yet encapsulated and released from tailored delivery materials [14,19,66,67,151–157]. Recently, Moon et al. reported a study using nitric oxide release to support the pulp–dentin regeneration [158]. However, in this case, release kinetics cannot be controlled or adjusted to the differentiation processes of the corresponding cells. Very few studies reported the application of specific drug encapsulation materials, mainly using hydrogels [63–65,159–161]. Hydrogels can easily be prepared using natural and artificial polymers (sometimes a combination of both classes). One of the most prominent groups of hydrogels is based on polysaccharides [149,162,163]. Furthermore, other polymers such as polyvinyl alcohol (PVA), polylactic acid and polyglycolic acid (PGA), polyacrylic acid (PAA), and polyethylene glycol (PEG) are intensively studied regarding their ability to form hydrogels used for controlled delivery [160,164]. Hydrogels offer various advantages; most importantly, they are tunable in their chemical structure resulting in controlled degradability. In a comprehensive review, Li et al. discussed various multiscale release kinetic mechanisms of hydrogels and classified them according to the structural interactions. Thus, the kinetics are significantly determined by the hydrogel mesh size, network degradation, swelling, and mechanical deformation. In addition, kinetics depend on various interactions of the hydrogel components such as conjugation, electrostatic interaction, and hydrophobic association [164].

For hard tissue such as bone, our group could recently show that it is possible to guide osteogenesis via purinergic receptor ligand release. Osteogenesis of mesenchymal stem cells is influenced by various purinergic receptors (P1, P2X, and P2Y) [122,124,165–168]. Thus, a release of specific agonists and/or antagonists enables tailoring of the corresponding receptor up- or downregulation. Furthermore, besides osteogenesis, purinergic receptors are also involved in angiogenesis—a process also required during tooth regeneration [68,169,170].

In a recently published paper, we reported the synthesis and testing of novel hybrid release materials based on hydroxyapatite and agarose used to improve the release kinetics of drugs applied for guided osteogenesis [171]. Scanning electron microscopy (SEM) revealed details regarding the influence of the drying treatment: lyophilized (LYO) versus supercritically-dried (SCD) gels were tested and compared. As shown in Figure 3, SEM confirmed a homogeneous distribution of the elements involved in the hybrid (carbon, calcium, and phosphorus). In addition to SEM, energy-dispersive X-ray spectroscopy (EDX) results are given in [171]).

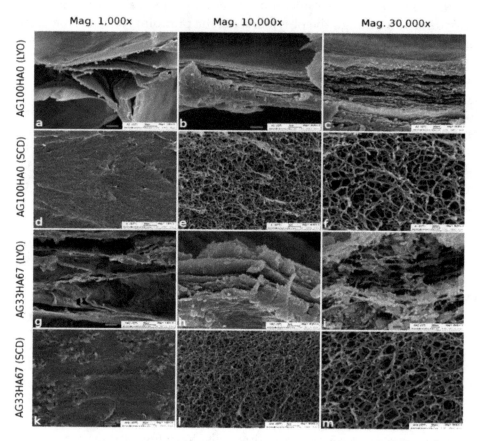

Figure 3. SEM images of agarose lyophilized (LYO) (**a–c**) and supercritically-dried (SCD) (**d–f**) and agarose/hydroxyapatite (33/76 w%) composite LYO (**g–i**) and SCD (**k–m**) at three different magnifications. The scale bars are 10 μm (left), 1 μm (middle), and 0.2 μm (right), respectively.

Hitherto, hybrid systems are mainly studied as release materials for hard tissue regeneration [67]. Here, sustained delivery is required for guided stem cell differentiation, a burst release is favorable to achieve anti-inflammatory and antibacterial effects. Since both processes are also relevant in tooth formation, hybrid materials would be promising candidates to be investigated as release materials to improve cascades, as shown in Figure 2. In previous studies, the HA/agarose hybrids were loaded with model drug compounds for guided differentiation of MSCs. Different release kinetic models were evaluated for adenosine 5'-triphosphate (ATP) and suramin (Figure 4) [171]. Although both drugs are highly water-soluble, the release could be slowed to four days, which is significantly longer than comparable systems reported in the literature [172].

Future efforts should be directed toward the development of tailored drug loading and/or encapsulation materials to be used for the controlled release of bioactive substances during tooth formation [157,173]. As shown in Figure 2 and Table 1, there are various signaling molecules and corresponding activators and suppressor molecules involved in the formation of enamel, dentin, and cementum. For a number of these substances, loading and controlled-release from non-cytotoxic materials already exist, as shown in Table 3. Release materials mainly consist of natural or artificial polymers, but also hybrids composed of organic and inorganic components. The focuses of the studies are release kinetics and corresponding mechanisms. However, some drugs are being successfully applied in vivo.

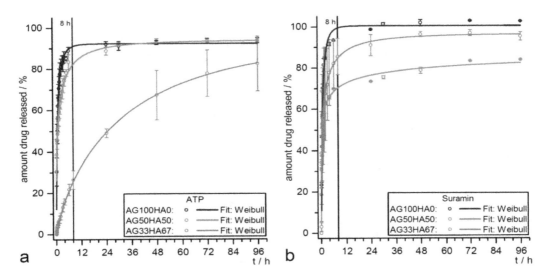

Figure 4. Release data of (**a**) adenosin triphosphate (ATP) and (**b**) suramin from agarose/hydroxyapatite (AG100HA0) (black), AG50HA50 (orange), and AG33HA67 (blue) scaffolds. Data fit: Weibull equation.

Table 3. Materials applicable for loading, encapsulation and drugs/signaling molecules release for promoting cell proliferation, and differentiation.

Signaling Molecule	Material for Drug Loading/Encapsulation and Release	Application	Release Efficiency/Kinetics Tested in	Reference
Amelogenin (EKR1/2 activator)	Self-assembled nanogels of cholesterol-bearing mannan as templates for hierarchical hybrid nanostructures	Amelogenin-releasing hydrogel for remineralization of enamel damage (artificial caries)	Cytotoxicity—in PDL fibroblasts; ex vivo enamel caries models of human molars	[174]
Purmorphamine (Hh activator/Smo agonist)	Glutaraldehyde (GA)-crosslinked gelatin type B matrix (for small molecules and proteins release)	In vitro delivery system for Wnt, Hh agonists and growth factors (e.g., FGF2, VEGF) beneficial for endochondral ossification	Release kinetics (burst vs. sustained release) studied without using cell culture; released molecules bioactivity verified in cell culture/biological assays	[175]
	Poly(propylene glycol–co-lactide) dimethacrylate (PPLM) adhesives for incorporating purmorphamine and TCP	Cell attachment and response to photocured, degradable bone adhesives containing TCP and purmorphamine	MC3T3-E1 (mouse pre-osteoblast cell line)	[176]
	PCL microspheres for encapsulating small molecules using a single emulsion oil-in-water method	Purmorphamine and retinoic acid-loaded microspheres for prolonged release during neural differentiation	Human iPSC aggregates differentiating into motor neurons	[177]
FGF	D-RADA16 peptide hydrogels coated on artificial bone composed of nanohydroxy-apatite/polyamide 66 (nHA/PA66) (for basic FGF release)	Porous growth factor-releasing structure for treating large bone defects	Female SD rat BM-MSCs; female SD rats with induced large bone defects	[178]
	Acetyl chitosan (chitin) gel (for binding and release of chitin binding peptide-FGF2 fusion protein)	Lysozyme-responsive (dose-dependent or activity-dependent) release of CBP-FGF2	Studies without using cell culture/biological assays	[179]
	Silk fibroin e-gel scaffolds (loaded with albumin = Fe3O4-bFGF conjugate)	Enhancing alkaline phosphatase, calcium deposition, and collagen synthesis during osteogenic differentiation	SaOS-2, osteogenic differentiation	[180]
BIO (Wnt/β-catenin activator)	Polymersomes (PMs) consisting of PEG-PCL block copolymer (approved for clinical use) loaded with BIO	BIO-loaded PMs for controlled activation of Wnt signaling and Runx2 during osteogenesis	Murine 3T3 Wnt reporter cells; Human BM-MSCs, osteogenic differentiation	[181]
	None	Local application of Wnt pathway modulators (BIO, CHIR, and Tegusib) to promote dentine regeneration	Wistar rats and CD1 mice molar damage	[121]
BMP2	Porous silica–calcium phosphate composite (SCPC50) (loaded with rhBMP2)	Sustained release of fhBMP2 for alveolar ridge augmentation in saddle-type defect	Mongrel dog with induced mandible defect	[182]
	Calcium phosphate (Ca-P)/poly(L-lactic acid) (PLLA) nanocomposites loaded with rhBMP2	3D Ca-P-PLLA scaffold sustainably releasing Ca^{2+} and rhBMP2 for enhanced osteogenesis	Human BM-MSCs, osteogenic differentiation	[183]
	Poly(lactic-co-glycolic acid)-multistage vector composite microspheres (PLGA-MSV) (for BMP2 release)	Controlled prolonged release of BMP2 for osteoinduction of rat BM-MSCs	Male SD rat BM-MSCs, osteogenic differentiation	[184]

Table 3. *Cont.*

Signaling Molecule	Material for Drug Loading/Encapsulation and Release	Application	Release Efficiency/Kinetics Tested in	Reference
TGF-β 1, 3	Poly(ethylene oxide terephthalate)/poly(butylene terephthalate) (PEOT/PBT) fibrous resins for loading the growth factors	Sustained delivery of growth factors (TGF-β1, PDGF-ββ, IGF-1) using a layer by layer assembly for supporting fibroblast attachment and proliferation	TK173 (human renal fibroblast cell line), neonatal rat dermal fibroblasts (nRDFs)	[185]
	Poly(vinylidene fluoride) (PVDF) nanofibers fabricated via electro-spinning method with/without chitosan nanoparticles (loaded with TGF-β1)	PVDF-TGF-β1 as a bio-functionalscaffold for enhancing smooth muscle cells (SMC) differentiation	AT-MSCs, SMC differentiation	[186]
	Alginate nanogel with cross-junction microchannels (encapsulating TGF-β3)	Controlled release of TGF-β3 from polymeric nanogel for enhanced chondrogenesis	Human MSCs, chondrogenic differentiation	[187]
ATP, suramin (P2XR activators)	Albumin nanoparticles (aNPs) of low polydispersity loaded with ATP and coated with erythrocyte membrane (EM)	EM-aNPs developed as a delivery vehicle for ATP to be used as an anticancer agent	HeLa, HEK-293 cell lines	[188]
	Hydroxyapatite (HA)/agarose hybrids for ATP and suramin release	ATP and suramin release for hard tissue formation	Release kinetic studies without cells (see Figure 4); biocompatibility test using AT-MSCs and MG-63 cell line	[171]

In detail, a sequential and on-demand release of multiple drugs (signaling molecules, activators, and suppressors) would be required to control and guide the signaling cascades of amelogenesis, odontogenesis, and cementogenesis [164]. Moreover, on-demand release systems usually require specific stimuli as reported for example for conductive polymer-based delivery devices [150]. Finally, theoretical modeling could provide a more fundamental understanding of release kinetics [189].

4. Whole Tooth Regeneration

The regeneration of a whole tooth as an organ replacement therapy is considered to be the ultimate goal of regenerative dentistry. For patients, this therapeutic option could represent a dream for the replacement of decayed or lost teeth to overcome prosthodontic or implantology treatment using artificial replacements. Whole-tooth generation could be performed as a hybrid strategy where, e.g., biologically created tissue compartments such as the periodontal ligament or a tooth crown would be combined with a metallic or ceramic implant or where a biological regenerated tooth root ("bio-root") would be combined with a prosthetic crown (see, e.g., [190–192]). In the following years, efforts in creating a whole tooth from only cells and tissues ("bio-tooth") will be very likely in the focus. However, despite all efforts and achieved results in basic and translational research, this approach is still challenging [48,58,69,193,194].

4.1. Reactivation the Odontogenic Potency

On the background of teeth evolution, a genetic approach to generate whole teeth may be an option in the far future. Teethed fishes, reptiles, or amphibians are polyphyodonty, which means that several tooth generations can be formed and erupted. This highly regenerative capacity was reduced during evolution. In mammals, many species including human are only diphyodont with the capacity to form a second dentition or even monophyodont such as the mouse [71,72,195]. Revitalizing the odontogenic potency for the lost tooth regeneration capacity may be an interesting approach to induce tooth formation in vivo in the adult. One prerequisite for tooth replacement is the existence of a successional dental lamina (SDL) carrying the capacity for inducing odontogenesis. Even in monophyodont animals, rudimentary SDL has been identified. In addition, in the human species, rudimentary laminae are preserved, which might be responsible for a third dentition but this, however, has been observed very seldomly. On a molecular level, tooth replacement is regulated by signaling pathways [71]. For example, in alligators or snakes, stem cells in the SDL express Sox2, which is initiated by the Wnt/β-catenin pathway an interacts with BMP signaling [195]. Dysregulation of Wnt-signaling is discussed to be important for the de-activation of rudimentary SDL as it occurs in the mouse. Therefore, the revitalization by stabilizing Wnt signaling by application of appropriate factors or genes could be a strategy for the induction of re-growing teeth in the future [195,196].

4.2. Tissue Recombination Approaches

The basic principle of this "classical" approach is to mimic the natural development and formation of a tooth and to recapitulate the signaling cascades regulating tissue interactions during odontogenesis. For over a hundred years, progress has been made in understanding tooth development in different species including human, identifying tissue interactions and factors involved on the morphological, cellular and molecular levels [18,58,65,71,193,195,197]. Classical tissue recombination experiments undertaken in developmental biology research have shown that mouse embryonic tooth germs can be dissociated and later re-aggregated. After temporary ectopic grafting of these cell aggregates, e.g., into the anterior eye chamber, subcutaneously, or under the renal capsule, tooth-like organs with mineralized tissues (dentin and enamel) could be grown (e.g., [198]). This method has been improved in the last years by using collagen drops for the organoid culture of 5–7 days or seeding the re-aggregated germ cells on biodegradable polymers [199–201]. The final goal of these experiments was to implant the constructs into the jaws of postnatal animals to generate a whole "bio-tooth". In line with this cultured rat tooth, bud cells seeded onto biodegradable scaffolds for 12 weeks formed tooth-like crowns consisting of pulp, dentin, enamel, and periodontal ligament after implantation into rat jaws [200].

A breakthrough came with experiments of the group of Ikeda, who could demonstrate that the implantation of re-aggregated autogenous germ cells into the extraction socket of pigs succeeded in the formation, development, and eruption of teeth, which could be brought into occlusion and fulfilled all functions of normal teeth [50]. Over half (56%) of the implanted constructs had erupted. Later, it was also possible to create a unit of a regrown tooth with surrounding alveolar bone [202]. Whole-tooth restoration using autologous bioengineered tooth germ transplantation was also successful in canines [51]. An allogeneic approach was undertaken by Wu and colleagues, who transplanted re-associated tooth germs into the jawbones of minipigs [203]. A xenogeneic approach was published by Wang and co-workers in 2018 [52]. Cells from unerupted deciduous molar germs of pigs were recombined and transplanted first in mouse renal capsules and finally in jawbones. However, problems are caused by the limited sources of tooth germ cells and risks of immune rejection when using allogeneic or xenogeneic cells. In humans, there are many hindrances, e.g., that tooth germs may not be easily accessible, but also ethical and legal constraints must be considered. An alternative could be the use of adult stem cells (see Section 4.2) or of iPSCs [53,54].

Different types of adult dental stem cells, e.g., from the pulp, or differentiated orofacial cells, e.g., from the gingiva, can be used as sources to create iPSCs with a similar epigenetic pattern. These cells show the ability to differentiate into epithelial or mesenchymal tooth germ cells [58,92]. Cai and co-workers generated iPS cells from cells out of human urine, which were differentiated to epithelial sheets and recombined with embryonic mouse dental mesenchyme [91]. Tooth-like structures were generated in which the epithelial cells differentiated into enamel-secreting ameloblasts. The formation of enamel, the hardest tissue of the body (see Section 2.2.1), is an important step in generating whole teeth, but also would be of importance for repair or regeneration of enamel loss in conservative dentistry. Thus, it is of major interest to find tissue sources able to generate dental epithelial cells which can be differentiated into enamel-secreting ameloblasts. Aside from iPSCs, examples for this are epithelial cells from the skin or gingiva as well as epithelial rests of Malassez, which can be found in the PDL, co-culture of these cells with different types of dental mesenchymal cells can lead to ameloblast differentiation or even formation of enamel-like structures [58,87,89].

4.3. Adult Stem Cell Approaches

The optimal method to create whole teeth would be the use of autogenous dental cells from patients demanding tooth regeneration. For whole tooth bioengineering, different strategies in the application of these cells have been developed. One idea was to combine adult stem cells with cells of the progenitor cells of embryonic tooth. Adult stem cells should have an odontogenic competence and should function as a "tooth inducer" when combined with mesenchymal cells or they should express a dental mesenchymal competence when combined with dental epithelium. Already in 2002, Young et al. cultured cells obtained from unerupted porcine tooth buds [199]. The aggregates were grown on biodegradable scaffolds in vitro or transplanted. This led to the formation of a primitive tooth crown with pulp, dentin, and enamel formation. Later, similar bioengineered tooth-like structures could be obtained by using rat and human cells [204,205]. In 2004, Ohazama and colleagues used non-dental adult MSCs in combination with inductive embryonic dental epithelium first transplanted under the renal capsule and transplanted them in adult jaws. Tooth formation including root occurred and the teeth erupted. In addition, bone was induced [206]. Adipose-derived MSCs alone were able to generate tooth bud-like structures in vitro [90]. Human gingival epithelial cells were used by Volponi Angelova and associates and combined with embryonic mouse tooth mesenchyme, which yielded an entire tooth outside of an embryo [207].

However, for all these experiments, relatively large amounts of adult cell populations were necessary that should be able to retain any odontogenic potential and, in addition, a large number of embryonic cells was needed as well. In a case of embryonic mouse tooth mesenchyme, a minimum cell number of 4×10^4 to 4×10^5 was sufficient according to the experiments of Hu et al. (2006) [208]. Therefore, to do so, cells from multiple embryos must be harvested. Another problem is the loss of the inductive capacity already after 24–48 h in culture, which makes the in vitro expansion of these cells using standard methods impossible [209]. This phenomenon can be explained by the fact that mesenchymal stem cells lose their dense packaging formed by cellular condensation and thus their linked cell contacts, which is a prerequisite for an inductive capacity in vivo. Ongoing research focuses therefore also on how an odontogenic potential can be maintained in vitro [194]. 3D micro-culture systems such as the hanging drop method in liquid media may allow the preservation of such signals. However, many cells are necessary for these methods [210]. Gene expression studies must be undertaken to identify signaling factors, which are lost in 2D cell cultures. In a study using postnatal dental pulp stem cells, Yang and collaborators could obtain "a rescue" in cultured cells due to the combination with uncultured mesenchymal tooth germ cells [57]. This rescue or community effect is responsible for the reactivation of inductive signals. Forthcoming, iPS cells (see Section 4.2) may be an appropriate cell substitute to overcome these biological problems.

In the future, research will presumably focus on using adult stem cells from dental and non-dental sources to test recombination or co-culturing for their effects on tooth development. Zhang and coworkers optimized such a method by recombinant 3D-tissue engineering of intact dental tissues and cell suspensions from postnatal porcine teeth and human third molars [211]. After osteogenic culturing and subcutaneous transplantation in athymic nude rat hosts, tooth-like constructs forming all dental hard substances could be harvested. Recently, tooth buds could be generated by co-culturing postnatal dental stem cells with human HUVEC cells encapsulated in gelatine hydrogel [56]. Only postnatal dental stem cells were used by Yang et al. (2016), who differentiated odontoblasts and osteoblasts from pig dental pulp stem cells and seeded them with gingival epithelium on a bioactive scaffold. Implantation into extraction sockets of 13.5-month-old pigs revealed the development of teeth in seven of eight animals. The regenerated molar teeth expressed dentin-matrix protein-1 and osteopontin [212].

4.4. Problems in Whole Tooth Regeneration

Despite the progress in some basic strategies for tooth regeneration, we still face a lot of problems [18,48]. An important condition for a proper functional occlusion in a dentition where teeth should be replaced by regeneration is the correct anatomical size and shape of the crown. Especially the relief of the occlusal surface with its specific pattern of fissures and cusps is relevant for a functional occlusion. The proper size and shape of a crown are determined by epithelial morphogenesis forming spatially regulated cellular condensations as signaling centers, called knots [71,197]. These knots (initiation knot, primary enamel knot, and secondary enamel knot) regulate crown development and cusp number, morphology, and pattern by expressing different factors such as FGF, BMP, Wnt, or Shh, as already mentioned. The number of tooth cusps in the mouse depends on the activity of Shh, EDA, and Activin A pathways [71,197]. The tooth size is independently regulated from the cusp number and is not only dependent on epithelial, but also mesenchymal influences. Therefore, it was suggested that the tooth size could be controlled by prolonging the activity of tooth epithelial stem cells and increasing the number of mesenchymal stem cells in recombination experiments [197]. The different tooth types such as molars or incisors have specific morphological features not only of the occlusal surface but also of the crown and root morphology. This will also be an important aspect for future tooth engineering [193]. The quality and the biomechanical loading of dental hard tissues are also important for occlusion and mastication. In already developed models of tooth regeneration, only a low level of enamel mineralization could be observed.

Tooth health is also dependent on proper vascularization and innervation. While vascularization occurs in different models already published [213], the question is whether this would be also sufficient for the long-lasting survival of regenerated teeth. Efforts have been made to induce neurogenesis and formation of nerve fibers, e.g., by using exogenous agents such as semaphorin 3 receptor inhibitors, by application of immunomodulation using cyclosporin A, or implication of bone marrow stromal cells [48]. Recently, Strub et al. recombined embryonic dental epithelium with a mixture of dental mesenchymal cells and bone marrow-derived cells and cultured and implanted these cells subcutaneously. The tooth-like tissues obtained were innervated with axons entering the newly formed pulp [214].

Other problems include the formation of a proper periodontium or infections occurring during or after transplantation. If whole tooth constructs can be implanted, the role of the tissue environment will play an important role in the success: How is the quality of the jawbone? How will the implantation be affected by age or systemic diseases of the patients? How resistant will the newly created tooth be against probable infections? Finally, the costs of creating a "bio-tooth" are also unpredictable yet [48,156].

5. Conclusions and Future Perspectives

Progress in regenerating whole teeth will need scientific research on different levels such as identification of appropriate cell sources with tooth inductive signals. For this further research on the feasibility of iPS cells for this approach is important. Furthermore, the identification of master genes in gene regulatory networks responsible for tooth induction and tooth formation is necessary for successful manipulation of, e.g., adult cells to form bioengineered dental tissues, and to control tooth crown, size and tooth identity.

Applying the acquired knowledge about signaling pathways shaping dental tissue genesis might stimulate novel cell culture techniques establishment and functionalized scaffolds development. Functionalized biomaterials will presumably play a central role in hard dental tissue regeneration such as dentin and cementum and probably the main role in enamel regeneration since this tissue is acellular and cannot be reproduced in vitro relying solely on a cell-based approach. Although several potentially appropriate biomaterials have already been investigated and tested, only very few examples were used in clinical studies until now. Future efforts in stem cell-based approaches will very likely be directed toward biomaterials that allow sequential and on-demand drug release of multiple drugs in order to tailor timely the different cascade processes during amelogenesis, dentinogenesis, and cementogenesis, respectively.

On the translational level, methods to improve 3D organogenesis, 3D printing applications, or the appropriate application of stimulatory molecules and drugs should be tested intensively. Solutions must be found for the proper mineralization of dental hard tissue formed by the regeneration process to ensure the natural properties of teeth in occlusion and mastication. Finally, there are considerable financial investment problems that should be taken into account. Then, but only then, whole biological tooth regeneration may even be a blueprint for the regeneration of other complex organs [70].

Author Contributions: Conceptualization and writing—original draft preparation, J.B., D.B., M.S., E.T., W.G.; writing—review and editing, E.T., M.S.; supervision and project administration, E.T., M.S.; funding acquisition, E.T., M.S., W.G. All authors have read and agree to the proofread version of the manuscript.

References

1. Petersen, P.E.; Bourgeois, D.; Ogawa, H.; Estupinan-Day, S.; Ndiaye, C. The global burden of oral diseases and risks to oral health. *Bull. World Health Organ.* **2005**, *83*, 661–669. [PubMed]

2. Nazir, M.A. Prevalence of periodontal disease, its association with systemic diseases and prevention. *Int. J. Health Sci. (Qassim)* **2017**, *11*, 72–80.

3. Conrads, G.; About, I. Pathophysiology of Dental Caries. In *Monographs in Oral Science*; S. Karger AG: Basel, Switzerland, 2018; Volume 27, pp. 1–10, ISBN 0077-0892.

4. Pan, W.; Wang, Q.; Chen, Q. The cytokine network involved in the host immune response to periodontitis. *Int. J. Oral Sci.* **2019**, *11*, 30. [CrossRef] [PubMed]

5. Kreiborg, S.; Jensen, B.L. Tooth formation and eruption—Lessons learnt from cleidocranial dysplasia. *Eur. J. Oral Sci.* **2018**, *126*, 72–80. [CrossRef] [PubMed]

6. Thesleff, I. Current understanding of the process of tooth formation: Transfer from the laboratory to the clinic. *Aust. Dent. J.* **2014**, *59*, 48–54. [CrossRef]

7. Smith, C.E.L.; Poulter, J.A.; Antanaviciute, A.; Kirkham, J.; Brookes, S.J.; Inglehearn, C.F.; Mighell, A.J. Amelogenesis Imperfecta; Genes, Proteins, and Pathways. *Front. Physiol.* **2017**, *8*, 435. [CrossRef]

8. Barron, M.J.; McDonnell, S.T.; Mackie, I.; Dixon, M.J. Hereditary dentine disorders: Dentinogenesis imperfecta and dentine dysplasia. *Orphanet J. Rare Dis.* **2008**, *3*, 31. [CrossRef]

9. Chen, F.-M.; Gao, L.-N.; Tian, B.-M.; Zhang, X.-Y.; Zhang, Y.-J.; Dong, G.-Y.; Lu, H.; Chu, Q.; Xu, J.; Yu, Y.; et al. Treatment of periodontal intrabony defects using autologous periodontal ligament stem cells: A randomized clinical trial. *Stem Cell Res. Ther.* **2016**, *7*, 33. [CrossRef]

10. Lacruz, R.S.; Habelitz, S.; Wright, J.T.; Paine, M.L. Dental Enamel Formation and Implications For Oral Health and Disease. *Physiol. Rev.* **2017**, *97*, 939–993. [CrossRef]

11. Tompkins, K. Molecular Mechanisms of Cytodifferentiation in Mammalian Tooth Development. *Connect. Tissue Res.* **2006**, *47*, 111–118. [CrossRef]

12. Goldberg, M.; Kulkarni, A.B.; Young, M.; Boskey, A. Dentin: Structure, composition and mineralization. *Front. Biosci. (Elite Ed.)* **2011**, *3*, 711–735. [CrossRef] [PubMed]

13. Yamamoto, T.; Hasegawa, T.; Yamamoto, T.; Hongo, H.; Amizuka, N. Histology of human cementum: Its structure, function, and development. *Jpn. Dent. Sci. Rev.* **2016**, *52*, 63–74. [CrossRef] [PubMed]

14. Yuan, Z.; Nie, H.; Wang, S.; Lee, C.H.; Li, A.; Fu, S.Y.; Zhou, H.; Chen, L.; Mao, J.J. Biomaterial selection for tooth regeneration. *Tissue Eng. Part B. Rev.* **2011**, *17*, 373–388. [CrossRef] [PubMed]

15. Nakata, A.; Kameda, T.; Nagai, H.; Ikegami, K.; Duan, Y.; Terada, K.; Sugiyama, T. Establishment and characterization of a spontaneously immortalized mouse ameloblast-lineage cell line. *Biochem. Biophys. Res. Commun.* **2003**, *308*, 834–839. [CrossRef]

16. Moussa, D.G.; Aparicio, C. Present and future of tissue engineering scaffolds for dentin-pulp complex regeneration. *J. Tissue Eng. Regen. Med.* **2019**, *13*, 58–75. [CrossRef]

17. Jung, C.; Kim, S.; Sun, T.; Cho, Y.-B.; Song, M. Pulp-dentin regeneration: Current approaches and challenges. *J. Tissue Eng.* **2019**, *10*. [CrossRef]

18. Thesleff, I. From understanding tooth development to bioengineering of teeth. *Eur. J. Oral Sci.* **2018**, *126*, 67–71. [CrossRef]

19. Shrestha, S.; Kishen, A. Bioactive Molecule Delivery Systems for Dentin-pulp Tissue Engineering. *J. Endod.* **2017**, *43*, 733–744. [CrossRef]

20. Katsura, K.A.; Horst, J.A.; Chandra, D.; Le, T.Q.; Nakano, Y.; Zhang, Y.; Horst, O.V.; Zhu, L.; Le, M.H.; DenBesten, P.K. WDR72 models of structure and function: A stage-specific regulator of enamel mineralization. *Matrix Biol.* **2014**, *38*, 48–58. [CrossRef]

21. Guo, F.; Feng, J.; Wang, F.; Li, W.; Gao, Q.; Chen, Z.; Shoff, L.; Donly, K.J.; Gluhak-Heinrich, J.; Chun, Y.H.P.; et al. Bmp2 deletion causes an amelogenesis imperfecta phenotype via regulating enamel gene expression. *J. Cell. Physiol.* **2015**, *230*, 1871–1882. [CrossRef]

22. Xie, X.; Liu, C.; Zhang, H.; Jani, P.H.; Lu, Y.; Wang, X.; Zhang, B.; Qin, C. Abrogation of epithelial BMP2 and BMP4 causes Amelogenesis Imperfecta by reducing MMP20 and KLK4 expression. *Sci. Rep.* **2016**, *6*, 25364. [CrossRef] [PubMed]

23. Shimo, T.; Koyama, E.; Kanayama, M.; Kurio, N.; Okui, T.; Yamamoto, D.; Hassan, N.M.M.; Sasaki, A. Sonic Hedgehog Positively Regulates Odontoblast Differentiation by a BMP2/4-dependent Mechanism. *J. Oral Tissue Eng.* **2009**, *7*, 26–37.

24. Li, S.; Shao, J.; Zhou, Y.; Friis, T.; Yao, J.; Shi, B.; Xiao, Y. The impact of Wnt signalling and hypoxia on osteogenic and cementogenic differentiation in human periodontal ligament cells. *Mol. Med. Rep.* **2016**, *14*, 4975–4982. [CrossRef] [PubMed]

25. Ngo, V.A.; Jung, J.-Y.; Koh, J.-T.; Oh, W.-M.; Hwang, Y.-C.; Lee, B.-N. Leptin Induces Odontogenic Differentiation and Angiogenesis in Human Dental Pulp Cells via Activation of the Mitogen-activated Protein Kinase Signaling Pathway. *J. Endod.* **2018**, *44*, 585–591. [CrossRef]

26. Sagomonyants, K.; Mina, M. Biphasic effects of FGF2 on odontoblast differentiation involve changes in the BMP and Wnt signaling pathways. *Connect. Tissue Res.* **2014**, *55* (Suppl. 1), 53–56. [CrossRef]

27. Sagomonyants, K.; Kalajzic, I.; Maye, P.; Mina, M. Enhanced Dentinogenesis of Pulp Progenitors by Early Exposure to FGF2. *J. Dent. Res.* **2015**, *94*, 1582–1590. [CrossRef]

28. Malik, Z.; Alexiou, M.; Hallgrimsson, B.; Economides, A.N.; Luder, H.U.; Graf, D. Bone Morphogenetic Protein 2 Coordinates Early Tooth Mineralization. *J. Dent. Res.* **2018**, *97*, 835–843. [CrossRef]

29. Kim, T.-H.; Bae, C.-H.; Lee, J.-Y.; Lee, J.-C.; Ko, S.-O.; Chai, Y.; Cho, E.-S. Temporo-spatial requirement of Smad4 in dentin formation. *Biochem. Biophys. Res. Commun.* **2015**, *459*, 706–712. [CrossRef]

30. Ishikawa, Y.; Nakatomi, M.; Ida-Yonemochi, H.; Ohshima, H. Quiescent adult stem cells in murine teeth are regulated by Shh signaling. *Cell Tissue Res.* **2017**, *369*, 497–512. [CrossRef]

31. Lv, H.; Yang, J.; Wang, C.; Yu, F.; Huang, D.; Ye, L. The WNT7B protein promotes the migration and differentiation of human dental pulp cells partly through WNT/beta-catenin and c-Jun N-terminal kinase signalling pathways. *Arch. Oral Biol.* **2018**, *87*, 54–61. [CrossRef]

32. Wang, W.; Yi, X.; Ren, Y.; Xie, Q. Effects of Adenosine Triphosphate on Proliferation and Odontoblastic Differentiation of Human Dental Pulp Cells. *J. Endod.* **2016**, *42*, 1483–1489. [CrossRef] [PubMed]

33. Tao, H.; Lin, H.; Sun, Z.; Pei, F.; Zhang, J.; Chen, S.; Liu, H.; Chen, Z. Klf4 Promotes Dentinogenesis and Odontoblastic Differentiation via Modulation of TGF-β Signaling Pathway and Interaction With Histone Acetylation. *J. Bone Miner. Res.* **2019**, *34*, 1502–1516. [CrossRef] [PubMed]

34. He, Y.D.; Sui, B.D.; Li, M.; Huang, J.; Chen, S.; Wu, L.A. Site-specific function and regulation of Osterix in tooth root formation. *Int. Endod. J.* **2016**, *49*, 1124–1131. [CrossRef] [PubMed]

35. Cao, Z.; Liu, R.; Zhang, H.; Liao, H.; Zhang, Y.; Hinton, R.J.; Feng, J.Q. Osterix controls cementoblast differentiation through downregulation of Wnt-signaling via enhancing DKK1 expression. *Int. J. Biol. Sci.* **2015**, *11*, 335–344. [CrossRef]

36. Choi, H.; Kim, T.-H.; Yang, S.; Lee, J.-C.; You, H.-K.; Cho, E.-S. A Reciprocal Interaction between β-Catenin and Osterix in Cementogenesis. *Sci. Rep.* **2017**, *7*, 8160. [CrossRef]

37. Choi, H.; Ahn, Y.-H.; Kim, T.-H.; Bae, C.-H.; Lee, J.-C.; You, H.-K.; Cho, E.-S. TGF-β Signaling Regulates Cementum Formation through Osterix Expression. *Sci. Rep.* **2016**, *6*, 26046. [CrossRef]

38. Zhang, H.; Tompkins, K.; Garrigues, J.; Snead, M.L.; Gibson, C.W.; Somerman, M.J. Full length amelogenin binds to cell surface LAMP-1 on tooth root/periodontium associated cells. *Arch. Oral Biol.* **2010**, *55*, 417–425. [CrossRef]

39. Hakki, S.S.; Bozkurt, S.B.; Türkay, E.; Dard, M.; Purali, N.; Götz, W. Recombinant amelogenin regulates the bioactivity of mouse cementoblasts in vitro. *Int. J. Oral Sci.* **2018**, *10*, 15. [CrossRef]

40. Olley, R.; Xavier, G.M.; Seppala, M.; Volponi, A.A.; Geoghegan, F.; Sharpe, P.T.; Cobourne, M.T. Expression analysis of candidate genes regulating successional tooth formation in the human embryo. *Front. Physiol.* **2014**, *5*, 445. [CrossRef]

41. Foster, B.L.; Nagatomo, K.J.; Nociti, F.H., Jr.; Fong, H.; Dunn, D.; Tran, A.B.; Wang, W.; Narisawa, S.; Millán, J.L.; Somerman, M.J. Central role of pyrophosphate in acellular cementum formation. *PLoS ONE* **2012**, *7*, e38393. [CrossRef]

42. Takahashi, S.; Kawashima, N.; Sakamoto, K.; Nakata, A.; Kameda, T.; Sugiyama, T.; Katsube, K.; Suda, H. Differentiation of an ameloblast-lineage cell line (ALC) is induced by Sonic hedgehog signaling. *Biochem. Biophys. Res. Commun.* **2007**, *353*, 405–411. [CrossRef] [PubMed]

43. Lee, H.-K.; Lee, D.-S.; Ryoo, H.-M.; Park, J.-T.; Park, S.-J.; Bae, H.-S.; Cho, M.-I.; Park, J.-C. The odontogenic ameloblast-associated protein (ODAM) cooperates with RUNX2 and modulates enamel mineralization via regulation of MMP-20. *J. Cell. Biochem.* **2010**, *111*, 755–767. [CrossRef] [PubMed]

44. Liu, X.; Xu, C.; Tian, Y.; Sun, Y.; Zhang, J.; Bai, J.; Pan, Z.; Feng, W.; Xu, M.; Li, C.; et al. RUNX2 contributes to TGF-β1-induced expression of Wdr72 in ameloblasts during enamel mineralization. *Biomed. Pharmacother.* **2019**, *118*, 109235. [CrossRef] [PubMed]

45. Aurrekoetxea, M.; Irastorza, I.; García-Gallastegui, P.; Jiménez-Rojo, L.; Nakamura, T.; Yamada, Y.; Ibarretxe, G.; Unda, F.J. Wnt/β-Catenin Regulates the Activity of Epiprofin/Sp6, SHH, FGF, and BMP to Coordinate the Stages of Odontogenesis. *Front. Cell Dev. Biol.* **2016**, *4*, 25. [CrossRef] [PubMed]

46. Fan, L.; Shijian, D.; Xin, S.; Mengmeng, L.; Cheng, S.; Wang, Y.; Gao, Y.; Chu, C.-H.; Zhan, Q. Constitutive activation of β-catenin in ameloblasts leads to incisor enamel hypomineralization. *J. Mol. Histol.* **2018**, *49*, 499–507. [CrossRef]

47. Okubo, M.; Chiba, R.; Karakida, T.; Yamazaki, H.; Yamamoto, R.; Kobayashi, S.; Niwa, T.; Margolis, H.C.; Nagano, T.; Yamakoshi, Y.; et al. Potential function of TGF-β isoforms in maturation-stage ameloblasts. *J. Oral Biosci.* **2019**, *61*, 43–54. [CrossRef]

48. Yelick, P.C.; Sharpe, P.T. Tooth Bioengineering and Regenerative Dentistry. *J. Dent. Res.* **2019**, *98*, 1173–1182. [CrossRef]

49. Dosedělová, H.; Dumková, J.; Lesot, H.; Glocová, K.; Kunová, M.; Tucker, A.S.; Veselá, I.; Krejčí, P.; Tichý, F.; Hampl, A.; et al. Fate of the molar dental lamina in the monophyodont mouse. *PLoS ONE* **2015**, *10*, e0127543. [CrossRef]

50. Ikeda, E.; Morita, R.; Nakao, K.; Ishida, K.; Nakamura, T.; Takano-Yamamoto, T.; Ogawa, M.; Mizuno, M.; Kasugai, S.; Tsuji, T. Fully functional bioengineered tooth replacement as an organ replacement therapy. *Proc. Natl. Acad. Sci. USA* **2009**, *106*, 13475–13480. [CrossRef]

51. Ono, M.; Oshima, M.; Ogawa, M.; Sonoyama, W.; Hara, E.S.; Oida, Y.; Shinkawa, S.; Nakajima, R.; Mine, A.; Hayano, S.; et al. Practical whole-tooth restoration utilizing autologous bioengineered tooth germ transplantation in a postnatal canine model. *Sci. Rep.* **2017**, *7*, 44522. [CrossRef]

52. Wang, F.; Wu, Z.; Fan, Z.; Wu, T.; Wang, J.; Zhang, C.; Wang, S. The cell re-association-based whole-tooth regeneration strategies in large animal, Sus scrofa. *Cell Prolif.* **2018**, *51*, e12479. [CrossRef] [PubMed]

53. Lee, J.-H.; Seo, S.-J. Biomedical Application of Dental Tissue-Derived Induced Pluripotent Stem Cells. *Stem Cells Int.* **2016**, *2016*, 9762465. [PubMed]

54. Morsczeck, C.; Reichert, T.E. Dental stem cells in tooth regeneration and repair in the future. *Expert Opin. Biol. Ther.* **2018**, *18*, 187–196. [CrossRef] [PubMed]

55. Zhang, W.; Vazquez, B.; Oreadi, D.; Yelick, P.C. Decellularized Tooth Bud Scaffolds for Tooth Regeneration. *J. Dent. Res.* **2017**, *96*, 516–523. [CrossRef] [PubMed]

56. Smith, E.E.; Angstadt, S.; Monteiro, N.; Zhang, W.; Khademhosseini, A.; Yelick, P.C. Bioengineered Tooth Buds Exhibit Features of Natural Tooth Buds. *J. Dent. Res.* **2018**, *97*, 1144–1151. [CrossRef]

57. Yang, L.; Angelova Volponi, A.; Pang, Y.; Sharpe, P.T. Mesenchymal Cell Community Effect in Whole Tooth Bioengineering. *J. Dent. Res.* **2016**, *96*, 186–191. [CrossRef]

58. Smith, E.E.; Yelick, P.C. Progress in Bioengineered Whole Tooth Research: From Bench to Dental Patient Chair. *Curr. Oral Health Rep.* **2016**, *3*, 302–308. [CrossRef]

59. Schulze, M.; Tobiasch, E. Artificial Scaffolds and Mesenchymal Stem Cells for Hard Tissues. In *Tissue Engineering III: Cell—Surface Interactions for Tissue Culture*; Kasper, C., Witte, F., Pörtner, R., Eds.; Springer: Berlin/Heidelberg, Germany, 2011; pp. 153–194.

60. Leiendecker, A.; Witzleben, S.; Tobiasch, M.S. and E. Template-Mediated Biomineralization for Bone Tissue Engineering. *Curr. Stem Cell Res. Ther.* **2017**, *12*, 103–123. [CrossRef]

61. El Khaldi-Hansen, B.; El-Sayed, F.; Tobiasch, E.; Witzleben, S.; Schulze, M. Functionalized 3D Scaffolds for Template- Mediated Biomineralization in Bone Regeneration. *Front. Stem Cell Regen. Med. Res.* **2017**, *4*, 3–58. [CrossRef]

62. Götz, W.; Tobiasch, E.; Witzleben, S.; Schulze, M. Effects of Silicon Compounds on Biomineralization, Osteogenesis, and Hard Tissue Formation. *Pharmaceutics* **2019**, *11*, 117. [CrossRef]

63. Witzler, M.; Alzagameem, A.; Bergs, M.; El Khaldi-Hansen, B.; Klein, S.; Hielscher, D.; Kamm, B.; Kreyenschmidt, J.; Tobiasch, E.; Schulze, M. Lignin-Derived Biomaterials for Drug Release and Tissue Engineering. *Molecules* **2018**, *23*, 1885. [CrossRef] [PubMed]

64. Zhang, K.; Wang, S.; Zhou, C.; Cheng, L.; Gao, X.; Xie, X.; Sun, J.; Wang, H.; Weir, M.D.; Reynolds, M.A.; et al. Advanced smart biomaterials and constructs for hard tissue engineering and regeneration. *Bone Res.* **2018**, *6*, 31. [CrossRef] [PubMed]

65. Huang, D.; Ren, J.; Li, R.; Guan, C.; Feng, Z.; Bao, B.; Wang, W.; Zhou, C. Tooth Regeneration: Insights from Tooth Development and Spatial-Temporal Control of Bioactive Drug Release. *Stem Cell Rev. Rep.* **2020**, *16*, 41–55. [CrossRef] [PubMed]

66. Patel, E.; Pradeep, P.; Kumar, P.; Choonara, Y.E.; Pillay, V. Oroactive dental biomaterials and their use in endodontic therapy. *J. Biomed. Mater. Res. Part B Appl. Biomater.* **2020**, *108*, 201–212. [CrossRef]

67. Sa, Y.; Gao, Y.; Wang, M.; Wang, T.; Feng, X.; Wang, Z.; Wang, Y.; Jiang, T. Bioactive calcium phosphate cement with excellent injectability, mineralization capacity and drug-delivery properties for dental biomimetic reconstruction and minimum intervention therapy. *RSC Adv.* **2016**, *6*, 27349–27359. [CrossRef]

68. Nosrati, H.; Pourmotabed, S.; Sharifi, E. A Review on Some Natural Biopolymers and Their Applications in Angiogenesis and Tissue Engineering. *J. Appl. Biotechnol. Rep.* **2018**, *5*, 81–91. [CrossRef]

69. Hu, L.; Liu, Y.; Wang, S. Stem cell-based tooth and periodontal regeneration. *Oral Dis.* **2018**, *24*, 696–705. [CrossRef]

70. Takeo, M.; Tsuji, T. Organ regeneration based on developmental biology: Past and future. *Curr. Opin. Genet. Dev.* **2018**, *52*, 42–47. [CrossRef]

71. Yu, T.; Klein, O.D. Molecular and cellular mechanisms of tooth development, homeostasis and repair. *Development* **2020**, *147*, dev184754. [CrossRef]

72. Calamari, Z.T.; Hu, J.K.-H.; Klein, O.D. Tissue Mechanical Forces and Evolutionary Developmental Changes Act Through Space and Time to Shape Tooth Morphology and Function. *Bioessays* **2018**, *40*, e1800140. [CrossRef]

73. Yuan, Y.; Chai, Y. Regulatory mechanisms of jaw bone and tooth development. *Curr. Top. Dev. Biol.* **2019**, *133*, 91–118. [PubMed]

74. Orchardson, R.; Cadden, S.W. An Update on the Physiology of the Dentine–Pulp Complex. *Dent. Update* **2001**, *28*, 200–209. [CrossRef] [PubMed]

75. Hyun, S.-Y.; Lee, J.-H.; Kang, K.-J.; Jang, Y.-J. Effect of FGF-2, TGF-β-1, and BMPs on Teno/Ligamentogenesis and Osteo/Cementogenesis of Human Periodontal Ligament Stem Cells. *Mol. Cells* **2017**, *40*, 550–557. [CrossRef] [PubMed]

76. Kawashima, N.; Okiji, T. Odontoblasts: Specialized hard-tissue-forming cells in the dentin-pulp complex. *Congenit. Anom. (Kyoto)* **2016**, *56*, 144–153. [CrossRef] [PubMed]

77. Seppala, M.; Fraser, G.J.; Birjandi, A.A.; Xavier, G.M.; Cobourne, M.T. Sonic Hedgehog Signaling and Development of the Dentition. *J. Dev. Biol.* **2017**, *5*, 6. [CrossRef]

78. Saygin, N.E.; Tokiyasu, Y.; Giannobile, W.V.; Somerman, M.J. Growth factors regulate expression of mineral associated genes in cementoblasts. *J. Periodontol.* **2000**, *71*, 1591–1600. [CrossRef]

79. Boabaid, F.; Gibson, C.W.; Kuehl, M.A.; Berry, J.E.; Snead, M.L.; Nociti, F.H., Jr.; Katchburian, E.; Somerman, M.J. Leucine-Rich Amelogenin Peptide: A Candidate Signaling Molecule During Cementogenesis. *J. Periodontol.* **2004**, *75*, 1126–1136. [CrossRef]

80. Casagrande, L.; Demarco, F.F.; Zhang, Z.; Araujo, F.B.; Shi, S.; Nör, J.E. Dentin-derived BMP-2 and Odontoblast Differentiation. *J. Dent. Res.* **2010**, *89*, 603–608. [CrossRef]

81. Yi, X.; Wang, W.; Xie, Q. Adenosine receptors enhance the ATP-induced odontoblastic differentiation of human dental pulp cells. *Biochem. Biophys. Res. Commun.* **2018**, *497*, 850–856. [CrossRef]

82. Zhang, S.; Ye, D.; Ma, L.; Ren, Y.; Dirksen, R.T.; Liu, X. Purinergic Signaling Modulates Survival/Proliferation of Human Dental Pulp Stem Cells. *J. Dent. Res.* **2019**, *98*, 242–249. [CrossRef]

83. Chen, D.; Yu, F.; Wu, F.; Bai, M.; Lou, F.; Liao, X.; Wang, C.; Ye, L. The role of Wnt7B in the mediation of dentinogenesis via the ERK1/2 pathway. *Arch. Oral Biol.* **2019**, *104*, 123–132. [CrossRef] [PubMed]

84. Shin, N.-Y.; Yamazaki, H.; Beniash, E.; Yang, X.; Margolis, S.; Pugach, M.; Simmer, J.; Margolis, H. Amelogenin phosphorylation regulates tooth enamel formation by stabilizing a transient amorphous mineral precursor. *J. Biol. Chem.* **2020**, *295*, jbc.RA119.010506. [CrossRef] [PubMed]

85. Zhang, J.; Wang, L.; Zhang, W.; Putnis, C. V Phosphorylated/Nonphosphorylated Motifs in Amelotin Turn Off/On the Acidic Amorphous Calcium Phosphate-to-Apatite Phase Transformation. *Langmuir* **2020**, *36*, 2102–2109. [CrossRef] [PubMed]

86. Green, D.R.; Schulte, F.; Lee, K.-H.; Pugach, M.K.; Hardt, M.; Bidlack, F.B. Mapping the Tooth Enamel Proteome and Amelogenin Phosphorylation Onto Mineralizing Porcine Tooth Crowns. *Front. Physiol.* **2019**, *10*, 925. [CrossRef] [PubMed]

87. Hu, X.; Lee, J.-W.; Zheng, X.; Zhang, J.; Lin, X.; Song, Y.; Wang, B.; Hu, X.; Chang, H.-H.; Chen, Y.; et al. Efficient induction of functional ameloblasts from human keratinocyte stem cells. *Stem Cell Res. Ther.* **2018**, *9*, 126. [CrossRef] [PubMed]

88. Shinmura, Y.; Tsuchiya, S.; Hata, K.; Honda, M.J. Quiescent epithelial cell rests of Malassez can differentiate into ameloblast-like cells. *J. Cell. Physiol.* **2008**, *217*, 728–738. [CrossRef] [PubMed]

89. Padma Priya, S.; Higuchi, A.; Abu Fanas, S.; Pooi Ling, M.; Kumari Neela, V.; Sunil, P.M.; Saraswathi, T.R.; Murugan, K.; Alarfaj, A.A.; Munusamy, M.A.; et al. Odontogenic epithelial stem cells: Hidden sources. *Lab. Investig.* **2015**, *95*, 1344–1352. [CrossRef]

90. Ferro, F.; Spelat, R.; Falini, G.; Gallelli, A.; D'Aurizio, F.; Puppato, E.; Pandolfi, M.; Beltrami, A.P.; Cesselli, D.; Beltrami, C.A.; et al. Adipose tissue-derived stem cell in vitro differentiation in a three-dimensional dental bud structure. *Am. J. Pathol.* **2011**, *178*, 2299–2310. [CrossRef]

91. Cai, J.; Zhang, Y.; Liu, P.; Chen, S.; Wu, X.; Sun, Y.; Li, A.; Huang, K.; Luo, R.; Wang, L.; et al. Generation of tooth-like structures from integration-free human urine induced pluripotent stem cells. *Cell Regen. (Lond. Engl.)* **2013**, *2*, 6. [CrossRef]

92. Kim, E.-J.; Yoon, K.-S.; Arakaki, M.; Otsu, K.; Fukumoto, S.; Harada, H.; Green, D.W.; Lee, J.-M.; Jung, H.-S. Effective Differentiation of Induced Pluripotent Stem Cells Into Dental Cells. *Dev. Dyn.* **2019**, *248*, 129–139. [CrossRef]

93. Hosoya, A.; Shalehin, N.; Takebe, H.; Shimo, T.; Irie, K. Sonic Hedgehog Signaling and Tooth Development. *Int. J. Mol. Sci.* **2020**, *21*. [CrossRef] [PubMed]

94. Shi, X.; Mao, J.; Liu, Y. Concise review: Pulp stem cells derived from human permanent and deciduous teeth: Biological characteristics and therapeutic applications. *Stem Cells Transl. Med.* **2020**, *9*, 445–464. [CrossRef] [PubMed]

95. Liu, X.; Gao, Y. Runx2 is involved in regulating amelotin promoter activity and gene expression in ameloblasts. In Proceedings of the 2013 ICME International Conference on Complex Medical Engineering, Beijing, China, 25–28 May 2013; pp. 91–96.

96. Cantù, C.; Pagella, P.; Shajiei, T.D.; Zimmerli, D.; Valenta, T.; Hausmann, G.; Basler, K.; Mitsiadis, T.A. A cytoplasmic role of Wnt/β-catenin transcriptional cofactors Bcl9, Bcl9l, and Pygopus in tooth enamel formation. *Sci. Signal.* **2017**, *10*, eaah4598. [CrossRef] [PubMed]

97. Guan, X.; Xu, M.; Millar, S.E.; Bartlett, J.D. Beta-catenin is essential for ameloblast movement during enamel development. *Eur. J. Oral Sci.* **2016**, *124*, 221–227. [CrossRef]

98. Järvinen, E.; Shimomura-Kuroki, J.; Balic, A.; Jussila, M.; Thesleff, I. Mesenchymal Wnt/β-catenin signaling limits tooth number. *Development* **2018**, *145*, dev158048. [CrossRef]

99. Yang, Y.; Li, Z.; Chen, G.; Li, J.; Li, H.; Yu, M.; Zhang, W.; Guo, W.; Tian, W. GSK3β regulates ameloblast differentiation via Wnt and TGF-β pathways. *J. Cell. Physiol.* **2018**, *233*, 5322–5333. [CrossRef]

100. Kobayashi-Kinoshita, S.; Yamakoshi, Y.; Onuma, K.; Yamamoto, R.; Asada, Y. TGF-β1 autocrine signalling and enamel matrix components. *Sci. Rep.* **2016**, *6*, 33644. [CrossRef]

101. Yang, J.; Ye, L.; Hui, T.-Q.; Yang, D.-M.; Huang, D.-M.; Zhou, X.-D.; Mao, J.J.; Wang, C.-L. Bone morphogenetic protein 2-induced human dental pulp cell differentiation involves p38 mitogen-activated protein kinase-activated canonical WNT pathway. *Int. J. Oral Sci.* **2015**, *7*, 95–102. [CrossRef]

102. Davies, O.G.; Cooper, P.R.; Shelton, R.M.; Smith, A.J.; Scheven, B.A. A comparison of the in vitro mineralisation and dentinogenic potential of mesenchymal stem cells derived from adipose tissue, bone marrow and dental pulp. *J. Bone Miner. Metab.* **2015**, *33*, 371–382. [CrossRef]

103. Xie, H.; Dubey, N.; Shim, W.; Ramachandra, C.J.A.; Min, K.S.; Cao, T.; Rosa, V. Functional Odontoblastic-Like Cells Derived from Human iPSCs. *J. Dent. Res.* **2017**, *97*, 77–83. [CrossRef]

104. Naihui, Y.; Shiting, L.; Yong, J.; Songbo, Q.; Yinghui, T. Amelogenin promotes odontoblast-like MDPC-23 cell differentiation via activation of ERK1/2 and p38 MAPK. *Mol. Cell. Biochem.* **2011**, *355*, 91–97. [CrossRef] [PubMed]

105. Martín-González, J.; Pérez-Pérez, A.; Cabanillas-Balsera, D.; Vilariño-García, T.; Sánchez-Margalet, V.; Segura-Egea, J.J. Leptin stimulates DMP-1 and DSPP expression in human dental pulp via MAPK 1/3 and PI3K signaling pathways. *Arch. Oral Biol.* **2019**, *98*, 126–131. [CrossRef] [PubMed]

106. Choi, S.-H.; Jang, J.-H.; Koh, J.-T.; Chang, H.-S.; Hwang, Y.-C.; Hwang, I.-N.; Lee, B.-N.; Oh, W.-M. Effect of Leptin on Odontoblastic Differentiation and Angiogenesis: An In Vivo Study. *J. Endod.* **2019**, *45*, 1332–1341. [CrossRef] [PubMed]

107. Åberg, T.; Wozney, J.; Thesleff, I. Expression patterns of bone morphogenetic proteins (Bmps) in the developing mouse tooth suggest roles in morphogenesis and cell differentiation. *Dev. Dyn.* **1997**, *210*, 383–396. [CrossRef]

108. Jani, P.; Liu, C.; Zhang, H.; Younes, K.; Benson, M.D.; Qin, C. The role of bone morphogenetic proteins 2 and 4 in mouse dentinogenesis. *Arch. Oral Biol.* **2018**, *90*, 33–39. [CrossRef]

109. Tucker, A.S.; Khamis, A.A.; Sharpe, P.T. Interactions between Bmp-4 and Msx-1 act to restrict gene expression to odontogenic mesenchyme. *Dev. Dyn.* **1998**, *212*, 533–539. [CrossRef]

110. Lu, X.; Yang, J.; Zhao, S.; Liu, S. Advances of Wnt signalling pathway in dental development and potential clinical application. *Organogenesis* **2019**, *15*, 101–110. [CrossRef]

111. Neves, V.C.M.; Babb, R.; Chandrasekaran, D.; Sharpe, P.T. Promotion of natural tooth repair by small molecule GSK3 antagonists. *Sci. Rep.* **2017**, *7*, 39654. [CrossRef]

112. Foster, B.L. Methods for studying tooth root cementum by light microscopy. *Int. J. Oral Sci.* **2012**, *4*, 119–128. [CrossRef]

113. Sowmya, S.; Chennazhi, K.P.; Arzate, H.; Jayachandran, P.; Nair, S.V.; Jayakumar, R. Periodontal Specific Differentiation of Dental Follicle Stem Cells into Osteoblast, Fibroblast, and Cementoblast. *Tissue Eng. Part C Methods* **2015**, *21*, 1044–1058. [CrossRef]

114. Duan, X.; Tu, Q.; Zhang, J.; Ye, J.; Sommer, C.; Mostoslavsky, G.; Kaplan, D.; Yang, P.; Chen, J. Application of induced pluripotent stem (iPS) cells in periodontal tissue regeneration. *J. Cell. Physiol.* **2011**, *226*, 150–157. [CrossRef]

115. Cao, Z.; Zhang, H.; Zhou, X.; Han, X.; Ren, Y.; Gao, T.; Xiao, Y.; de Crombrugghe, B.; Somerman, M.J.; Feng, J.Q. Genetic evidence for the vital function of Osterix in cementogenesis. *J. Bone Miner. Res.* **2012**, *27*, 1080–1092. [CrossRef] [PubMed]

116. Nagayasu-Tanaka, T.; Anzai, J.; Takaki, S.; Shiraishi, N.; Terashima, A.; Asano, T.; Nozaki, T.; Kitamura, M.; Murakami, S. Action Mechanism of Fibroblast Growth Factor-2 (FGF-2) in the Promotion of Periodontal Regeneration in Beagle Dogs. *PLoS ONE* **2015**, *10*, e0131870. [CrossRef] [PubMed]

117. Liu, L.; Liu, Y.-F.; Zhang, J.; Duan, Y.-Z.; Jin, Y. Ameloblasts serum-free conditioned medium: Bone morphogenic protein 4-induced odontogenic differentiation of mouse induced pluripotent stem cells. *J. Tissue Eng. Regen. Med.* **2016**, *10*, 466–474. [CrossRef] [PubMed]

118. Li, J.; Chatzeli, L.; Panousopoulou, E.; Tucker, A.S.; Green, J.B.A. Epithelial stratification and placode invagination are separable functions in early morphogenesis of the molar tooth. *Development* **2016**, *143*, 670–681. [CrossRef]

119. Liu, Z.; Chen, T.; Bai, D.; Tian, W.; Chen, Y. Smad7 Regulates Dental Epithelial Proliferation during Tooth Development. *J. Dent. Res.* **2019**, *98*, 1376–1385. [CrossRef] [PubMed]

120. Yuan, X.; Liu, M.; Cao, X.; Yang, S. Ciliary IFT80 regulates dental pulp stem cells differentiation by FGF/FGFR1 and Hh/BMP2 signaling. *Int. J. Biol. Sci.* **2019**, *15*, 2087–2099. [CrossRef]

121. Zaugg, L.K.; Banu, A.; Walther, A.R.; Chandrasekaran, D.; Babb, R.C.; Salzlechner, C.; Hedegaard, M.A.B.; Gentleman, E.; Sharpe, P.T. Translation Approach for Dentine Regeneration Using GSK-3 Antagonists. *J. Dent. Res.* **2020**, 0022034520908593. [CrossRef]

122. Zippel, N.; Schulze, M.; Tobiasch, E. Biomaterials and Mesenchymal Stem Cells for Regenerative Medicine. *Recent Pat. Biotechnol.* **2010**, *4*, 1–22. [CrossRef]

123. Tonk, C.; Witzler, M.; Schulze, M.; Tobiasch, E. Mesenchymal Stem Cells. In *Essential Current Concepts in Stem Cell Biology*; Brand-Saberi, B., Ed.; Springer: Berlin, Germany, 2020; pp. 21–39, ISBN 978-3-030-33922-7.

124. Ottensmeyer, P.F.; Witzler, M.; Schulze, M.; Tobiasch, E. Small Molecules Enhance Scaffold-Based Bone Grafts via Purinergic Receptor Signaling in Stem Cells. *Int. J. Mol. Sci.* **2018**, *19*, 3601. [CrossRef]

125. Zabrovsky, A.; Beyth, N.; Pietrokovski, Y.; Ben-Gal, G.; Houri-Haddad, Y. 5-Biocompatibility and functionality of dental restorative materials. In *Woodhead Publishing Series in Biomaterials*; Shelton, R., Ed.; Woodhead Publishing: Sawston/Cambridge, UK, 2017; pp. 63–75, ISBN 978-0-08-100884-3.

126. El Gezawi, M.; Wölfle, U.C.; Haridy, R.; Fliefel, R.; Kaisarly, D. Remineralization, Regeneration, and Repair of Natural Tooth Structure: Influences on the Future of Restorative Dentistry Practice. *ACS Biomater. Sci. Eng.* **2019**, *5*, 4899–4919. [CrossRef]

127. Jazayeri, H.E.; Lee, S.-M.; Kuhn, L.; Fahimipour, F.; Tahriri, M.; Tayebi, L. Polymeric scaffolds for dental pulp tissue engineering: A review. *Dental Mater.* **2020**, *36*, e47–e58. [CrossRef] [PubMed]

128. Haugen, J.H.; Basu, P.; Sukul, M.; Mano, F.J.; Reseland, E.J. Injectable Biomaterials for Dental Tissue Regeneration. *Int. J. Mol. Sci.* **2020**, *21*, 3442. [CrossRef] [PubMed]

129. Pandya, M.; Diekwisch, T. Enamel biomimetics—Fiction or future of dentistry. *Int. J. Oral Sci.* **2019**, *11*, 1–9. [CrossRef]

130. Yu, H.-P.; Zhu, Y.-J.; Lu, B.-Q. Dental Enamel-Mimetic Large-Sized Multi-Scale Ordered Architecture Built by a Well Controlled Bottom-Up Strategy. *Chem. Eng. J.* **2018**, *360*, 1633–1645. [CrossRef]

131. Zheng, W.; Ding, L.; Wang, Y.; Han, S.; Zheng, S.; Guo, Q.; Li, W.; Zhou, X.; Zhang, L. The effects of 8DSS peptide on remineralization in a rat model of enamel caries evaluated by two nondestructive techniques. *J. Appl. Biomater. Funct. Mater.* **2019**, *17*, 2280800019827798. [CrossRef]

132. Zhou, Y.; Zhou, Y.; Gao, L.; Wu, C.; Chang, J. Synthesis of Artificial Dental Enamel by Elastin-like Polypeptide Assisted Biomimetic Approach. *J. Mater. Chem. B* **2018**, *6*, 844–853. [CrossRef]

133. Fan, M.; Zhang, M.; Xu, H.H.K.; Tao, S.; Yu, Z.; Yang, J.; Yuan, H.; Zhou, X.; Liang, K.; Li, J. Remineralization effectiveness of the PAMAM dendrimer with different terminal groups on artificial initial enamel caries in vitro. *Dent. Mater.* **2020**, *36*, 210–220. [CrossRef]

134. Gao, Y.; Liang, K.; Weir, M.D.; Gao, J.; Imazato, S.; Tay, F.R.; Lynch, C.D.; Oates, T.W.; Li, J.; Xu, H.H.K. Enamel remineralization via poly(amido amine) and adhesive resin containing calcium phosphate nanoparticles. *J. Dent.* **2020**, *92*, 103262. [CrossRef]

135. Raddall, G.; Mello, I.; Leung, B.M. Biomaterials and Scaffold Design Strategies for Regenerative Endodontic Therapy. *Front. Bioeng. Biotechnol.* **2019**, *7*, 317. [CrossRef]

136. Metlerska, J.; Fagogeni, I.; Nowicka, A. Efficacy of Autologous Platelet Concentrates in Regenerative Endodontic Treatment: A Systematic Review of Human Studies. *J. Endod.* **2019**, *45*, 20–30. [CrossRef] [PubMed]

137. Shimizu, E.; Ricucci, D.; Albert, J.; Alobaid, A.S.; Gibbs, J.L.; Huang, G.T.-J.; Lin, L.M. Clinical, Radiographic, and Histological Observation of a Human Immature Permanent Tooth with Chronic Apical Abscess after Revitalization Treatment. *J. Endod.* **2013**, *39*, 1078–1083. [CrossRef] [PubMed]

138. Mandakhbayar, N.; El-Fiqi, A.; Lee, J.-H.; Kim, H.-W. Evaluation of Strontium-Doped Nanobioactive Glass Cement for Dentin–Pulp Complex Regeneration Therapy. *ACS Biomater. Sci. Eng.* **2019**, *5*, 6117–6126. [CrossRef]

139. Moonesi Rad, R.; Atila, D.; Akgün, E.E.; Evis, Z.; Keskin, D.; Tezcaner, A. Evaluation of human dental pulp stem cells behavior on a novel nanobiocomposite scaffold prepared for regenerative endodontics. *Mater. Sci. Eng. C* **2019**, *100*, 928–948. [CrossRef]

140. Pankajakshan, D.; Voytik-Harbin, S.L.; Nör, J.E.; Bottino, M.C. Injectable Highly Tunable Oligomeric Collagen Matrices for Dental Tissue Regeneration. *ACS Appl. Bio Mater.* **2020**, *3*, 859–868. [CrossRef]

141. Goudouri, O.-M.; Kontonasaki, E.; Boccaccini, A.R. 17-Layered scaffolds for periodontal regeneration. In *Biomaterials for Oral and Dental Tissue Engineering*; Tayebi, L., Moharamzadeh, K., Eds.; Woodhead Publishing: Sawston/Cambridge, UK, 2017; pp. 279–295, ISBN 978-0-08-100961-1.

142. Iwasaki, K.; Washio, K.; Meinzer, W.; Tsumanuma, Y.; Yano, K.; Ishikawa, I. Application of cell-sheet engineering for new formation of cementum around dental implants. *Heliyon* **2019**, *5*, e01991. [CrossRef]

143. Fakheran, O.; Birang, R.; Schmidlin, P.R.; Razavi, S.M.; Behfarnia, P. Retro MTA and tricalcium phosphate/retro MTA for guided tissue regeneration of periodontal dehiscence defects in a dog model: A pilot study. *Biomater. Res.* **2019**, *23*, 14. [CrossRef]

144. Wei, L.; Teng, F.; Deng, L.; Liu, G.; Luan, M.; Jiang, J.; Liu, Z.; Liu, Y. Periodontal regeneration using bone morphogenetic protein 2 incorporated biomimetic calcium phosphate in conjunction with barrier membrane: A pre-clinical study in dogs. *J. Clin. Periodontol.* **2019**, *46*, 1254–1263. [CrossRef]

145. Wang, B.; Mastrogiacomo, S.; Yang, F.; Shao, J.; Ong, M.; Chanchareonsook, N.; Jansen, J.; Walboomers, X.; Yu, N. Application of BMP-Bone Cement and FGF-Gel on Periodontal Tissue Regeneration in Nonhuman Primates. *Tissue Eng. Part C Methods* **2019**, *25*, 748–756. [CrossRef]

146. Oortgiesen, D.A.W.; Walboomers, X.F.; Bronckers, A.L.J.J.; Meijer, G.J.; Jansen, J.A. Periodontal regeneration using an injectable bone cement combined with BMP-2 or FGF-2. *J. Tissue Eng. Regen. Med.* **2014**, *8*, 202–209. [CrossRef]

147. Vaquette, C.; Saifzadeh, S.; Farag, A.; Hutmacher, D.W.; Ivanovski, S. Periodontal Tissue Engineering with a Multiphasic Construct and Cell Sheets. *J. Dent. Res.* **2019**, *98*, 673–681. [CrossRef] [PubMed]

148. Fredenberg, S.; Wahlgren, M.; Reslow, M.; Axelsson, A. The mechanisms of drug release in poly(lactic-co-glycolic acid)-based drug delivery systems—A review. *Int. J. Pharm.* **2011**, *415*, 34–52. [CrossRef] [PubMed]

149. Witzler, M.; Büchner, D.; Shoushrah, S.H.; Babczyk, P.; Baranova, J.; Witzleben, S.; Tobiasch, E.; Schulze, M. Polysaccharide-Based Systems for Targeted Stem Cell Differentiation and Bone Regeneration. *Biomolecules* **2019**, *9*, 840. [CrossRef] [PubMed]

150. Chapman, C.A.R.; Cuttaz, E.A.; Goding, J.A.; Green, R.A. Actively controlled local drug delivery using conductive polymer-based devices. *Appl. Phys. Lett.* **2020**, *116*, 10501. [CrossRef]

151. Kikuchi, N.; Kitamura, C.; Morotomi, T.; Inuyama, Y.; Ishimatsu, H.; Tabata, Y.; Nishihara, T.; Terashita, M. Formation of Dentin-like Particles in Dentin Defects above Exposed Pulp by Controlled Release of Fibroblast Growth Factor 2 from Gelatin Hydrogels. *J. Endod.* **2007**, *33*, 1198–1202. [CrossRef] [PubMed]

152. Moioli, E.K.; Clark, P.A.; Xin, X.; Lal, S.; Mao, J.J. Matrices and scaffolds for drug delivery in dental, oral and craniofacial tissue engineering. *Adv. Drug Deliv. Rev.* **2007**, *59*, 308–324. [CrossRef] [PubMed]

153. Kitamura, C.; Nishihara, T.; Terashita, M.; Tabata, Y.; Washio, A. Local regeneration of dentin-pulp complex using controlled release of fgf-2 and naturally derived sponge-like scaffolds. *Int. J. Dent.* **2012**, *2012*, 190561. [CrossRef]

154. Piva, E.; Silva, A.F.; Nör, J.E. Functionalized scaffolds to control dental pulp stem cell fate. *J. Endod.* **2014**, *40*, S33–S40. [CrossRef]

155. Zhang, N.; Weir, M.D.; Chen, C.; Melo, M.A.S.; Bai, Y.; Xu, H.H.K. Orthodontic cement with protein-repellent and antibacterial properties and the release of calcium and phosphate ions. *J. Dent.* **2016**, *50*, 51–59. [CrossRef]

156. Monteiro, N.; Yelick, P.C. Advances and perspectives in tooth tissue engineering. *J. Tissue Eng. Regen. Med.* **2017**, *11*, 2443–2461. [CrossRef]

157. Tabatabaei, F.; Torres, R.; Tayebi, L. Biomedical Materials in Dentistry. In *Applications of Biomedical Engineering in Dentistry*; Tayebi, L., Ed.; Springer: Cham, Switzerland, 2020; pp. 3–20, ISBN 978-3-030-21582-8.

158. Moon, C.-Y.; Nam, O.H.; Kim, M.; Lee, H.-S.; Kaushik, S.N.; Cruz Walma, D.A.; Jun, H.-W.; Cheon, K.; Choi, S.C. Effects of the nitric oxide releasing biomimetic nanomatrix gel on pulp-dentin regeneration: Pilot study. *PLoS ONE* **2018**, *13*, e0205534. [CrossRef] [PubMed]

159. Ishimatsu, H.; Kitamura, C.; Morotomi, T.; Tabata, Y.; Nishihara, T.; Chen, K.-K.; Terashita, M. Formation of Dentinal Bridge on Surface of Regenerated Dental Pulp in Dentin Defects by Controlled Release of Fibroblast Growth Factor–2 From Gelatin Hydrogels. *J. Endod.* **2009**, *35*, 858–865. [CrossRef] [PubMed]

160. Tutar, R.; Motealleh, A.; Khademhosseini, A.; Kehr, N.S. Functional Nanomaterials on 2D Surfaces and in 3D Nanocomposite Hydrogels for Biomedical Applications. *Adv. Funct. Mater.* **2019**, *29*, 1904344. [CrossRef]

161. Inoue, B.S.; Streit, S.; dos Santos Schneider, A.L.; Meier, M.M. Bioactive bacterial cellulose membrane with prolonged release of chlorhexidine for dental medical application. *Int. J. Biol. Macromol.* **2020**, *148*, 1098–1108. [CrossRef]

162. Gericke, M.; Heinze, T. Homogeneous tosylation of agarose as an approach toward novel functional polysaccharide materials. *Carbohydr. Polym.* **2015**, *127*, 236–245. [CrossRef]

163. Kostag, M.; Gericke, M.; Heinze, T.; El Seoud, O.A. Twenty-five years of cellulose chemistry: Innovations in the dissolution of the biopolymer and its transformation into esters and ethers. *Cellulose* **2019**, *26*, 139–184. [CrossRef]

164. Li, J.; Mooney, D.J. Designing hydrogels for controlled drug delivery. *Nat. Rev. Mater.* **2016**, *1*, 16071. [CrossRef]

165. Zippel, N.; Limbach, C.A.; Ratajski, N.; Urban, C.; Luparello, C.; Pansky, A.; Kassack, M.U.; Tobiasch, E. Purinergic Receptors Influence the Differentiation of Human Mesenchymal Stem Cells. *Stem Cells Dev.* **2011**, *21*, 884–900. [CrossRef]

166. Kaebisch, C.; Schipper, D.; Babczyk, P.; Tobiasch, E. The role of purinergic receptors in stem cell differentiation. *Comput. Struct. Biotechnol. J.* **2014**, *13*, 75–84. [CrossRef]

167. Schipper, D.; Babczyk, P.; Elsayed, F.; Klein, S.E.; Schulze, M.; Tobiasch, E. The effect of nanostructured surfaces on stem cell fate. In *Nanostructures for Novel Therapy. Synthesis, Characterization and Applications*; Grumezescu, F., Ed.; Elsevier: Amsterdam, The Netherlands, 2017; pp. 567–589, ISBN 978-0-323-46142-9.

168. Zhang, Y.; Lau, P.; Pansky, A.; Kassack, M.; Hemmersbach, R.; Tobiasch, E. The Influence of Simulated Microgravity on Purinergic Signaling Is Different between Individual Culture and Endothelial and Smooth Muscle Cell Coculture. *Biomed. Res. Int.* **2014**, *2014*, 413708. [CrossRef]

169. Babczyk, P.; Conzendorf, C.; Klose, J.; Schulze, M.; Harre, K.; Tobiasch, E. Stem Cells on Biomaterials for Synthetic Grafts to Promote Vascular Healing. *J. Clin. Med.* **2014**, *3*, 39–87. [CrossRef] [PubMed]

170. Grotheer, V.; Schulze, M.; Tobiasch, E. Trends in Bone Tissue Engineering: Proteins for Osteogenic Differentiation and the Respective Scaffolding. In *Protein Purification: Principles and Trends*; iConcept Press: Hong Kong, China, 2014; ISBN 978-1-922227-40-9.

171. Witzler, M.; Ottensmeyer, P.F.; Gericke, M.; Heinze, T.; Tobiasch, E.; Schulze, M. Non-Cytotoxic Agarose/Hydroxyapatite Composite Scaffolds for Drug Release. *Int. J. Mol. Sci.* **2019**, *20*, 3565. [CrossRef] [PubMed]

172. Kolanthai, E.; Dikeshwar Colon, V.S.; Sindu, P.A.; Chandra, V.S.; Karthikeyan, K.R.; Babu, M.S.; Sundaram, S.M.; Palanichamy, M.; Kalkura, S.N. Effect of solvent; enhancing the wettability and engineering the porous structure of a calcium phosphate/agarose composite for drug delivery. *RSC Adv.* **2015**, *5*, 18301–18311. [CrossRef]

173. Izadi, Z.; Derakhshankhah, H.; Alaei, L.; Karkazis, E.; Jafari, S.; Tayebi, L. Recent Advances in Nanodentistry. In *Applications of Biomedical Engineering in Dentistry*; Tayebi, L., Ed.; Springer: Cham, Switzerland, 2020; pp. 263–287.

174. Fan, Y.; Wen, Z.; Liao, S.; Lallier, T.; Hagan, J.; Twomley, J.; Zhang, J.-F.; Sun, Z.; Xu, X. Novel amelogenin-releasing hydrogel for remineralization of enamel artificial caries. *J. Bioact. Compat. Polym.* **2012**, *27*, 585–603. [CrossRef] [PubMed]

175. Ahrens, L.A.J.; Vonwil, D.; Christensen, J.; Shastri, V.P. Gelatin device for the delivery of growth factors involved in endochondral ossification. *PLoS ONE* **2017**, *12*, e0175095. [CrossRef] [PubMed]

176. Gellynck, K.; Abou Neel, E.A.; Li, H.; Mardas, N.; Donos, N.; Buxton, P.; Young, A.M. Cell attachment and response to photocured, degradable bone adhesives containing tricalcium phosphate and purmorphamine. *Acta Biomater.* **2011**, *7*, 2672–2677. [CrossRef]

177. De la Vega, L.; Karmirian, K.; Willerth, S.M. Engineering Neural Tissue from Human Pluripotent Stem Cells Using Novel Small Molecule Releasing Microspheres. *Adv. Biosyst.* **2018**, *2*, 1800133. [CrossRef]

178. Zhao, W.; Li, Y.; Zhou, A.; Chen, X.; Li, K.; Chen, S.; Qiao, B.; Jiang, D. Controlled release of basic fibroblast growth factor from a peptide biomaterial for bone regeneration. *R. Soc. Open Sci.* **2020**, *7*, 191830. [CrossRef]

179. Tachibana, A.; Yasuma, D.; Takahashi, R.; Tanabe, T. Chitin degradation enzyme-responsive system for controlled release of fibroblast growth factor-2. *J. Biosci. Bioeng.* **2020**, *129*, 116–120. [CrossRef]

180. Karahaliloğlu, Z.; Yalçın, E.; Demirbilek, M.; Denkbaş, E.B. Magnetic silk fibroin e-gel scaffolds for bone tissue engineering applications. *J. Bioact. Compat. Polym.* **2017**, *32*, 596–614. [CrossRef]

181. Scarpa, E.; Janeczek, A.A.; Hailes, A.; de Andrés, M.C.; De Grazia, A.; Oreffo, R.O.C.; Newman, T.A.; Evans, N.D. Polymersome nanoparticles for delivery of Wnt-activating small molecules. *Nanomed. Nanotechnol. Biol. Med.* **2018**, *14*, 1267–1277. [CrossRef] [PubMed]

182. Fahmy, R.A.; Mahmoud, N.; Soliman, S.; Nouh, S.R.; Cunningham, L.; El-Ghannam, A. Acceleration of Alveolar Ridge Augmentation Using a Low Dose of Recombinant Human Bone Morphogenetic Protein-2 Loaded on a Resorbable Bioactive Ceramic. *J. Oral Maxillofac. Surg.* **2015**, *73*, 2257–2272. [CrossRef]

183. Wang, C.; Zhao, Q.; Wang, M. Cryogenic 3D printing for producing hierarchical porous and rhBMP-2-loaded Ca-P/PLLA nanocomposite scaffolds for bone tissue engineering. *Biofabrication* **2017**, *9*, 25031. [CrossRef] [PubMed]

Tooth Formation: Are the Hardest Tissues of Human Body Hard...

65

184. Minardi, S.; Fernandez-Moure, S.J.; Fan, D.; Murphy, B.M.; Yazdi, K.I.; Liu, X.; Weiner, K.B.; Tasciotti, E. Biocompatible PLGA-Mesoporous Silicon Microspheres for the Controlled Release of BMP-2 for Bone Augmentation. *Pharmaceutics* **2020**, *12*, 118. [CrossRef] [PubMed]

185. Damanik, F.F.R.; Brunelli, M.; Pastorino, L.; Ruggiero, C.; van Blitterswijk, C.; Rotmans, J.; Moroni, L. Sustained delivery of growth factors with high loading efficiency in a layer by layer assembly. *Biomater. Sci.* **2020**, *8*, 174–188. [CrossRef] [PubMed]

186. Ardeshirylajimi, A.; Ghaderian, S.M.-H.; Omrani, M.D.; Moradi, S.L. Biomimetic scaffold containing PVDF nanofibers with sustained TGF-β release in combination with AT-MSCs for bladder tissue engineering. *Gene* **2018**, *676*, 195–201. [CrossRef]

187. Mahmoudi, Z.; Mohammadnejad, J.; Razavi Bazaz, S.; Abouei Mehrizi, A.; Saidijam, M.; Dinarvand, R.; Ebrahimi Warkiani, M.; Soleimani, M. Promoted chondrogenesis of hMCSs with controlled release of TGF-β3 via microfluidics synthesized alginate nanogels. *Carbohydr. Polym.* **2020**, *229*, 115551. [CrossRef]

188. Díaz-Saldívar, P.; Huidobro-Toro, J.P. ATP-loaded biomimetic nanoparticles as controlled release system for extracellular drugs in cancer applications. *Int. J. Nanomed.* **2019**, *14*, 2433–2447. [CrossRef]

189. Mohammadi Amirabad, L.; Zarrintaj, P.; Lindemuth, A.; Tayebi, L. Whole Tooth Engineering. In *Applications of Biomedical Engineering in Dentistry*; Tayebi, L., Ed.; Springer: Cham, Switzerland, 2020; pp. 443–462, ISBN 978-3-030-21582-8.

190. Sonoyama, W.; Liu, Y.; Fang, D.; Yamaza, T.; Seo, B.-M.; Zhang, C.; Liu, H.; Gronthos, S.; Wang, C.-Y.; Wang, S.; et al. Mesenchymal stem cell-mediated functional tooth regeneration in swine. *PLoS ONE* **2006**, *1*, e79. [CrossRef]

191. Oshima, M.; Inoue, K.; Nakajima, K.; Tachikawa, T.; Yamazaki, H.; Isobe, T.; Sugawara, A.; Ogawa, M.; Tanaka, C.; Saito, M.; et al. Functional tooth restoration by next-generation bio-hybrid implant as a bio-hybrid artificial organ replacement therapy. *Sci. Rep.* **2014**, *4*, 6044. [CrossRef]

192. Gao, Z.H.; Hu, L.; Liu, G.L.; Wei, F.L.; Liu, Y.; Liu, Z.H.; Fan, Z.P.; Zhang, C.M.; Wang, J.S.; Wang, S.L. Bio-Root and Implant-Based Restoration as a Tooth Replacement Alternative. *J. Dent. Res.* **2016**, *95*, 642–649. [CrossRef] [PubMed]

193. Oshima, M.; Tsuji, T. Whole Tooth Regeneration as a Future Dental Treatment. *Adv. Exp. Med. Biol.* **2015**, *881*, 255–269. [PubMed]

194. Angelova Volponi, A.; Zaugg, L.K.; Neves, V.; Liu, Y.; Sharpe, P.T. Tooth Repair and Regeneration. *Curr. Oral Health Rep.* **2018**, *5*, 295–303. [CrossRef] [PubMed]

195. Li, L.; Tang, Q.; Wang, A.; Chen, Y. Regrowing a tooth: In vitro and in vivo approaches. *Curr. Opin. Cell Biol.* **2019**, *61*, 126–131. [CrossRef] [PubMed]

196. Popa, E.M.; Buchtova, M.; Tucker, A.S. Revitalising the rudimentary replacement dentition in the mouse. *Development* **2019**, *146*, dev171363. [CrossRef] [PubMed]

197. Balic, A. Biology Explaining Tooth Repair and Regeneration: A Mini-Review. *Gerontology* **2018**, *64*, 382–388. [CrossRef] [PubMed]

198. Mina, M.; Kollar, E.J. The induction of odontogenesis in non-dental mesenchyme combined with early murine mandibular arch epithelium. *Arch. Oral Biol.* **1987**, *32*, 123–127. [CrossRef]

199. Young, C.S.; Terada, S.; Vacanti, J.P.; Honda, M.; Bartlett, J.D.; Yelick, P.C. Tissue Engineering of Complex Tooth Structures on Biodegradable Polymer Scaffolds. *J. Dent. Res.* **2002**, *81*, 695–700. [CrossRef]

200. Duailibi, M.T.; Duailibi, S.E.; Young, C.S.; Bartlett, J.D.; Vacanti, J.P.; Yelick, P.C. Bioengineered Teeth from Cultured Rat Tooth Bud Cells. *J. Dent. Res.* **2004**, *83*, 523–528. [CrossRef]

201. Nakao, K.; Morita, R.; Saji, Y.; Ishida, K.; Tomita, Y.; Ogawa, M.; Saitoh, M.; Tomooka, Y.; Tsuji, T. The development of a bioengineered organ germ method. *Nat. Methods* **2007**, *4*, 227–230. [CrossRef]

202. Oshima, M.; Mizuno, M.; Imamura, A.; Ogawa, M.; Yasukawa, M.; Yamazaki, H.; Morita, R.; Ikeda, E.; Nakao, K.; Takano-Yamamoto, T.; et al. Functional tooth regeneration using a bioengineered tooth unit as a mature organ replacement regenerative therapy. *PLoS ONE* **2011**, *6*, e21531. [CrossRef] [PubMed]

203. Wu, Z.; Wang, F.; Fan, Z.; Wu, T.; He, J.; Wang, J.; Zhang, C.; Wang, S. Whole-Tooth Regeneration by Allogeneic Cell Reassociation in Pig Jawbone. *Tissue Eng. Part A* **2019**, *25*, 1202–1212. [CrossRef] [PubMed]

204. Young, C.S.; Abukawa, H.; Asrican, R.; Ravens, M.; Troulis, M.J.; Kaban, L.B.; Vacanti, J.P.; Yelick, P.C. Tissue-Engineered Hybrid Tooth and Bone. *Tissue Eng.* **2005**, *11*, 1599–1610. [CrossRef] [PubMed]

205. Young, C.S.; Kim, S.-W.; Qin, C.; Baba, O.; Butler, W.T.; Taylor, R.R.; Bartlett, J.D.; Vacanti, J.P.; Yelick, P.C. Developmental analysis and computer modelling of bioengineered teeth. *Arch. Oral Biol.* **2005**, *50*, 259–265. [CrossRef]

206. Ohazama, A.; Modino, S.A.C.; Miletich, I.; Sharpe, P.T. Stem-cell-based Tissue Engineering of Murine Teeth. *J. Dent. Res.* **2004**, *83*, 518–522. [CrossRef]

207. Angelova Volponi, A.; Kawasaki, M.; Sharpe, P.T. Adult human gingival epithelial cells as a source for whole-tooth bioengineering. *J. Dent. Res.* **2013**, *92*, 329–334. [CrossRef]

208. Hu, B.; Nadiri, A.; Kuchler-Bopp, S.; perrin-schmitt, F.; Peters, H.; Lesot, H. Tissue Engineering of Tooth Crown, Root, and Periodontium. *Tissue Eng.* **2006**, *12*, 2069–2075. [CrossRef]

209. Zheng, Y.; Cai, J.; Hutchins, A.P.; Jia, L.; Liu, P.; Yang, D.; Chen, S.; Ge, L.; Pei, D.; Wei, S. Remission for Loss of Odontogenic Potential in a New Micromilieu In Vitro. *PLoS ONE* **2016**, *11*, e0152893. [CrossRef]

210. Kuchler-Bopp, S.; Bécavin, T.; Kökten, T.; Weickert, J.L.; Keller, L.; Lesot, H.; Deveaux, E.; Benkirane-Jessel, N. Three-dimensional Micro-culture System for Tooth Tissue Engineering. *J. Dent. Res.* **2016**, *95*, 657–664. [CrossRef]

211. Zhang, W.; Vázquez, B.; Yelick, P.C. Bioengineered post-natal recombinant tooth bud models. *J. Tissue Eng. Regen. Med.* **2017**, *11*, 658–668. [CrossRef]

212. Yang, K.-C.; Kitamura, Y.; Wu, C.-C.; Chang, H.-H.; Ling, T.-Y.; Kuo, T.-F. Tooth Germ-Like Construct Transplantation for Whole-Tooth Regeneration: An In Vivo Study in the Miniature Pig. *Artif. Organs* **2016**, *40*, E39–E50. [CrossRef] [PubMed]

213. Nait Lechguer, A.; Kuchler-Bopp, S.; Hu, B.; Haïkel, Y.; Lesot, H. Vascularization of Engineered Teeth. *J. Dent. Res.* **2008**, *87*, 1138–1143. [CrossRef] [PubMed]

214. Strub, M.; Keller, L.; Idoux-Gillet, Y.; Lesot, H.; Clauss, F.; Benkirane-Jessel, N.; Kuchler-Bopp, S. Bone Marrow Stromal Cells Promote Innervation of Bioengineered Teeth. *J. Dent. Res.* **2018**, *97*, 1152–1159. [CrossRef] [PubMed]

4

Tissue-Specific Decellularization Methods: Rationale and Strategies to Achieve Regenerative Compounds

Unai Mendibil [1,2], Raquel Ruiz-Hernandez [1,†], Sugoi Retegi-Carrion [1,†], Nerea Garcia-Urquia [2,†], Beatriz Olalde-Graells [2] and Ander Abarrategi [1,3,*]

1   Center for Cooperative Research in Biomaterials (CIC biomaGUNE), Basque Research and Technology Alliance (BRTA), 20014 Donostia-San Sebastian, Spain; umendibil@cicbiomagune.es (U.M.); rruiz@cicbiomagune.es (R.R.-H.); sretegi@cicbiomagune.es (S.R.-C.)
2   TECNALIA, Basque Research and Technology Alliance (BRTA), 20009 Donostia-San Sebastian, Spain; nerea.garcia@tecnalia.com (N.G.-U.); beatriz.olalde@tecnalia.com (B.O.-G.)
3   Ikerbasque, Basque Foundation for Science, 48013 Bilbao, Spain
*   Correspondence: aabarrategi@cicbiomagune.es
†   These authors contribute equally to the work.

Abstract: The extracellular matrix (ECM) is a complex network with multiple functions, including specific functions during tissue regeneration. Precisely, the properties of the ECM have been thoroughly used in tissue engineering and regenerative medicine research, aiming to restore the function of damaged or dysfunctional tissues. Tissue decellularization is gaining momentum as a technique to obtain potentially implantable decellularized extracellular matrix (dECM) with well-preserved key components. Interestingly, the tissue-specific dECM is becoming a feasible option to carry out regenerative medicine research, with multiple advantages compared to other approaches. This review provides an overview of the most common methods used to obtain the dECM and summarizes the strategies adopted to decellularize specific tissues, aiming to provide a helpful guide for future research development.

Keywords: extracellular matrix; decellularization; regenerative medicine

1. Introduction

In many adult animal tissues, the main component in terms of volume is not the cells, but the cell-secreted three-dimensional (3D) structure known as the extracellular matrix (ECM). Structural and specialized proteins and proteoglycans are some of the ECM's macromolecular components, and they interact in this network with multiple key roles and functions. During homeostasis, the ECM provides structural integrity and mechanical support for tissues and organ architecture. In parallel, the ECM is the reservoir and the place for the active exchange of ions, nutrients, waters, metabolites, and signals [1]. In this way, the ECM serves as the environment in which tissue-resident cells attach, communicate, and interact, thereby regulating cell dynamics and behavior, and contributing to the maintenance of tissue-specific cell phenotypes and functions (Figure 1). Notably, the ECM provides a niche for tissue-resident stem cells and drives their fate decisions, a property particularly relevant during homeostasis and tissue repair–regeneration processes.

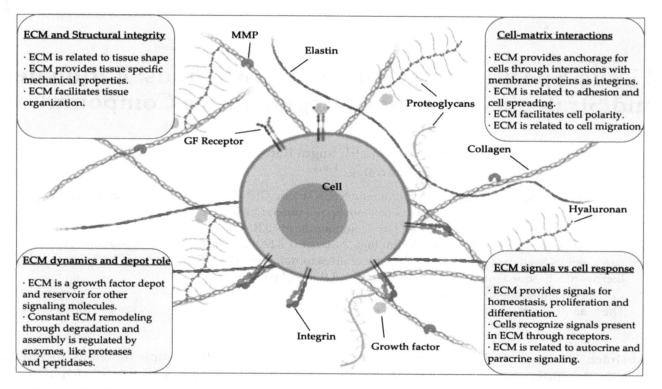

Figure 1. Structure, components, and functions of extracellular matrix (ECM) (MMP, matrix metalloprotease; GF, growth factor).

The properties of the ECM have been thoroughly used in tissue engineering and regenerative medicine research, aiming to restore the function of damaged or dysfunctional tissues [2]. In this context, applications of ECM-derived components are multiple, from in vitro stem cell basic research to clinical settings. For example, ECM-derived components have been used as surface coatings for cell adhesion purposes; gel matrices for establishing organoid cultures; 3D environments for cell seeding and growth factor delivery approaches; and as biocompatible and biomimetic implantable allograft or xenograft scaffolds with in vivo tissue regeneration properties. Often, strategies are based on specific ECM components, such as collagens, fibrin, hyaluronic acid, or even cell-culture-derived ECM. Using these materials, multiple fabrication techniques have been implemented to generate successful ECM-derived biomimetic structures such as hydrogels, or even more sophisticated engineering approaches such as micropatterned surfaces, electrospinning, 3D printing, and bioprinting-designed scaffolds, among others [3].

Interestingly, tissue decellularization is becoming a common technique to obtain decellularized extracellular matrix (dECM) [4] and it is a research field gaining momentum in recent years (Figure 2). The rationale for decellularization is related to the adverse response that cell waste may induce when tissue-derived material is used for implantation procedures, including immune reaction and inflammation, leading to implant rejection. Therefore, dECM is usually obtained by chemical, enzymatic, and/or physical decellularization methods, developed to eliminate the cells and their waste, mainly DNA [5]. These procedures yield decellularized materials formed by the multiple ECM components, which are maintained similar to the original tissue in composition, even in architecture, if required. dECM-related advantages are often associated with better performance and applicability as implants for tissue repair, and also with better mechanical/biochemical properties for the intended use. Initially, research in this field was focused precisely on developing proper decellularization methods and techniques, while more recently, the field is moving to implement approaches related to bioengineering and which tackle applied research aims.

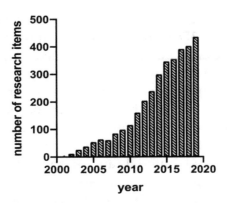

Figure 2. Research items per year with the words "decellularization" or "decellularized" in the title (Source: https://scholar.google.es "allintitle: decellularized OR decellularization").

Notably, regenerative medicine research can now take advantage of approaches based on the use of the targeted tissue-specific dECM [6,7]. The ECM is different from one tissue to another, and therefore, decellularization methods and techniques to obtain dECM have been extensively studied and tissue-specifically improved, aiming to preserve the ECM molecules and structures relevant for the intended use. In this sense, research has been mainly based on empirical testing of tissue-specific decellularization methods intended to achieve specific aims or applications.

In this review, we provide an overview of the methods used to obtain the dECM, and we summarize the most common strategies adopted to decellularize specific tissues, aiming to provide a helpful guide for future research development.

2. Organ Decellularization and Tissue Decellularization Approaches for Biomedical Applications

Whole tissue or intact tissue pieces or sections are a common starting point for decellularization, especially when the final purpose is whole organ decellularization for bioengineering purposes. In these cases, the resulting dECM is a tissue scaffold generally created to keep its structure as intact as possible. Note that dECM tissue scaffold decellularization processes tend to be long, due to the need to be sure about all the reagents reaching the target cells in order to achieve complete decellularization [5]. The bigger the tissue pieces, the longer it takes to make sure that they are completely decellularized. Moreover, the longer the period of chemicals and enzymatic reactions, the higher the chance of damaging the ECM components.

In tissue pieces, it is easy to assess by histology the decellularization and integrity of the remaining ECM. However, even if there are no histologically visible nuclei in the tissue, it is still important to quantify the DNA content by molecular techniques, with the safe limit of DNA content in a decellularized tissue established as below 50 pg DNA per milligram of dry tissue. Regarding macromolecules, they need to be assessed both in quantity and quality by histology, spectrophotometry, and other techniques, while the other ECM components as growth factors may need to be assessed and quantified as well [8]. In some cases, tissue-specific tests are also required to characterize those properties key to the intended biomedical application. For example, a mechanical stress test is required to assess the mechanical/elastic properties after the decellularization of tendons, muscles, and cartilage tissues [9].

Keeping the ECM as intact as possible can be a problem when dECM tissue scaffolds are meant to be used for cell colonization purposes [10]. The ECM is grown by and around the cells, providing physical support, and therefore keeping intact the intricate ECM net, which may impede proper cell seeding of decellularized material. Moreover, potential implantation may be physically limited by the dECM structure and form. Therefore, the decellularization strategy is often designed to degrade some of the ECM components, aiming to maximize the further cell seeding strategy [11].

Tissue decellularization is achieved using as starting material tissues treated with mechanical or chemical methods for tissue grinding, pulverization, or homogenization before decellularization.

This approach is gaining relevance, especially in strategies aiming to use the dECM in postprocessing fabrication approaches (hydrogels, 3D printing, electrospinning, and similar). The outcome of decellularization in these cases is dECM powder, an intermediate product mainly used to artificially generate further ECM coatings or 3D structures [12]. Note that tissue powder processing yields dECM powder with multiple components, but does not keep the tissue architecture and affects the structure of the ECM macromolecules. This is because the aim of approaches using this processing method is not to keep the structural proteins untouched, but rather to use the properties of the relevant ECM components to improve biocompatibility, adhesion, differentiation, and/or other properties or purposes.

The applications of decellularized materials and matrices in regenerative medicine context are multiple, including clinically used implantable materials, and continue to expand. For example, whole decellularized pieces are mainly used as scaffolds for transplantation purposes; dECM processed to form sheets and/or patches is useful in soft tissue and cardiac repair; Powder of demineralized bone matrix can be resuspended and be used to fill and heal bone defects; dECM-derived hydrogels are useful as injectable materials with regenerative properties; Hydrogels can be processed too, to generate inks and bioinks useful in 3D printing and electrospun-based strategies; dECM-derived scaffolds can be used as cell carriers for in vitro modeling or in vivo regenerative purposes [2,13–19].

In any case, the decellularization procedure selected and the further characterization needs depend on the final aim and approach (Figure 3). Whole organ, tissue pieces, or powdered ECM are the most common starting materials and multiple possible decellularization methods can be applied, all of them with advantages and disadvantages to be taken into account in light of the specific aim and context [18].

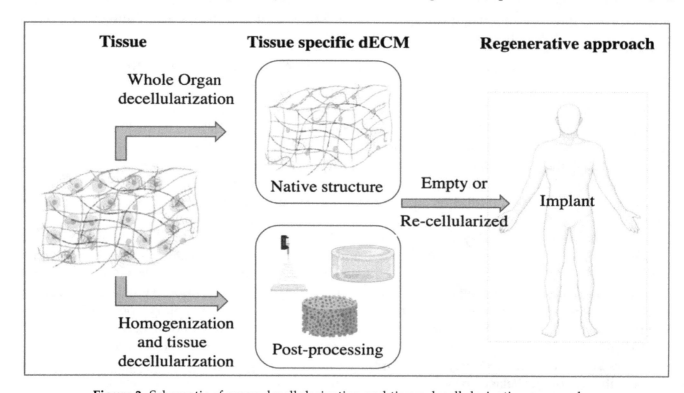

Figure 3. Schematic of organ decellularization and tissue decellularization approaches.

3. ECM Decellularization and Sterilization Methods

Most protocols describe combinatorial and sequential use of different physical, chemical, and enzymatic techniques in order to achieve tissue-specific decellularization. In general, chemical and enzymatic techniques are mainly responsible for successful decellularization in most protocols. Physical techniques are generally used to complement chemical and enzymatic techniques and therefore increase the decellularization effects (Table 1). Physical techniques can produce damage in the matrix, while chemical techniques can produce reactions that change the chemical composition of the ECM [20–22]. For this reason, setting up the decellularization protocol is of paramount importance in each specific approach.

3.1. Chemical and Enzymatic Methods for Decellularization

Detergents are chemical agents used to solubilize cell membranes and to dissociate their inner structure. Among them, Triton X-100 is the most commonly used detergent in decellularization processes. It targets the lipid–lipid and lipid–protein interactions, but it leaves the protein–protein interaction intact [9,23]. It is a very useful agent in those tissues where the key matrix components are primarily proteins. It is an effective detergent to eliminate cells from many tissues, but it is generally avoided in tissues with glycosaminoglycans (GAGs) as a key component in their matrix.

Side by side with the Triton X-100, sodium dodecyl sulfate (SDS) is the other most commonly used detergent in decellularization procedures. SDS solubilizes both the external and nuclear membranes, but also tends to denature proteins and may alter the native structure of the matrix [24,25]. For these reasons, short time SDS treatment is the most common, aiming to minimize the possible damage to proteins and the overall matrix structure [26]. Nevertheless, it is very efficient in removing nuclear and cytoplasmic waste.

Other detergents are useful for specific tissues and applications in a decellularization context. CHAPS reagent has properties of both ionic and nonionic detergents, and therefore, it targets lipid–lipid and lipid–protein interactions, while also solubilizes membranes [27]. CHAPS is not good as a permeating agent, and is therefore mainly used to decellularize thin tissues, for which it is more effective. Triton X-200 is another detergent, less used than its X-100 counterpart because it is more prone to alter the ECM's structure, but it is highly effective for tissues such as neural tissue [28].

Enzymes are also used in most decellularization protocols, mainly to eliminate cell waste and other undesirable components of the ECM. However enzymatic treatments can often lead to additional problems related to enzyme removal [20], a problem usually tackled with further steps using nonenzymatic agents or detergent treatment. Among such enzymes, trypsin is the most commonly used in decellularization procedures [29]. Trypsin targets the C-side bonds in arginine and lysine amino acids and is mostly used combined with EDTA, a chemical agent able to break cell–matrix interactions. Of note, the prolonged exposure to trypsin–EDTA treatment can significantly alter the structure of the matrix, destroy laminin, and remove GAGs, resulting in severe mechanical weakness of the tissue [30]. Pepsin is another useful enzyme in decellularization processes. It is a highly aggressive protease commonly found in the stomach and, like trypsin, it targets the bounds between peptides. Thus, it may cause damage to the ECM if used under long exposition times [31].

Hypotonic and hypertonic solutions [32] use osmotic properties to make the cells explode. The osmotic shock kills the cells, but it does not remove the cell waste that it releases to the matrix, which should be taken into account in the design of a decellularization procedure. Moreover, the treatment of the DNA waste is of paramount relevance in all decellularization processes, due to the tendency of the nuclear material to remain stuck to ECM proteins. In this sense, endonucleases and exonucleases are other kinds of enzymes that are of great use to eliminate the waste of nuclear components [8,30].

Finally, chemical acid–base [33] and TBP [32] treatments are rarely used, because they are very aggressive toward the proteins of the matrix. Specifically, acid–base solutions damage collagen and TBP disrupts protein–protein bonds.

3.2. Physical Methods for Decellularization

As previously mentioned, physical techniques are not enough to decellularize the tissue, but they can help in combination with chemical and enzymatic processes. For example, when big tissue pieces or whole organs are the target of decellularization, perfusion is recommended in order to better reach all tissue areas [21,34].

Table 1. Methods used in decellularization processes.

Methods	Mechanism	Side Effects on the ECM	References
Chemical			
Acid; Base	Solubilizes cytoplasmic components, disrupts nucleic acids	Damages collagen and GAG	[32]
Triton X-100	Breaks lipid–lipid and lipid–protein unions, while leaving the protein interactions untouched	Not recommended for ECM where the lipids and GAG are important components	[9,23,26,44,45]
SDS	Liquefies the internal and external cell membranes	Tends to denaturalize proteins and may induce nuclear and cytoplasmic waste in the remaining matrix	[24–26,46–48]
Triton X-200	Similar to its X-100 counterpart. Very effective in some tissues	Needs to be combined with a zwitterionic detergent to be effective. Damages the matrix in a similar way that SDS does.	[28]
CHAPS	Properties of ionic and nonionic detergents	Similar damage level compared to Triton X-100	[27]
TBP	Disrupts protein–protein interactions	Variable results, collagen degradation but keeping the mechanical properties	[25,49]
Hypertonic and hypotonic solutions	Osmotic pressure to make the membrane explode	High amount of cell waste in the remaining matrix	[11,33,50]
Enzymes			
EDTA, EGTA	Breaks cell adhesion to matrix. It is usually combined with trypsin	Does not actually kill the cells	[29,30,46,50,51]
Trypsin	Digestion of membrane proteins leading to cell dead	Can damage the proteins in the ECM, in particular laminin and GAG	[29,30,43,52]
Pepsin	It targets peptide bounds	Causes high damage in the ECM proteins if left for too long	[31]
Endonucleases and Exonucleases	Degradation of the nuclear material inside and outside of the nucleus	Further cleaning and enzyme removal is required, as they may promote immune response	[6,29,53–57]
Physical			
Freezing	Crystals created in the freezing process destroy the cell membrane	The overall protein structure of the ECM may be compromised	[24,35–38,58]
Force	Mechanical pressure can be enough to induce the lysis in some tissues	Limited to tissues with hard structures, as it can greatly damage the ECM structure	[39]
Agitation	Commonly used to facilitate chemical agent infiltration and to induce cell lysis	Aggressive processes like sonication can greatly damage the ECM	[40,41]
Vacuum-assisted decellularization (VAD)	Enables chemical agents to reach the more inner parts of the tissue	It is not a decellularization method but a facilitator	[42]
Hydrostatic pressure	Applies high pressure to the tissue and induces cell lysis	Excessive pressure can damage the structure	[43,59]

The most commonly used physical technique is snap freezing, or freeze–thawing, as the first step of a decellularization process. By freezing tissue, intracellular ice crystals are formed, thereby disrupting cellular membranes and causing cell lysis. Thus, freezing is a common and effective method for cell lysis and it eases further uniform decellularization. Protocols using this approach have to carefully control the rate of temperature change to control the size of the formed ice crystals, therefore preventing excessive damage to the ECM [35–38].

Cells can be lysed by applying direct pressure to tissue, but this method is only effective for tissues or organs that are not characterized by a densely organized ECM (e.g., liver and lung). Mechanical force has also been used to delaminate layers of tissue from organs that are characterized by natural planes of dissection, such as the small intestine and the urinary bladder. These methods are effective and cause minimal disruption to the three-dimensional architecture of the ECM within these tissues [39].

Mechanical agitation and sonication are useful in combination with a chemical treatment to assist in cell lysis and the removal of cellular debris [40,41]. Mechanical agitation can be applied by using a magnetic stir plate, an orbital shaker, or a low-profile roller. There are no studies to determine the optimal magnitude or frequency of sonication for the disruption of cells. However, the standard ultrasonic cleaner appears to be as effective at removing cellular material as the movement of an orbital shaker. In all of these procedures, the optimal speed, the volume of reagent, and the length of mechanical agitation are dependent on the composition, volume, and density of the tissue.

Vacuum-assisted decellularization (VAD) cannot decellularize, but it is highly effective in enabling chemical agents to reach the whole tissue [42]. Hydrostatic pressure, on the other hand, is an effective decellularization method, but it is usually combined with enzymes such as DNases to achieve complete decellularization [43].

3.3. Sterilization Methods

The dECM is commonly used in in vivo implantable approaches. Therefore, in order to prevent the transmission of pathogens, the dECM needs to be sterilized to eliminate any microorganisms and to prevent infections [60]. There are some useful physical and chemical dECM sterilization techniques, although their suitability depends on multiple factors, which may need to be considered in each specific approach; for example, humidity, time of exposure, temperature, and the nature/load of the bioburden are some of the factors to be taken into account.

Among the physical methods, two groups are differentiated—those using heat and those using radiation. Heat techniques have very limited use in a dECM context, as dECM products are usually thermosensitive and high temperatures may denaturalize important proteins, as well as disrupt the structure [61,62]. Irradiation effectively eliminates microorganisms, and UV-irradiation is a common sterilization method used in research settings, mainly due to easy accessibility in terms of research labs and cell culture facilities [63]. On the other hand, gamma-irradiation is the preferred sterilization method for many pharmaceutical and clinical products, due to its high penetration, good assurance of sterilization, and feasible temperature during the sterilization process. Conversely, radiation may affect structural proteins such as collagen, reducing the strength of the treated material [64–66]. Therefore, gamma-irradiation is often selected as the sterilization method for dECM products, but the irradiation dosage has to be optimized for each specific case, and the properties must be assessed after treatment.

Liquid chemical wash can be used as a sterilization method. Alcohol-based sterilization is common in laboratories, as it is cheap and easy to access. This method is more effective at killing microbes in aqueous solutions, but may also affect the protein structure in dECM pieces. Phenols act by disrupting membranes, precipitating proteins, and inactivating enzymes. They are bactericidal, fungicidal, and mycobactericidal, but are ineffective against spores and most viruses. Aldehydes are alkylating agents that damage nucleic acids and kill all microorganisms, including spores. In contrast to alcohols, which are volatile, phenols and aldehydes are generally toxic, corrosive, and/or irritating. Liquid chemicals must be removed after the sterilization process due to their potential toxicity in further in vivo uses [60].

Chemical methods are also useful to sterilize dECM materials, and are an alternative to physical methods. Ozone and hydrogen peroxide are traditional sterilization gases, but ethylene oxide is more commonly used in a dECM context, as the ultrastructure and the mechanical properties of the dECM are usually not altered under such treatment [64,67]. After the sterilization process, it is important to ensure the elimination of residual sterilizing agents and other possible volatile residues.

4. Decellularization Methods by Tissue

In order to choose a decellularization method, the tissue itself and the strategy to be followed have to be taken into consideration. Since each tissue has a different structure and composition, the decellularization methods have to be specifically selected and empirically tested. Moreover, if the aim is to keep the tissue structure as intact as possible, chemical methods have to be chosen carefully because of the damage that they may cause to the structural macromolecules of the matrix. Therefore, the literature shows trends in the use of specific combinatorial physical and chemical approaches for each specific tissue and application, as follows.

4.1. Bone Tissue

Bones are the main components of the skeletal system. They give support to the body, allow movement, and produce red and white cells in their marrow. Bone is a connective tissue and its ECM is formed by different key proteins that confer tensile strength, such as collagen type I, noncollagenous glycoproteins, and proteoglycans. Calcium hydroxyapatite (HA) in the ECM serves to store minerals and provides specific properties to bone tissue, such as resistance and hardness [68], while a series of signaling molecules, such as bone morphogenetic proteins (BMPs), are also part of a bone's ECM compartment.

The properties of bone's ECM as regenerative material have been thoroughly described in the literature, and that is why diseased and damaged bones are frequently approached using bone grafts. Bone autografts are the best option in order to avoid rejection, while allografts and xenografts are often used [46,69–72]. Implantable ceramic scaffolds are also frequent in clinical settings as osteoconductive materials, and some of these products are bovine or porcine bone HA calcareous matrices obtained after heat-treating bones in a muffle furnace to remove all organic compounds, including cells [68,73].

The first use of decalcified bone as bone implantable material was described as early as 1889 by Senn [74], when he used muriatic acid as a decalcification agent, followed by washing and alcohol sterilization before implantation in human bone defects. However, the experimental evidence of the demineralized bone matrix (DBM) as an osteoinductive material was established by Urist in 1965 [52]. At present, osteoinductive activity in the bone matrix is largely related to some of the BMPs present in bone's ECM, and therefore, it is known that demineralization processing has to be done with the aim to preserve BMPs' biological activity. In any case, the DBM obtained by different decalcification methods is commercially available and extensively reported in clinical settings as an osteoinductive implantable material suitable for treating bone defects [70,75,76]. Hydrochloric acid (HCl) and EDTA are common decalcifying agents, while chloroform and methanol can be used for lipid extraction. Then, the DBM can be snap frozen, lyophilized, or kept at $-20\ °C$ until necessary [71,77].

The decellularized bone matrix (DecBM) is frequently achieved by chemical methods, such as EDTA in combination with trypsin or SDS, along with ammonium hydroxide [46,70]. Alternatively, thermal shock can be used, together with Triton X-100, for effective osseous tissue decellularization [44]. Additionally, high-hydrostatic pressurization, a physical method, has been used with good results regarding bone decellularization [60]. Nucleases and dehydrated alcohol are used as complementary and final steps in order to remove waste nuclear acids and other cellular remains [53,54].

Some authors have described specifically the serial decalcification and decellularization steps in their protocols toward the generation of cell-free demineralized implantable materials [71]. In this sense, hydrogels made from the ECM of decalcified and decellularized bone are quite common due to its versatility and osteoconductivity [53,70].

4.2. Cartilage Tissue

Cartilage in adult animals is a connective, smooth, and resilient tissue. Hyaline cartilage is avascular, and it is present in the stress points of skeletal tissue, such as bone heads, where it provides flexibility and prevents abrasion and damage [77]. It is also present in the rib cage, nose, larynx, and trachea, while its extracellular matrix is composed of collagens, mainly collagen type II, glycosaminoglycans (GAGs), and laminin [78]. GAGs are closely related to cartilage mechanical properties and help the tissue to cope with sudden external forces [24]. Elastic cartilage is a supportive structure for tissues such as the outer ear and epiglottis, and histologically, it is similar to hyaline cartilage, but with much more elastic fibers [21]. On the other hand, fibrocartilage is the only cartilage with collagen type I in its structure, because it is a mixture of fibrous tissue and cartilaginous tissue, with unique toughness and elasticity properties, and is present in intervertebral discs and menisci, among others.

Hyaline cartilage is the most commonly targeted cartilage in regenerative treatment—in particular, the one covering the end of bones in articulations. Damage to articular cartilage is usually related to trauma or pathology, and can cause pain, osteoarthritis, or even loss of functionality [78]. Surgical interventions attempt to solve these conditions, but at present, they are largely temporary solutions, pushing research toward searching for new regenerative approaches. The lack of vascularization greatly limits the number of nutrients and oxygen that can reach the inner parts of the tissue. This condition makes cartilage regrowth particularly challenging [79]. Therefore, tackling articular cartilage regeneration is a challenging goal. The literature is extensive in terms of approaches aiming to regenerate articular cartilage tissue, using different kinds of implantable scaffolds [80,81]. In this sense, cartilage dECM derivatives have been used for coating other implantable materials or as a 3D-printed structure for knee and meniscal regeneration, among others.

The basics for articular cartilage decellularization are the same for the different types of cartilages, and in general, decellularization aims to keep in the matrix as many GAGs as possible. Of note, as the different cartilages have different permeability properties, it is paramount to set up proper conditions for optimum decellularization in each case.

Tissue snap freezing or freeze–thawing is a common pretreatment in many decellularization protocols [24,58]. Similarly, pretreatment with hypotonic and hypertonic solutions is also a popular method, used to induce apoptosis by osmotic pressure [11]. These methods do not decellularize the structure by itself, but they help further decellularization methods to work better and to reach the inner parts of the tissue. It has been proved that snap freezing does not affect the matrix component, and it has no significant negative impact on the structure [82]. Conversely, initial tissue homogenization, suggested in some articles, leads to the significant loss of GAGs and structural proteins [83].

Further to physical pretreatments, decellularization has been assayed with enzymatic or chemical detergent methods. The enzyme trypsin–EDTA is the most common decellularization approach reported [83]. Trypsin–EDTA breaks both the proteins that hold the cell in place inside the matrix, as well as the cell membrane proteins. Based on protease activity, the main setback of using this technique is the degradation of the proteins of potential interest in the final dECM. In order to prevent this, the exposure time to trypsin–EDTA needs to be highly controlled, and it is usually limited to 6–24 h [51].

Regarding the detergents used as decellularization reagents, Triton X-100 and SDS are the most commonly used ones. Both cause some extent of damage to the structure of the matrix, but the mechanical integrity of the matrix maintains at an acceptable level. Note that both incur damage to GAG content and integrity, being Triton X-100 the worse in this sense [26,45].

An additional issue to be taken into account regarding cartilage tissue is the need for enzymatic treatment with nuclease activity to prevent waste nucleic acid material from sticking to the matrix [11]. In elastic cartilage, the nuclease time may be necessarily longer to ensure decellularization. In articular cartilage, on the other hand, being a less dense structure, the risk of having DNA material stick to the protein matrix is lower.

The digestion of GAGs using chondroitinase ABC (ChABC) has been reported as a cartilage decellularization method. It is not a commonly used technique; although it facilitates the removal of native chondrocytes, it reduces the mechanical properties of the tissue as well [11].

4.3. Adipose Tissue

White adipose tissue is defined as a connective tissue that stores energy in the form of lipids (triglycerides), insulates the body, and provides cushioning and support for subcutaneous tissues and internal organs. It is composed of clusters of fat-storing cells (adipocytes) surrounded by a reticular fiber network and interspersed small blood vessels. The key ECM proteins of adipose tissue are collagen type I, collagen type IV, and laminin. Collagen with laminin provides anchoring sites and barrier functions for adipocytes. Collagen types IV and VII and laminin are major components of the basement membrane [84].

Reconstruction of soft tissue defects is needed after certain tumor resections, external injury, or due to congenital malformations, and presents a major challenge in plastic and reconstructive surgery. At present, the main complications related to adipose tissue reconstruction include capsular contracture, necrosis and donor site morbidity, and immune rejection, and therefore, new clinical approaches are required to improve the success rate. The subcutaneous adipose tissue discarded from surgical operations represents an abundant and easy-to-collect human tissue source, processable by dECM biomaterial [85]. In this sense, allograft and xenograft dECM biomimetic scaffolds have proved to be effective tools for promoting tissue repair and regeneration in numerous preclinical and clinical studies [86–88].

The optimal adipose tissue decellularization includes the extraction of lipids (delipidation), followed by the extraction of cells and cell components, thereby maintaining key proteins and the 3D structure. Human and porcine are the most common sources of adipose tissue extraction, and there are two different kinds of initial adipose tissue samples useful for decellularization purposes. Such samples can be solid tissue derived from resection surgery, usually performed in the abdominal area, which has to be cut into small pieces for decellularization. On the other hand, liposuction-derived samples are gel-like tissues that require homogenization and centrifugation as the initial step for separation of the lipid phase. After initial processing, decellularization can be achieved using detergent-based or detergent-free protocols [89].

A detergent-free method for adipose tissue decellularization was described by Flynn et al. in 2010, in which the dECM was produced with a combination of multiple physical and chemical strategies, such as freeze–thaw cycles in hypotonic buffer to loosen the ECM, isopropanol to remove lipids, and enzymatic digestion with trypsin–EDTA, DNase, RNase, and lipase to remove cells and lipids. The resulting dECM conserves the collagen architecture and provides a microenvironment for the differentiation of human adipose stem cells [90,91]. A further published similar protocol demonstrated, by immunohistochemical staining, that laminin and collagen type IV remain abundant in the decellularized matrix. In vitro and in vivo models with microporous foams and hydrogel scaffolds with cells both demonstrate strong support for adipogenesis and induce an angiogenic response and formation of new adipose tissue [55,92]. dECM scaffolds generated by similar detergent-free methods, with a combination of isopropanol, trypsin, EDTA, and DNase–RNAse in gel-like liposuction-derived samples, served as support for human adipose-derived stem cells and adipose regeneration [93].

In adipose tissue, the use of detergents for decellularization seems to increase the risk of matrix protein denaturalization and degradation. Wang et al. reported a method with multiple sequential physical and chemical steps, including a polar solvent extraction and Triton X-100 treatment, which resulted in the maintenance of collagens but the absence of laminin in the final dECM [94]. Similar results were obtained when SDS was used during the decellularization process [95]. Note that although laminin was absent in Triton-X-100-treated samples, in vivo studies have confirmed that the dECM undergoes vascularization and adipose tissue regeneration at Day 30 of implantation, which is consistent with other reports on the adipose tissue-derived matrix.

4.4. Skeletal Muscle and Tendons

Muscles are connective tissues formed by contractile fibers. Skeletal muscles are responsible for voluntary movement and homeostasis, and they are attached to bones by collagen fibrillar structures called tendons. Skeletal muscle is divided into several innervated and vascularized subtypes. Given the complex structure of skeletal muscle, it is difficult to pinpoint the exact distribution and composition of the ECM. Collagen is the most common component, as it contains collagen types I, II, III, IV, V, VI, XI, XII, XIV, XV, and XVIII. GAGs are ubiquitous in the ECM, while interactions between proteins and glycans are particularly important to regulate protein distribution. Moreover, ECM glycoproteins and cell membrane–protein interactions transmit the mechanical force in the muscle and are active during muscle injury regeneration [96].

Skeletal muscle loss is often the result of a traumatic injury. In this sense, reconstruction surgery may be required to recover functionality [97]. The first option is always an autologous transfer from nearby tissue, but this implicates a partial loss of functionality or volume. Among other approaches, the use of the muscle's decellularized ECM is a promising treatment, due to several reasons. Decellularized muscle xenografts are feasible, as a muscle's ECM is similar among different species, thereby minimizing the risk of the immune response [98,99]. Moreover, the dECM from skeletal muscle shows good integration in vivo, promoting vascularization, remodulation, and differentiation.

The initial skeletal muscle decellularization protocols included physical methods such as freeze–thawing or proteases, which have been further proven too aggressive for proper muscle tissue decellularization. At present, and aiming to preserve the matrix content and tissue structure, some less aggressive detergents, salt solutions, and nucleus-specific enzymes are the preferred decellularization methods. Most of the protocols use weak acids or detergents, such as sodium deoxycholate or Triton X-100 and SDS, respectively, followed by DNase treatment—all of them at low concentrations and exposition times and with multiple repeated cycles [100,101]. In some cases, trypsin is used in low concentrations and for short times, ensuring it does not damage in excess the protein structure.

Tendon tissue is a highly fibroelastic structure that connects muscles to bones. A tendon's ECM is mainly formed of collagen type I, elastin, and proteoglycans, and it provides mechanical and elastic capabilities. Collagen, in particular, constitutes up to 80% of the dry mass of tendons. These proteins are organized by creating fibers, fascicles, and, finally, the tendon itself. Other than that, there is a huge network of proteoglycans and other elastic macromolecules [49].

Tendons have a natural healing capacity, but they can be damaged if the injury goes beyond this healing capacity. Damage to tendons can be the result of severe trauma or the result of continuously repeated injuries during recovery processes [47]. When using a material to repair possible damage, the main properties required are mechanical and elastic capacities. In this context, dECM derivate materials show good regenerative properties, particularly the ones created with decellularized tendons, as they are assimilated easily and promote new tissue formation.

In order to decellularize tendons, detergents are the most commonly used reagents. Triton X-100 and SDS have been tested and compared independently as tendon decellularization methods, with and without a previous freeze–thawing cycle [102]. Triton X-100 treatment shows less efficiency in cell lysis, with no significant cell removal, and induces damage to the tendon structure. On the other hand, SDS is more effective as a tendon decellularization agent, with less damage to the ECM components and collagen structure [102]. Tri-n-butyl phosphate (TnBP) detergent has also been used for tendon decellularization purposes, with improved results compared to previous methods. Specifically, TnBP treatment results in a significant decrease in cell density, without disruption of the collagen matrix, even when used in relatively high concentrations [47].

4.5. Cardiovascular Tissue

The heart is a muscular organ whose function is mainly the pumping of blood through the circulatory system. The ECM of the heart is quite specific and is composed predominantly of collagens (types I, II, and III), fibrillin, hyaluronan, laminin, fibronectin, and proteoglycans [34,103,104]. Due to its composition, it shows great strength, flexibility, and durability [105].

Cardiovascular diseases are the target of multiple regenerative medicine approaches. Stem cell-related approaches have been tested as promising therapies for myocardial infarction, chronic ischemic myocardial dysfunction, and nonischemic dilated cardiomyopathy, among others. Moreover, regenerative and repair strategies have also been investigated as alternatives to heart transplant procedures. On the other hand, research efforts are directed toward defining regenerative or reparative approaches as a substitute to heart valve replacement procedures, a clinically required procedure in certain patients of valvular heart disease (VHD). In all of these cardiovascular regenerative research contexts, decellularized cardiac tissue is now thoroughly assayed, sometimes as a biocompatible and cytokine-carrying implantable scaffold, and in other cases, as a recellularized carrier of therapeutic cells [106,107]. Indeed, whole tissue decellularization, followed by recellularization, is a strategy that many labs are working on [108].

In comparison to other tissues, which are hard to decellularize, current methods yield decellularized heart tissue that retains a great number of its original properties, including its elasticity. The most frequent method to achieve heart decellularization is the use of specific detergents combined with other decellularizing agents. Detergents such as SDS, sodium deoxycholate, PEG, or Triton X-100 work very well as decellularization agents in valves, tissue pieces, and in a whole heart decellularization context. These detergents are often used together [76], but they have also been individually compared and optimized. For example, for whole porcine heart decellularization, Ferng et al. suggested 3% SDS as the optimal detergent, especially when perfused at a pressure between 90 and 120 mmHg, while discouraging the use of CHAPS or OGP due to their inability to successfully decellularize the tissue. Physical and enzymatic methods are often used before or after the detergent-based decellularization step. For example, osmotic shock before decellularization induces loosening of the ECM, while pretreatment with trypsin–EDTA also improves further decellularization steps [30,109,110]. After the detergent-based decellularization step(s), and in order to achieve more porosity in the scaffold, hearts are often freeze-dried [109–111], and sucrose may be added to the freezing media in order to avoid damaging or denaturizing biomolecules in the heart scaffold [111]. Focused on heart valve decellularization, postdetergent treatment, enzymes at low concentrations have been used in order to remove the waste of nucleic acids [112], a treatment that does not affect collagen valves or elastane and is compatible with further cell seeding strategies [30].

Some authors claim the most common detergents used as decellularizing agents for hearts may cause ECM denaturation and further loss of mechanical properties. That is why some physical and enzymatic treatments have been assayed with no detergent step for heart decellularization. For example, Seo et al. based their study on the supercritical fluid technology using the $scCO_2$–EtOH cosolvent system to decellularize hearts, showing that the supercritical carbon dioxide method maintains higher GAG and collagen levels in the remaining decellularized scaffolds [113]. For heart valves, in terms of enzymatic methods, an accutase solution, followed by nuclease treatment, has been reported [112], while other proteolytic and collagenolytic enzymes combined with nucleases are also able to effectively remove nearly all nucleic acids [112]. Other methods have been assayed, but seem more controversial in terms of the achieved nucleic acid removal, such as the use of a combination of proteases and chelating agents, e.g., trypsin–ethylenediaminetetraacetic acid [112].

4.6. Vascular Tissue

Vascular tissue is related to the transport of nutrients, oxygen, CO_2, hormones, and blood cells through the body. Essentially, it comprises arteries, arterioles, and veins, while only arteries are the common target for decellularization approaches. Arteries are elastic blood vessels that carry oxygenated blood from the heart to the whole body. The main arteries are composed of three tissue layers, from inside out: intima, media, and adventitia. The tunica intima's ECM contains mainly laminin and collagen type IV, while the medium layer is principally composed of collagen type III, elastin, glycoproteins, and GAGs. In contrast, the ECM of the outermost tunica consists primarily of collagen type I and elastin, but it also contains proteoglycans such as biglycan and decorin, as well as thrombospondin-2 [114]. Although arteries have a very complex structure, it is very important to maintain the ECM components in the interest of keeping the ECM properties intact, such as elasticity and resistance.

Arterial diseases such as pulmonary arterial hypertension, restenosis, and peripheral arterial diseases are now targeted with experimental and promising stem cell therapies [115–118]. On the other hand, peripheral and coronary artery bypasses are clinical procedures often based on artery replacement by autologous graft transplantation. The use of natural or synthetic biopolymers as grafting materials is a clinically feasible option [119,120], but decellularized artery grafts are gaining increased research attention as artery bypass grafting materials due to their proper molecular and mechanical properties and their reduced immunogenicity [121,122].

Before the artery decellularization technique starts, there are different steps that are recommended for better results. Some protocols suggest to lyse blood cells by washing the arteries in distilled water while shaking [120], while others include three freeze–thaw cycles with EDTA [50]. As in many other tissues, decellularization is mainly based on the activity of detergents such as SDS, EDTA, SDC, CHAPS, Triton X-100, or DOC [27,50,56,120,123,124], used individually or in combination [50]. Some protocols report trypsin and hypo/hypertonic solutions used together with these detergents [27,50,56], while the use of enzymatic DNA and RNA removal as one of the final steps is also recommended [27,56]. All of these chemical methods report quite good results, both in vitro and in vivo, implying the use of detergents as a valuable approach to vascular decellularization, while improvements related to avoiding immunogenicity and cytotoxicity are required [121].

Recently, arterial decellularization mediated by supercritical and pressurized CO_2 has been described. It briefly consists of a high-pressure syringe pump that delivers liquid CO_2 and ethanol or limonene (as cosolvents) through a preheated extraction vessel. Samples are treated with $scCO_2$ and endonucleases to remove residual cosolvent and DNA. This approach yields a nearly intact decellularized tissue free of cells, lipids, and nucleic acids, proposing an alternative to traditional decellularization methods. Nevertheless, further in vitro and in vivo analyses need to be completed [125].

4.7. Dermal Tissue

The skin's main function is to protect, but it is also in charge of thermoregulation and perception. Dermal tissue is complex, with different layers, such as the epidermis, dermis, and subdermis, each with different compositions, properties, and functions. The epidermis is the outermost layer, and its main function is protection. Behind it, the dermis softens stress and strain, while also provides a sense of touch, elasticity, and heat. The inner layer is the subdermis and it is in charge of insulation. Dermal tissue, as many tissues, is composed of different collagens (85% of the dermal tissue's ECM), matricellular proteins, elastin, fibrillin, and also other fiber-forming proteins, such as vitronectin and fibronectin, which are necessary for wound healing. The ECM of dermal tissue is also formed by proteoglycans and GAGs, with functions related to hydration and osmosis balance. Examples of these are hyaluronic acid, decorin, and versican [126].

Dermal conditions such as cutaneous burns and scars require regenerative medicine approaches to restore dermal function. These therapeutic targets comprise stem cell transplants, growth factors, and tissue engineering. When scars, burns, and wounds are not able to heal on their own, they require replacement of the dermal barrier. It is very common to find skin equivalent and reconstruction research in the literature using primary human fibroblasts and keratinocytes, regularly supplemented with a collagen type I matrix or ascorbate [127,128]. In terms of decellularized approaches, acellular dermal matrices (ADMs) are widely used in clinical regenerative medicine approaches because of their biological and structural organization. ADMs can be of both animal and human origin, and they have a lot of different applications, such as the regeneration of skin tissue in burn, wound, and scar reconstructions, among others.

The majority of commercial ADMs are based on patented or proprietary decellularization protocols. Among the experimental research examples, the first step is always the mechanical isolation of the dermal layer, which is obtained by individual or combined chemical, physical, and biological methods supplemented with agitation. Dermal tissue can be incubated in hypotonic buffer for cell lysing [129] before the dermal decellularization step, which is usually carried out with detergents such as Triton X-100, DOC, N-lauroylsarcosinate (NLS), or SDS, and often in combination with trypsin, BSA, EDTA, and/or dispase [25,76,124,129–132]. Detergents can be also combined with acids and bases for the hydrolytic degradation of residual nucleic acids or even hair, but they can damage the ECM [109]. Some protocols describe the use of a further step in endonucleases treatment to remove residual genetic material [129,132].

It is worth noting that there are some other methods that have been reported that are detergent-free, such as osmotic shock and latrunculin B treatment, but they are not mainstream in the dermal tissue decellularization field [132].

4.8. Tissues Related to Respiratory System

Taking oxygen and expelling carbon dioxide is a function of the respiratory system, which is formed by multiple organs. Some of them, such as the trachea, lungs, and even diaphragm, have been tested as raw materials for decellularization purposes [133–135].

The trachea is related to essential physiological functions, such as airway protection, phonation, and breathing. It is composed of hyaline cartilage, fibrous tissue, respiratory epithelium, and smooth muscle, with cartilage being the most prominent [134]. Trachea damage requiring replacement surgery can be a result of trauma, neoplastic diseases, or congenital stenosis. Tracheal reconstruction in these cases is still a great challenge, while currently, autologous tissue and cell transplantation, with or without additional grafting material, seems the best solution [136,137]. In this sense, decellularized allograft and xenograft trachea is a material that has been tested in in vivo settings. While hyaline cartilage is part of this tissue, its decellularization protocols are slightly different to the ones described for cartilage, being tissue freeze-drying, followed by detergent treatment, the most common [138]. Some protocols add DNases into the detergent treatment step to ensure the elimination of the nuclear material. Commonly, mild reagents are used with several cycles of repeated treatment. According to the reports, this procedure yields mild disruption of the mucosa layer, with preservation of the majority of the remaining tissue structure [139].

The function of the lungs is related to gas exchange from the environment to the bloodstream. The structure of the bronchi in the lungs is similar to the trachea, while air circulates from the bronchi to the bronchioles on its way to alveolar airspace, where gas exchange occurs. The regenerative capacity of the lungs is low, and a broad spectrum of severe lung diseases, such as obstructive diseases, fibrosis, and sarcoidosis, may require a lung transplantation procedure [140]. In this context, decellularized lung tissue has been tested as an experimental alternative to transplantation. Perfusion of detergents such as SDS and Triton X-100 is a technique useful to decellularize mouse and rat cadaveric lung tissue, thereby preserving the vascular and airway structures of the tissue [141–143]. Due to the complexity of lung tissue and the need for instant functionality of used implants, a lung's dECM requires recellularization with epithelial and endothelial cells via cell infusion and bioreactors for ex vivo generation, maturation, and maintenance of the so-called bioartificial lungs. In vivo implantation of these organs in rats yields anastomosis, but long-term success is still to be achieved [143,144].

4.9. Tissues Related to Gastrointestinal Tract

The gastrointestinal tract is a complex microenvironment with different parts and multiple related organs. It is in charge of digesting food to extract energy and nutrients and to absorb them. Damage to the gastrointestinal tract or the related organs can be caused by stress, injury, or diseases that affect one or several tissues of the gastrointestinal tract, for example, trauma, surgeries, neoplasia, cancers, fibrosis, inflammatory bowel disease, esophagectomy, and congenital or acquired defects [145]. The complex anatomy of the gastrointestinal tract makes the use of bioengineered scaffolds a difficult task, requiring the use of multilayered structures and the seed of different types of cells, depending on the tissue anatomy [146]. The esophagus and intestines have been the main target of decellularization in gastrointestinal tissue, along with the related organs such as the liver and pancreas, and decellularization has often been complemented with the seeding of functional cells [147]. Research of the esophagus and pancreas is still in the initial stages of development [148]; therefore, here, we focus on intestine and liver tissue decellularization.

The intestine has a complex cellular and matrix architecture with multiple gradients, and it is therefore difficult to replicate using simple scaffolds. The three-dimensional architecture of the intestine is maintained by the ECM, composed of an intricate network of fibrous structural proteins (proteoglycans and glycoproteins), along with fibronectin, laminin isoforms, collagens, and heparin sulfate proteoglycans (HSPGs). Furthermore, multiple cell phenotypes are present in the intestine, including stem cells, pericryptal myofibroblasts, fibroblasts, endothelial cells, pericytes, immune cells, neural cells, and smooth muscle cells [149]. The intestinal stem cell niche is a well-known dynamic environment located at the base of crypts and embedded within a specific ECM in which the intestinal stem cells (ISCs) reside and control proliferation, differentiation, and tissue homeostasis. That is why multiple research efforts have been conducted to bioengineer the intestinal stem cell niche, including the use of decellularized tissue [150]. For intestine decellularization, a combination of chemical and enzymatic solutions (perfusion of sodium deoxycholate, use of DNase, immersion in a hypotonic solution, etc.) is used to remove cells in this tissue, while maintaining the 3D structure [151,152]. Recent studies have shown that the integrin effector protein focal adhesion kinase (FAK) is essential for intestinal regeneration, and thus the preservation of FAK in decellularized tissue scaffolds is essential for regenerative purposes [153].

Hydrogels derived from decellularized intestine are useful for generating endoderm-derived human organoids, such as gastric, hepatic, pancreatic, and small intestine organoids [154]. Moreover, decellularized intestine scaffolds have been used for regenerative purposes in many other tissues, due to their intrinsic ability to induce site-specific in vivo cellular repopulation and regeneration without the need for an in vitro recellularization step. For example, decellularized intestine has been used for regeneration purposes in vascular [155], cardiac [156], dura mater [157], abdominal wall [158], bladder [159], bowel [160], corneal [161], esophagus [162], tendon [163], ligament, cartilage (meniscus) [164], dermal [165], and bone tissues [166].

In recent decades, several methods have been implemented for liver decellularization, aiming to preserve the liver's major ECM components such as laminin, elastin, fibronectin, collagen types I and IV, and sulfated glycosaminoglycan (sGAG) [167]. The most promising of such methods is the perfusion decellularization method, in which detergents (SDS and/or Triton X-100), hyperosmotic (NaCl), and enzymatic (DNase) solutions are injected intravascularly [168]. Additionally, liver fragments can be immersed into the detergent solutions under mechanical agitation, and the previously described detergent–enzymatic method can be combined with high G-force oscillation treatment to reduce processing times. In such decellularized livers, the structural properties and the protein composition of the ECM are maintained, while they show good biocompatibility and neovascularization in vivo [168,169].

4.10. Nervous Tissue

The nervous system is a complex system of specialized cells that connect the parts of the body and coordinate it using signals. The main cells that transmit said signals are neurons, while they require a network of nonneural supportive cells, i.e., glia cells. In this context, the nervous tissue's ECM is indeed created by glia cells, which protect, isolate, and feed neurons, thereby allowing synapsis. The main structural components of the ECM are collagen (types II and IV) and laminin. Fibronectin guides axon growth, and acetylcholinesterase helps to regulate neural signals [170].

Damage of the nervous system can be caused by multiple conditions, diseases, and injuries, while symptoms related to tissue damage can be multiple, from mild to severe, including the loss of motor functions [171]. The regeneration of damaged nerve connections is a long-time pursued aim and it has been widely assayed using different approaches. There is consensus in defining the properties of the ideal graft for nerve regeneration, which should be of a flexible, thin, neuro-inductive, conductive, and biocompatible structure, capable of promoting axon proliferation and of guiding its growth toward the reconnection of damaged nervous edges [172,173]. To do this, the structure should be able to be molded by Schwann cells, which are specific glia cells related to the guidance of neural regeneration processes. In this context, the usefulness of dECM-derived nerve scaffolds as implantable grafts has been assessed in in vitro and in vivo animal trials, with some promising results [174]. Some studies have shown the capacity of the dECM to promote axon growth and the regeneration of peripheral nerve connections in rats [69,172,175]. The regeneration process requires weeks, and the subjects are in need of rehabilitation in order to begin proper recovery.

For decellularization of nervous tissue, the most commonly used approach is washing with detergents. Note that in the peripheral nerves, the main DNA source is that of Schwann cells, which protect the axons that carry signals. The detergent that shows the best results in this tissue is usually Triton X-200 [28]. A low concentration and prolonged Triton X-200 treatment has been proven successful in eliminating both the axon and Schwann cells, as well as myelin waste. The elimination causes slight damage to the ECM proteins, but it retains good condition of the structure. This treatment is commonly used in combination with an osmotic cell burst, which breaks the cells, facilitating cell waste removal by Triton X-200. Other detergents, such as SDS or Triton-100, are used without osmotic shock [33,171]. Other than these, nucleases have been used in combination with detergents and osmotic methods to ensure that the DNA is properly removed.

4.11. Cornea

The cornea is a transparent, avascular, and highly innervated connective tissue that acts as the primary structural barrier to infections, and is the first lens of the eye optical system. The human cornea is organized in five layers, three of which are cellular (i.e., the epithelium, stroma, and endothelium), and two are considered interphases (i.e., the Bowman membrane and the Descemet membrane). This highly organized structure contributes to the cornea's transparency and mechanical strength, while disruptions to this pattern disturb said transparency and result in loss of vision [176,177]. The cornea's ECM is composed of water, inorganic salts, proteoglycans, glycoproteins, and collagens. The stromal lamellar collagen fibrils are heterotypic hybrids of types I and V, with significant amounts of collagen types VI, XII, and XIV. A high concentration of small leucine-rich proteoglycans, including decorin, lumican, and keratocan, decorated with dermatan sulfate and keratan sulfate are present in the lamellae, credited with maintaining the interfibrillar spacing required for transparency, and contributing to the regulation of corneal hydration.

Many corneal disorders require a corneal transplant, while obviously there is limited availability of donor tissue. As an alternative to cadaveric corneas, among others, the dECM from acellular porcine and bovine cornea and decellularized amniotic membrane have been combined with different cell types to form full-thickness corneas with stroma, epithelium, and endothelium layers [178–180]. This dECM replicates the structure and functional requirements of the native cornea, with the maintenance of the collagen fibril organization, transparency, biocompatibility, suitable mechanical toughness, and low immunogenicity [181,182].

Detergents such as SDS and Triton X-100 were commonly used in the pioneering cornea decellularization methods. Du et al. used a 24 h SDS (0.5% or 1%) treatment to generate a decellularized porcine cornea matrix, which was opaque and swollen after the decellularization [48]. Transparency was restored after soaking in sterile glycerol for one hour, but implantation in a rabbit model showed stromal edema and worsening of corneal opacity throughout the 28-day observation period. Another comparative study used NaCl, 0.05% SDS, or 1% Triton-X100 to decellularize human corneas, and observed that NaCl did not affect transparency, while Triton-X-treated corneas experiencing tissue clouding; meanwhile, SDS-treated corneas appeared the most cloudy/opaque after decellularization [183,184]. Conversely, SDS treatment was combined with benzonase (nuclease) and protease inhibitors in human corneal sheets; in this case, when the dECM was recellularized and implanted in a rabbit model, the implanted tissue maintained complete transparency for three months [185]. A conceptually similar decellularization approach combines sodium N-lauroyl glutamate (SLG) surfactant with supernuclease (a nuclease homologous to benzonase), and also provides adequate transparency and good biocompatibility without degradation 28 days after transplantation [57].

Benzonase endonuclease is often used as the main decellularizing agent in a detergent-free approach, based on its ability to quickly infiltrate the corneal stroma, combined with its easy removal by repeated washes. This approach minimizes the destruction of the ECM, with minimal loss of optical transparency and proper results in animal transplantation assays [186].

4.12. Thymus

The thymus is an innervated organ part of the lymphatic and endocrine systems. The function of the thymus is to allow the development and maturation of the T-cell repertoire, and therefore, it has a main role in the immune response. Specifically, T-cell precursors are generated in the bone marrow and migrate to the thymus to become thymocytes, ultimately maturing immunocompetent T-cells. Endothelial and epithelial cells are the main cellular components of the thymus and, along with thymocytes, contribute to creating specific ECMs and microenvironment. The complex interaction network in the thymus includes cytokines, chemokines, matrix metalloproteases, laminin, collagen type IV, and multiple isoforms of fibronectin and glycoproteins, among others, with specific roles and precisely tuned toward the T-cell development process.

Thymus organ cultures are achieved via serial disaggregation and reaggregation of the tissue, and they are useful for ex vivo study of thymus function and complex cell interactions [187]. The rationale of thymus decellularization is mainly related to modeling thymus development, as well as the generation of potential regenerative or therapeutic approaches for in vivo immune response modulation [188]. To this aim, decellularization should be soft enough to keep intact the key ECM components, and should allow further proper recellularization with thymic epithelial cells and endothelial cells. Specifically, the thymus's dECM-derived bioengineered structure has to be able to reproduce T-cell differentiation and maturation processes. Freeze–thawing, followed by SDS and Triton X-100 detergent treatments, is a common decellularization technique [189]. Thymic epithelial cell-seeded dECM scaffolds, also called thymic reconstructed organoids, have been implanted in immunocompromised mice, yielding the development of populations of mature T-cells overwise absent in these animals [190].

5. The Clinical Outcome and Market of the dECM

Translational research is already a reality for some dECM-derived approaches, including several ongoing clinical trials and products on the market (see Table 2). The most common products are decellularized tissue pieces, serving as implantable materials for tissue formation, with proprietary- or patented-specific decellularization procedures [191–193].

Decellularized products on the market are generally issued with the ISO standard for biological medical devices (ISO10993-1, the standard for biological evaluation of medical devices), while recently, a specific standard for the evaluation of decellularized products has become available (ASTM F3354-19, Standard Guide for Evaluating Extracellular Matrix Decellularization Processes) [8]. Characterization includes in vitro and in vivo studies to provide data related to the removal of donor DNA and to the safety of implantable commercial products [194,195].

For some specific tissues, there are multiple decellularized products, competing for the same application market and claiming different properties due to differences in decellularization treatments. Comparative clinical case studies are common, and they provide useful information related to clinical success and outcomes of the different commercial dECMs available for each specific application [196–

199]. In this sense, there is a lack of standardized tissue-specific decellularization methods, which would serve as the standard control for comparative purposes [5,200]. Such standardized controls would be useful not only for the assessment of products already on the market, but also to perform more efficient, comparable, and reliable experimental research studies [201].

Table 2. Some examples of commercially available tissue-derived ECM products provided by tissue source.

Tissue Source	Application	Examples of Commercial Products
Bone/cartilage tissues	Grafting material for tissue regeneration and orthopedic surgery	-AlloWedge® Bicortical Allograft Bone (RTI Surgical) -Chondrofix® Osteochondral Allograft (Zimmer Inc.) -BioAdapt® DBM (RTI Surgical)
Adipose tissue	Aesthetic soft tissue reconstruction. Multiple tissues.	-Leneva® Allograft adipose matrix (MTF Biologics) -Adipose allograft matrix (AAM) (Musculoskeletal Transplant Found.)
Muscle and tendons	Graft tissue for pelvic organ prolapse	-Suspend® (Coloplast Corp.)
Cardiovascular tissue: heart valve, Pericardium	Graft for valve replacement and aneurysm reconstruction	-Hancock® II, Mosaic® and Freestyle® (Medtronic Inc.) -Prima® Plus and Perimount® (Edwards Lifesciences LLC) -Epic® and SJM Biocor® (St. Jude Medical Inc.)
Vascular tissue: Descending aorta, carotid artery, mesenteric vein, femoral artery.	Xenografting material for arterial replacement, bypass, aneurysm reconstruction, and path graft	-Artegraft® (Artegraft Inc.) -CryoGraft® and CryoArtery® (CryoLife Inc.) -ProCol® (LeMaitre Vascular Inc.)
Nerve tissue	Surgical repair of peripheral nerve discontinuities.	-Avance nerve allograft (Axogen corporation)
Dermal tissue	Grafting matrix for damaged tissue repair	-Dermacell® AWM (LifeNet Health Inc) -Alloderm® RTM (BioHorizons) -AlloPatch HD® (MTF Biologics)
Gastrointestinal tract: small intestine	Xenograft for cardiac tissue repair	-CorMatrix ECM® (CorMatrix® Cardiovascular Inc.)
Others: amniotic membrane, peritoneum	Grafting matrix for damaged tissue repair	-Biovance® (Celgene Cellular Therapeutics) -Meso BioMatrix® Surgical Mesh (MTF Biologics)

6. Concluding Remarks

Decellularization is a great technique to generate tissue-specific ECM-derived products with multiple applications, including tissue regeneration in clinical settings. Decellularization can be achieved from many tissues, but it has to be designed in accordance with the properties of the target tissue and the intended approach, aiming to preserve specific ECM components. The literature is extensive, but mostly related to empirical experimental research data. As a consequence, a variety of decellularization protocols have been described for each one of the targeted tissues.

Therefore, the challenge remains in defining broadly acceptable standardized decellularization and characterization procedures for each specific tissue that would ease the selection of standard controls and the development of future research, ultimately helping in the transfer of knowledge to clinical settings. At present, tissues' dECM scaffolds are the core of most clinical products, while research efforts are now strongly moving toward the development of postprocessing-related products, such as bioink-related 3D structures. Therefore, we anticipate rapid growth in the number of tissue-specific dECM postprocessing-related clinical products for the coming years.

Author Contributions: Conceptualization, U.M. and A.A.; writing—original draft preparation, U.M. and R.R.-H., S.R.-C., N.G.-U., and A.A.; writing—review and editing, A.A.; visualization, B.O.-G.; supervision, A.A. All authors have read and agreed to the published version of the manuscript.

Acknowledgments: This work was performed under the Maria de Maeztu Units of Excellence Programme–Grant No. MDM-2017-0720 Ministry of Science, Innovation and Universities. And Basque Government Elkartek program (KK-2019/00093).

Abbreviations

ECM	Extracellular matrix
dECM	Decellularized extracellular matrix
3D	Three-dimensional
DNA	Deoxyribonucleic acid
GAG	Glycosaminoglycans
SDS	Sodium dodecyl sulfate
CHAPS	3-[(3-cholamidopropyl)dimethylammonio]-1-propanesulfonate
EDTA	Ethylenediaminetetraacetic acid
VAD	Vacuum-assisted decellularization
HA	Calcium hydroxyapatite
MMP	Matrix metalloprotease
BMPs	Bone morphogenetic proteins
DBM	Demineralized bone matrix
DecBM	Decellularized bone matrix
HCl	Hydrochloric acid
ChABC	Chondroitinase ABC
TnBP	Tri-n-butyl phosphate
VHD	Valvular heart disease
PEG	Polyethylene glycol
OGP	n-Octyl-β-D-glucopyranosid
DOC	Deoxycholic acid
SDC	Sodium deoxycholate
ADMs	Acellular dermal matrices
NLS	N-lauroyl sarcosinate
HSPGs	Heparin sulfate proteoglycans
ISCs	Intestinal stem cells
FAK	Focal adhesion kinase
sGAG	Sulfated glycosaminoglycan
SLG	Sodium N-lauroyl glutamate
ISO	International Organization for Standardization
ASTM	American Society of Testing and Materials

References

1. Gattazzo, F.; Urciuolo, A.; Bonaldo, P. Extracellular matrix: A dynamic microenvironment for stem cell niche. *Biochim. Biophys. Acta Gen. Subj.* **2014**, *1840*, 2506–2519. [CrossRef] [PubMed]

2. Hussey, G.S.; Dziki, J.L.; Badylak, S.F. Extracellular matrix-based materials for regenerative medicine. *Nat. Rev. Mater.* **2018**, *3*, 159–173. [CrossRef]

3. Ahmed, E.M. Hydrogel: Preparation, characterization, and applications: A review. *J. Adv. Res.* **2015**, *6*, 105–121. [CrossRef] [PubMed]

4. Porzionato, A.; Stocco, E.; Barbon, S.; Grandi, F.; Macchi, V.; De Caro, R. Tissue-engineered grafts from human decellularized extracellular matrices: A systematic review and future perspectives. *Int. J. Mol. Sci.* **2018**, *19*, 4117. [CrossRef] [PubMed]

5. Crapo, P.M.; Gilbert, T.W.; Badylak, S.F. An overview of tissue and whole organ decellularization processes. *Biomaterials* **2012**, *32*, 3233–3243. [CrossRef]

6. Parmaksiz, M.; Dogan, A.; Odabas, S.; El, A.E.; El, Y.M. Clinical applications of decellularized extracellular matrices for tissue engineering and regenerative medicine. *Biomed. Mater.* **2016**, *11*, 22003. [CrossRef]

7. Nakamura, N.; Kimura, T.; Kishida, A. Overview of the Development, Applications, and Future Perspectives of Decellularized Tissues and Organs. *Biomaterials* **2017**, *3*, 1236–1244. [CrossRef]

8. ASTM International. *ASTM F3354-19, Standard Guide for Evaluating Extracellular Matrix Decellularization Processes*; ASTM International: West Conshohocken, PA, USA, 2019.

9. Woods, T.; Gratzer, P.F. Effectiveness of three extraction techniques in the development of a decellularized bone-anterior cruciate ligament-bone graft. *Biomaterials* **2005**, *26*, 7339–7349. [CrossRef]

10. Choi, J.S.; Kim, B.S.; Kim, J.Y.; Kim, J.D.; Choi, Y.C.; Yang, H.J.; Park, K.; Lee, H.Y.; Cho, Y.W. Decellularized extracellular matrix derived from human adipose tissue as a potential scaffold for allograft tissue engineering. *J. Biomed. Mater. Res. Part A* **2011**, *97*, 292–299. [CrossRef]

11. Bautista, C.A.; Park, H.J.; Mazur, C.M.; Aaron, R.K.; Bilgen, B. Effects of Chondroitinase ABC-Mediated Proteoglycan Digestion on Decellularization and Recellularization of Articular Cartilage. *PLoS ONE* **2016**, *11*, e0158976. [CrossRef]

12. Yu, Y.; Hua, S.; Yang, M.; Fu, Z.; Teng, S.; Niu, K.; Zhao, Q.; Yi, C. Fabrication and characterization of electrospinning/3D printing bone tissue engineering scaffold. *RSC Adv.* **2016**, *6*, 110557–110565. [CrossRef]

13. Heath, D.E. A Review of Decellularized Extracellular Matrix Biomaterials for Regenerative. *Regen. Med. Transl. Med.* **2019**, *5*, 155–166. [CrossRef]

14. Kabirian, F.; Mozafari, M. Decellularized ECM-derived bioinks: Prospects for the future. *Methods* **2019**, *171*, 108–118. [CrossRef] [PubMed]

15. Taylor, D.A.; Sampaio, L.C.; Ferdous, Z.; Gobin, A.S.; Taite, L.J. Decellularized matrices in regenerative medicine. *Acta Biomater.* **2018**, *74*, 74–89. [CrossRef] [PubMed]

16. Choudhury, D.; Tun, H.W.; Wang, T.; Naing, M.W. Organ-Derived Decellularized Extracellular Matrix: A Game Changer for Bioink Manufacturing? *Trends Biotechnol.* **2018**, *36*, 787–805. [CrossRef] [PubMed]

17. Gilpin, A.; Yang, Y. Decellularization Strategies for Regenerative Medicine: From Processing Techniques to Applications. *Biomed Res. Int.* **2017**, *2017*, 9831534. [CrossRef] [PubMed]

18. Robb, K.P.; Shridhar, A.; Flynn, L.E. Decellularized Matrices as Cell-Instructive Scaffolds to Guide Tissue-Specific Regeneration. *ACS Biomater. Sci. Eng.* **2018**, *4*, 3627–3643. [CrossRef]

19. Spang, M.T.; Christman, K.L. Extracellular matrix hydrogel therapies: In vivo applications and development. *Acta Biomater.* **2017**, *68*, 1–14. [CrossRef]

20. Keane, T.J.; Swinehart, I.T.; Badylak, S.F. Methods of tissue decellularization used for preparation of biologic scaffolds and in vivo relevance. *Methods* **2015**, *84*, 25–34. [CrossRef]

21. Crapo, P.M.; Gilbert, T.W.; Badylak, S.F. An overview of tissue and whole organ decellularization processes. *Biomaterials* **2011**, *32*, 3233–3243. [CrossRef]

22. White, L.J.; Taylor, A.J.; Faulk, D.M.; Keane, T.J.; Saldin, L.T.; Reing, J.E.; Swinehart, I.T.; Turner, N.J.; Ratner, B.D.; Stephen, F. The impact of detergents on the tissue decellularization process: A ToF-SIMS study. *Acta Biomater.* **2017**, 207–219. [CrossRef] [PubMed]

23. Cartmell, J.S.; Dunn, M.G. Effect of chemical treatments on tendon cellularity and mechanical properties. *J. Biomed. Mater. Res.* **2000**, *49*, 134–140. [CrossRef]

24. Elder, B.D.; Kim, D.H.; Athanasiou, K.A. Developing an articular cartilage decellularization process toward facet joint cartilage replacement. *Neurosurgery* **2010**, *66*, 722–727. [CrossRef] [PubMed]
25. Chen, R.N.; Ho, H.O.; Tsai, Y.T.; Sheu, M.T. Process development of an acellular dermal matrix (ADM) for biomedical applications. *Biomaterials* **2004**, *25*, 2679–2686. [CrossRef] [PubMed]
26. Tavassoli, A.; Matin, M.M.; Niaki, M.A.; Mahdavi-Shahri, N.; Shahabipour, F. Mesenchymal stem cells can survive on the extracellular matrix-derived decellularized bovine articular cartilage scaffold. *Iran. J. Basic Med. Sci.* **2015**, *18*, 1221. [PubMed]
27. Dahl, S.L.M.; Koh, J.; Prabhakar, V.; Niklason, L.E. Decellularized native and engineered arterial scaffolds for transplantation. *Cell Transplant.* **2003**, *12*, 659–666. [CrossRef]
28. Hudson, T.W.; Liu, S.Y.; Schmidt, C.E. Engineering an improved acellular nerve graft via optimized chemical processing. *Tissue Eng.* **2004**, *10*, 1346–1358. [CrossRef]
29. McFetridge, P.S.; Daniel, J.W.; Bodamyali, T.; Horrocks, M.; Chaudhuri, J.B. Preparation of porcine carotid arteries for vascular tissue engineering applications. *J. Biomed. Mater. Res. Part A* **2004**, *70*, 224–234. [CrossRef]
30. Rieder, E.; Kasimir, M.T.; Silberhumer, G.; Seebacher, G.; Wolner, E.; Simon, P.; Weigel, G. Decellularization protocols of porcine heart valves differ importantly in efficiency of cell removal and susceptibility of the matrix to recellularization with human vascular cells. *J. Thorac. Cardiovasc. Surg.* **2004**, *127*, 399–405. [CrossRef]
31. Poon, C.J.; Pereira, M.V.; Cotta, E.; Sinha, S.; Palmer, J.A.; Woods, A.A.; Morrison, W.A.; Abberton, K.M. Preparation of an adipogenic hydrogel from subcutaneous adipose tissue. *Acta Biomater.* **2013**, *9*, 5609–5620. [CrossRef]
32. Reing, J.E.; Brown, B.N.; Daly, K.A.; Freund, J.M.; Gilbert, T.W.; Hsiong, S.X.; Huber, A.; Kullas, K.E.; Tottey, S.; Wolf, M.T.; et al. The effects of processing methods upon mechanical and biologic properties of porcine dermal extracellular matrix scaffolds. *Biomaterials* **2010**, *31*, 8626–8633. [CrossRef] [PubMed]
33. Cornelison, R.C.; Wellman, S.M.; Park, J.H.; Porvasnik, S.L.; Song, Y.H.; Wachs, R.A.; Schmidt, C.E. Development of an apoptosis-assisted decellularization method for maximal preservation of nerve tissue structure. *Acta Biomater.* **2018**, *77*, 116–126. [CrossRef] [PubMed]
34. Ott, H.C.; Matthiesen, T.S.; Goh, S.; Black, L.D.; Kren, S.M.; Netoff, T.I.; Taylor, D.A. Perfusion-decellularized matrix: Using nature's platform to engineer a bioartificial heart. *Nat. Med.* **2008**, *14*, 213–221. [CrossRef]
35. Burk, J.; Erbe, I.; Berner, D.; Kacza, J.; Kasper, C.; Pfeiffer, B.; Winter, K.; Brehm, W. Freeze-Thaw cycles enhance decellularization of large tendons. *Tissue Eng. Part C Methods* **2014**, *20*, 276–284. [CrossRef] [PubMed]
36. Jackson, D.W.; Windler, G.E.; Simon, T.M. Intraarticular reaction associated with the use of freeze-dried, ethylene oxide-sterilized bone-patella tendon-bone allografts in the reconstruction of the anterior cruciate ligament. *Am. J. Sports Med.* **1990**, *18*, 1–11. [CrossRef] [PubMed]
37. Roth, S.P.; Glauche, S.M.; Plenge, A.; Erbe, I.; Heller, S.; Burk, J. Automated freeze-thaw cycles for decellularization of tendon tissue—A pilot study. *BMC Biotechnol.* **2017**, *17*, 1–10. [CrossRef] [PubMed]
38. Hung, S.H.; Su, C.H.; Lee, F.P.; Tseng, H. Larynx Decellularization: Combining Freeze-Drying and Sonication as an Effective Method. *J. Voice* **2013**, *27*, 289–294. [CrossRef]
39. Lin, P.; Chan, W.C.; Badylak, S.F.; Bhatia, S.N. Assessing Porcine Liver-Derived Biomatrix for Hepatic Tissue Engineering. *Tissue Eng.* **2004**, *10*, 1042–1053. [CrossRef]
40. Starnecker, F.; König, F.; Hagl, C.; Thierfelder, N. Tissue-engineering acellular scaffolds—The significant influence of physical and procedural decellularization factors. *J. Biomed. Mater. Res. Part B Appl. Biomater.* **2018**, *106*, 153–162. [CrossRef]
41. Azhim, A.; Ono, T.; Fukui, Y.; Morimoto, Y.; Furukawa, K.; Ushida, T. Preparation of Decellularized Meniscal Scaffolds Using Sonication Treatment for Tissue Engineering. In Proceedings of the 2013 35th Annual International Conference of the IEEE Engineering in Medicine and Biology Society (EMBC), Osaka, Japan, 3–7 July 2013; pp. 6953–6956.
42. Butler, C.R.; Hynds, R.E.; Crowley, C.; Gowers, K.H.; Partington, L.; Hamilton, N.J.; Carvalho, C.; Platé, M.; Samuel, E.R.; Burns, A.J.; et al. Vacuum-assisted decellularization: An accelerated protocol to generate tissue-intestine in mice. *Biomaterials* **2017**, *124*, 95–105. [CrossRef]
43. Waletzko, J.; Dau, M.; Seyfarth, A.; Springer, A.; Frank, M.; Bader, R.; Jonitz-heincke, A. Devitalizing Effect of High Hydrostatic Pressure on Human Cells—Influence on Cell Death in Osteoblasts and Chondrocytes. *Int. J. Mol. Sci.* **2020**, *21*, 3836. [CrossRef] [PubMed]

44.   Gardin, C.; Ricci, S.; Ferroni, L.; Guazzo, R.; Sbricoli, L.; De Benedictis, G.; Finotti, L.; Isola, M.; Bressan, E.; Zavan, B. Decellularization and delipidation protocols of bovine bone and pericardium for bone grafting and guided bone regeneration procedures. *PLoS ONE* **2015**, *10*, e0132344. [CrossRef] [PubMed]

45.   Rothrauff, B.B.; Yang, G.; Tuan, R.S. Tissue-specific bioactivity of soluble tendon-derived and cartilage-derived extracellular matrices on adult mesenchymal stem cells. *Stem Cell Res. Ther.* **2017**, *8*, 1–17. [CrossRef] [PubMed]

46.   Lee, D.J.; Diachina, S.; Lee, Y.T.; Zhao, L.; Zou, R.; Tang, N.; Han, H.; Chen, X.; Ko, C.C. Decellularized bone matrix grafts for calvaria regeneration. *J. Tissue Eng.* **2016**, *7*, 2041731416680306. [CrossRef] [PubMed]

47.   Yang, G.; Rothrauff, B.B.; Tuan, R.S. Tendon and ligament regeneration and repair: Clinical relevance and developmental paradigm. *Birth Defects Res. Part C Embryo Today Rev.* **2013**, *99*, 203–222. [CrossRef]

48.   Du, L.; Wu, X. Development and characterization of a full-thickness acellular porcine cornea matrix for tissue engineering. *Artif. Organs* **2011**, *35*, 691–705. [CrossRef]

49.   Kannus, P. Structure of the tendon connective tissue. *Scand. J. Med. Sci. Sport.* **2000**, *10*, 312–320. [CrossRef]

50.   Cheng, J.; Wang, C.; Gu, Y. Combination of freeze-thaw with detergents: A promising approach to the decellularization of porcine carotid arteries. *Biomed. Mater. Eng.* **2019**, *30*, 191–205. [CrossRef]

51.   Chan, B.P.; Leong, K.W. Scaffolding in tissue engineering: General approaches and tissue-specific considerations. *Eur. Spine J.* **2008**, *17*, 467–479. [CrossRef]

52.   Urist, M.R. Bone: Formation by Autoinduction. *Science* **1965**, *150*, 893–899. [CrossRef]

53.   Rothrauff, B.B.; Tuan, R.S. Decellularized bone extracellular matrix in skeletal tissue engineering. *Biochem. Soc. Trans.* **2020**, *48*, 755–764. [CrossRef] [PubMed]

54.   Chen, G.; Lv, Y. Decellularized Bone Matrix Scaffold for Bone Regeneration. *Methods Mol. Biol.* **2017**, *1577*, 239–254. [CrossRef]

55.   Cheung, H.K.; Han, T.T.Y.; Marecak, D.M.; Watkins, J.F.; Amsden, B.G.; Flynn, L.E. Composite hydrogel scaffolds incorporating decellularized adipose tissue for soft tissue engineering with adipose-derived stem cells. *Biomaterials* **2014**, *35*, 1914–1923. [CrossRef]

56.   Lin, C.H.; Hsia, K.; Tsai, C.H.; Ma, H.; Lu, J.H.; Tsay, R.Y. Decellularized porcine coronary artery with adipose stem cells for vascular tissue engineering. *Biomed. Mater.* **2019**, *14*. [CrossRef] [PubMed]

57.   Dong, M.; Zhao, L.; Wang, F.; Hu, X.; Li, H.; Liu, T.; Zhou, Q.; Shi, W. Rapid porcine corneal decellularization through the use of sodium N-lauroyl glutamate and supernuclease. *J. Tissue Eng.* **2019**, *10*, 2041731419875876. [CrossRef] [PubMed]

58.   Sutherland, A.J.; Detamore, M.S. Bioactive Microsphere-Based Scaffolds Containing Decellularized Cartilage. *Macromol. Biosci.* **2015**, *15*, 979–989. [CrossRef] [PubMed]

59.   Hashimoto, Y.; Funamoto, S.; Kimura, T.; Nam, K.; Fujisato, T.; Kishida, A. The effect of decellularized bone/bone marrow produced by high-hydrostatic pressurization on the osteogenic differentiation of mesenchymal stem cells. *Biomaterials* **2011**, *32*, 7060–7067. [CrossRef]

60.   Rutala, W.A.; Weber, D.J. Disinfection, sterilization, and antisepsis: An overview. *Am. J. Infect. Control* **2016**, *44*, e1–e6. [CrossRef]

61.   Rogers, W.J. *Steam and Dry Heat Sterilization of Biomaterials and Medical Devices*; Elsevier Masson SAS: Issy les Moulineaux, France, 2012; pp. 20–55.

62.   Gosztyla, C.; Ladd, M.R.; Werts, A.; Fulton, W.; Johnson, B.; Sodhi, C.; Hackam, D.J.; Physiology, C. A comparison of sterilization techniques for production of decellularized engineered human tracheal scaffolds. *Biomaterials* **2017**, *124*, 95–105. [CrossRef]

63.   Singh, R.; Singh, D.; Singh, A. Radiation sterilization of tissue allografts: A review. *World J. Radiol.* **2016**, *8*, 355. [CrossRef]

64.   White, L.J.; Keane, T.J.; Smoulder, A.; Zhang, L.; Castleton, A.A.; Reing, J.E.; Turner, N.J.; Dearth, C.L.; Badylak, S.F. The impact of sterilization upon extracellular matrix hydrogel structure and function. *J. Immunol. Regen. Med.* **2018**, *2*, 11–20. [CrossRef]

65.   Van Nooten, G.; Somers, P.; Cuvelier, C.A.; De Somer, F.; Cornelissen, M.; Cox, E.; Verloo, M.; Chiers, K. Gamma radiation alters the ultrastructure in tissue-engineered heart valve scaffolds. *Tissue Eng. Part A* **2009**, *15*, 3597–3604. [CrossRef]

66.   Harrell, C.R.; Djonov, V.; Fellabaum, C.; Volarevic, V. Risks of using sterilization by gamma radiation: The other side of the coin. *Int. J. Med. Sci.* **2018**, *15*, 274–279. [CrossRef] [PubMed]

67. Hodde, J.; Janis, A.; Ernst, D.; Zopf, D.; Sherman, D.; Johnson, C. Effects of sterilization on an extracellular matrix scaffold: Part I. Composition and matrix architecture. *J. Mater. Sci. Mater. Med.* **2007**, *18*, 537–543. [CrossRef] [PubMed]

68. Galeano, S.; García-Lorenzo, M.L. Bone mineral change during experimental calcination: An X-ray diffraction study. *J. Forensic Sci.* **2014**, *59*, 1602–1606. [CrossRef] [PubMed]

69. Lin, T.; Liu, S.; Chen, S.; Qiu, S.; Rao, Z.; Liu, J.; Zhu, S.; Yan, L.; Mao, H.; Zhu, Q.; et al. Hydrogel derived from porcine decellularized nerve tissue as a promising biomaterial for repairing peripheral nerve defects. *Acta Biomater.* **2018**, *73*, 326–338. [CrossRef] [PubMed]

70. Sawkins, M.J.; Bowen, W.; Dhadda, P.; Markides, H.; Sidney, L.E.; Taylor, A.J.; Rose, F.R.A.J.; Badylak, S.F.; Shakesheff, K.M.; White, L.J. Hydrogels derived from demineralized and decellularized bone extracellular matrix. *Acta Biomater.* **2013**, *9*, 7865–7873. [CrossRef]

71. Bracey, D.N.; Seyler, T.M.; Jinnah, A.H.; Lively, M.O.; Willey, J.S.; Smith, T.L.; Van Dyke, M.E.; Whitlock, P.W. A decellularized porcine xenograft-derived bone scaffold for clinical use as a bone graft substitute: A critical evaluation of processing and structure. *J. Funct. Biomater.* **2018**, *9*, 45. [CrossRef]

72. Kim, Y.S.; Majid, M.; Melchiorri, A.J.; Mikos, A.G. Applications of decellularized extracellular matrix in bone and cartilage tissue engineering. *Bioeng. Transl. Med.* **2019**, *4*, 83–95. [CrossRef]

73. Khalpey, Z. Acellular porcine heart matrices: Whole organ decellularization with 3D-Bioscaffold & vascular preservation. *J. Clin. Transl. Res.* **2017**, *3*, 260–270. [CrossRef]

74. Senn, N. On the healing of aseptic bone cavities by implantation of antiseptic decalcified bone. *Am. J. Med. Sci.* **1889**, *3*, 219. [CrossRef]

75. Gruskin, E.; Doll, B.A.; Futrell, F.W.; Schmitz, J.P.; Hollinger, J.O. Demineralized bone matrix in bone repair: History and use. *Adv. Drug Deliv. Rev.* **2012**, *64*, 1063–1077. [CrossRef] [PubMed]

76. Saldin, L.T.; Cramer, M.C.; Velankar, S.S.; White, L.J.; Badylak, S.F. Extracellular matrix hydrogels from decellularized tissues: Structure and function. *Acta Biomater.* **2017**, *49*, 1–15. [CrossRef] [PubMed]

77. Kim, D.K.; In Kim, J.; Sim, B.R.; Khang, G. Bioengineered porous composite curcumin/silk scaffolds for cartilage regeneration. *Mater. Sci. Eng. C* **2017**, *78*, 571–578. [CrossRef] [PubMed]

78. Elder, B.D.; Vigneswaran, K.; Athanasiou, K.A.; Kim, D.H. Biomechanical, Biochemical, and Histological Characterization of Canine Lumbar Facet Joint Cartilage. *Neurosurgery* **2010**, *66*, 722–727. [CrossRef] [PubMed]

79. Reddi, A.H. Cartilage morphogenetic proteins: Role in joint development, homoeostasis, and regeneration. *Ann. Rheum. Dis.* **2003**, *62*, 73–78. [CrossRef]

80. Yang, Z.; Shi, Y.; Wei, X.; He, J.; Yang, S.; Dickson, G.; Tang, J.; Xiang, J.; Song, C.; Li, G. Fabrication and repair of cartilage defects with a novel acellular cartilage matrix scaffold. *Tissue Eng. Part C Methods* **2010**, *16*, 865–876. [CrossRef]

81. Nie, X.; Jin, Y.; Zhu, W.; He, P.; Peck, Y.; Wang, D. Decellularized tissue engineered hyaline cartilage graft for articular cartilage repair. *Biomaterials* **2020**, *235*, 119821. [CrossRef]

82. Shin, M.; Vacanti, J. Tissue engineering. *Emerg. Technol. Surg.* **2007**, *10*, 133–151.

83. Benders, K.E.M.; Boot, W.; Cokelaere, S.M.; Van Weeren, P.R.; Gawlitta, D.; Bergman, H.J.; Saris, D.B.F.; Dhert, W.J.A.; Malda, J. Multipotent Stromal Cells Outperform Chondrocytes on Cartilage-Derived Matrix Scaffolds. *Cartilage* **2014**, *5*, 221–230. [CrossRef]

84. Costa, A.; Naranjo, J.D.; Londono, R.; Badylak, S.F. Biologic scaffolds. *Cold Spring Harb. Perspect. Med.* **2017**, *7*, 1–24. [CrossRef] [PubMed]

85. Ruderman, N.; Chisholm, D.; Pi-Sunyer, X.; Schneider, S. The metabolically obese, normal-weight individual revisited. *Diabetes* **1998**, *47*, 699–713. [CrossRef] [PubMed]

86. Hemmrich, K.; von Heimburg, D. Biomaterials for adipose tissue engineering. *Expert Rev. Med. Devices* **2006**, *3*, 635–645. [CrossRef] [PubMed]

87. Girandon, L.; Kregar-Velikonja, N.; Božikov, K.; Barlič, A. In vitro models for adipose tissue engineering with adipose-derived stem cells using different scaffolds of natural origin. *Folia Biol.* **2011**, *57*, 47–56. [PubMed]

88. Onnely, E.D.; Riffin, M.G.; Utler, P.E.B. Breast Reconstruction with a Tissue Engineering and Regenerative Medicine Approach (Systematic Review). *Ann. Biomed. Eng.* **2020**, *48*, 9–25. [CrossRef]

89. Banyard, D.A.; Sarantopoulos, C.; Tassey, J.; Ziegler, M.; Chnari, E.; Evans, G.R.D.; Widgerow, A.D. Preparation, Characterization, and Clinical Implications of Human Decellularized Adipose Tissue Extracellular Matrix. In *Regenerative Medicine and Plastic Surgery*; Springer International Publishing: New York City, NY, USA, 2019; pp. 71–89.

90. Flynn, L.E. The use of decellularized adipose tissue to provide an inductive microenvironment for the adipogenic differentiation of human adipose-derived stem cells. *Biomaterials* **2010**, *31*, 4715–4724. [CrossRef]

91. Yu, C.; Kornmuller, A.; Brown, C.; Hoare, T.; Flynn, L.E. Decellularized adipose tissue microcarriers as a dynamic culture platform for human adipose-derived stem/stromal cell expansion. *Biomaterials* **2017**, *120*, 66–80. [CrossRef]

92. Yu, C.; Bianco, J.; Brown, C.; Fuetterer, L.; Watkins, J.F.; Samani, A.; Flynn, L.E. Porous decellularized adipose tissue foams for soft tissue regeneration. *Biomaterials* **2013**, *34*, 3290–3302. [CrossRef]

93. Song, M.; Liu, Y.; Hui, L. Preparation and characterization of acellular adipose tissue matrix using a combination of physical and chemical treatments. *Mol. Med. Rep.* **2018**, *17*, 138–146. [CrossRef]

94. Wang, L.; Johnson, J.A.; Zhang, Q.; Beahm, E.K. Combining decellularized human adipose tissue extracellular matrix and adipose-derived stem cells for adipose tissue engineering. *Acta Biomater.* **2013**, *9*, 8921–8931. [CrossRef]

95. Choi, Y.C.; Choi, J.S.; Kim, B.S.; Kim, J.D.; Yoon, H.I.; Cho, Y.W. Decellularized extracellular matrix derived from porcine adipose tissue as a xenogeneic biomaterial for tissue engineering. *Tissue Eng. Part C Methods* **2012**, *18*, 866–876. [CrossRef] [PubMed]

96. Gillies, A.R.; Lieber, R.L. Structure and function of the skeletal muscle extracellular matrix. *Muscle Nerve* **2011**, *44*, 318–331. [CrossRef]

97. Gefen, A.; van Nierop, B.; Bader, D.L.; Oomens, C.W. Strain-time cell-death threshold for skeletal muscle in a tissue-engineered model system for deep tissue injury. *J. Biomech.* **2008**, *41*, 2003–2012. [CrossRef] [PubMed]

98. Porzionato, A.; Sfriso, M.M.; Pontini, A.; Macchi, V.; Petrelli, L.; Pavan, P.G.; Natali, A.N.; Bassetto, F.; Vindigni, V.; De Caro, R. Decellularized human skeletal muscle as biologic scaffold for reconstructive surgery. *Int. J. Mol. Sci.* **2015**, *16*, 14808–14831. [CrossRef] [PubMed]

99. Zhang, J.; Qian, Z.; Turner, N.J.; Feng, S.; Yue, W.; Yang, H.; Zhang, L.; Wei, H.; Wang, Q.; Badylak, S.F. Perfusion-decellularized skeletal muscle as a three-dimensional scaffold with a vascular network template. *Biomaterials* **2016**, *89*, 114–126. [CrossRef] [PubMed]

100. Gillies, A.R.; Smith, L.R.; Lieber, R.L.; Varghese, S. Method for decellularizing skeletal muscle without detergents or proteolytic enzymes. *Tissue Eng. Part C Methods* **2011**, *17*, 383–389. [CrossRef]

101. Wilson, K.; Terlouw, A.; Roberts, K.; Wolchok, J.C.; Program, M.B. The Characterization of Decellularized Human Skeletal Muscle as a Blueprint for Mimetic Scaffolds. *J. Mater. Sci. Mater. Med.* **2016**, *27*, 1–29. [CrossRef]

102. Liu, Y.; Ramanath, H.S.; Wang, D.A. Tendon tissue engineering using scaffold enhancing strategies. *Trends Biotechnol.* **2008**, *26*, 201–209. [CrossRef]

103. Lockhart, M.; Wirrig, E.; Phelps, A.; Wessels, A. Extracellular Matrix and Heart Development. *Birth Defects Res. Part A Clin. Mol. Teratol.* **2011**, *91*, 535–550. [CrossRef]

104. Bayomy, A.F.; Bauer, M.; Qiu, Y.; Liao, R. Regeneration in heart disease—Is ECM the key? *Life Sci.* **2012**, *91*, 823–827. [CrossRef]

105. Eitan, Y.; Sarig, U.; Dahan, N.; Machluf, M. Acellular Cardiac Extracellular Matrix as a Scaffold for Tissue Engineering: In Vitro Cell Support, Remodeling, and Biocompatibility. *Tissue Eng. Part C Methods* **2010**, *16*. [CrossRef] [PubMed]

106. Prat-vidal, C.; Bayes-genis, A. Decellularized pericardial extracellular matrix: The preferred porous scaffold for regenerative medicine. *Xenotransplantation* **2020**, *27*, e12580. [CrossRef]

107. Di Meglio, F.; Nurzynska, D.; Romano, V.; Miraglia, R.; Belviso, I.; Sacco, A.M.; Barbato, V.; Di Gennaro, M.; Granato, G.; Maiello, C.; et al. Optimization of human myocardium decellularization method for the construction of implantable patches. *Tissue Eng.* **2017**, *23*, 525–539. [CrossRef] [PubMed]

108. Guyette, J.P.; Charest, J.M.; Mills, R.W.; Jank, B.J.; Moser, P.T.; Gilpin, S.E.; Gershlak, J.R.; Okamoto, T.; Gonzalez, G.; Milan, D.J.; et al. Bioengineering Human Myocardium on Native Extracellular Matrix. *Circ. Res.* **2015**, *118*, 56–72. [CrossRef]

109. Goldfracht, I.; Efraim, Y.; Shinnawi, R.; Kovalev, E.; Huber, I.; Gepstein, A.; Arbel, G.; Shaheen, N.; Tiburcy, M.; Zimmermann, W.H.; et al. Acta Biomaterialia Engineered heart tissue models from hiPSC-derived cardiomyocytes and cardiac ECM for disease modeling and drug testing applications. *Acta Biomater.* **2019**, *92*, 145–159. [CrossRef]

110. Efraim, Y.; Sarig, H.; Cohen, N.; Sarig, U.; de Berardinis, E.; Chaw, S. Acta Biomaterialia Biohybrid cardiac ECM-based hydrogels improve long term cardiac function post myocardial infarction. *Acta Biomater.* **2017**, *50*, 220–233. [CrossRef] [PubMed]

111. Wang, S.; Goecke, T.; Meixner, C.; Haverich, A.; Hilfiker, A.; Wolkers, W.F. Freeze-dried heart valve scaffolds. *Tissue Eng. Part C Methods* **2012**, *18*, 517–525. [CrossRef] [PubMed]

112. Haupt, J.; Lutter, G.; Gorb, S.N.; Simionescu, D.T.; Frank, D.; Seiler, J.; Paur, A.; Haben, I. Detergent-based decellularization strategy preserves macro- and microstructure of heart valves. *Interact. Cardio Vasc. Thorac. Surg.* **2018**, *26*, 230–236. [CrossRef] [PubMed]

113. Seo, Y.; Jung, Y.; Hyun, S. Biomaterialia Decellularized heart ECM hydrogel using supercritical carbon dioxide for improved angiogenesis. *Acta Biomater.* **2018**, *67*, 270–281. [CrossRef] [PubMed]

114. Cavinato, C.; Badel, P.; Krasny, W.; Avril, S.; Morin, C. Experimental Characterization of Adventitial Collagen Fiber Kinematics Using Second-Harmonic Generation Imaging Microscopy: Similarities and Differences Across Arteries, Species and Testing Conditions. *Tissue Eng. Biomater.* **2020**, *23*, 123–164. [CrossRef]

115. Suen, C.M.; Stewart, D.J.; Montroy, J.; Welsh, C.; Levac, B.; Wesch, N.; Zhai, A.; Fergusson, D.; McIntyre, L.; Lalu, M.M. Regenerative cell therapy for pulmonary arterial hypertension in animal models: A systematic review. *Stem Cell Res. Ther.* **2019**, *10*, 1–14. [CrossRef] [PubMed]

116. Homma, J.; Sekine, H.; Matsuura, K.; Kobayashi, E.; Shimizu, T. Mesenchymal Stem Cell Sheets Exert Antistenotic Effects in a Rat Arterial Injury Model. *Tissue Eng. Part A* **2018**, *24*, 1545–1553. [CrossRef] [PubMed]

117. Botham, C.M.; Bennett, W.L.; Cooke, J.P. Clinical trials of adult stem cell therapy for peripheral artery disease. *Methodist Debakey Cardiovasc. J.* **2013**, *9*, 201–205. [CrossRef]

118. Raval, Z.; Losordo, D.W. Cell therapy of peripheral arterial disease: From experimental findings to clinical trials. *Circ. Res.* **2013**, *112*, 1288–1302. [CrossRef] [PubMed]

119. Iwasaki, K.; Kojima, K.; Kodama, S.; Paz, A.C.; Chambers, M.; Umezu, M.; Vacanti, C.A. Bioengineered three-layered robust and elastic artery using hemodynamically-equivalent pulsatile bioreactor. *Circulation* **2008**, *118*, 52–57. [CrossRef]

120. Cai, Z.; Gu, Y.; Cheng, J.; Li, J.; Xu, Z.; Xing, Y.; Wang, C.; Wang, Z. Decellularization, cross-linking and heparin immobilization of porcine carotid arteries for tissue engineering vascular grafts. *Cell Tissue Bank.* **2019**, *20*, 569–578. [CrossRef]

121. Schneider, K.H.; Enayati, M.; Grasl, C.; Walter, I.; Budinsky, L.; Zebic, G.; Kaun, C.; Wagner, A.; Kratochwill, K.; Redl, H.; et al. Acellular vascular matrix grafts from human placenta chorion: Impact of ECM preservation on graft characteristics, protein composition and in vivo performance. *Biomaterials* **2018**, *177*, 14–26. [CrossRef]

122. Hany, K.; Park, K.; Yu, L.; Song, S.; Woo, H.; Kwak, H. Vascular reconstruction: A major challenge in developing a functional whole solid organ graft from decellularized organs. *Acta Biomater.* **2020**, *103*, 68–80. [CrossRef]

123. Kajbafzadeh, A.M.; Khorramirouz, R.; Kameli, S.M.; Hashemi, J.; Bagheri, A. Decellularization of Human Internal Mammary Artery: Biomechanical Properties and Histopathological Evaluation. *Biores. Open Access* **2017**, *6*, 74–84. [CrossRef]

124. Mayorca-Guiliani, A.E.; Willacy, O.; Madsen, C.D.; Rafaeva, M.; Elisabeth Heumüller, S.; Bock, F.; Sengle, G.; Koch, M.; Imhof, T.; Zaucke, F.; et al. Decellularization and antibody staining of mouse tissues to map native extracellular matrix structures in 3D. *Nat. Protoc.* **2019**, *14*, 3395–3425. [CrossRef]

125. Gil-Ramírez, A.; Rosmark, O.; Spégel, P.; Swärd, K.; Westergren-Thorsson, G.; Larsson-Callerfelt, A.K.; Rodríguez-Meizoso, I. Pressurized carbon dioxide as a potential tool for decellularization of pulmonary arteries for transplant purposes. *Sci. Rep.* **2020**, *10*, 1–12. [CrossRef] [PubMed]

126. Tracy, L.E.; Minasian, R.A.; Caterson, E.J. Extracellular Matrix and Dermal Fibroblast Function in the Healing Wound. *Adv. Wound Care* **2016**, *5*, 119–136. [CrossRef]

127. Jevtić, M.; Löwa, A.; Nováčková, A.; Kováčik, A.; Kaessmeyer, S.; Erdmann, G.; Vávrová, K.; Hedtrich, S. Impact of intercellular crosstalk between epidermal keratinocytes and dermal fibroblasts on skin homeostasis. *Biochim. Biophys. Acta Mol. Cell Res.* **2020**, *1867*, 118722. [CrossRef] [PubMed]

128. Pouliot, R.; Larouche, D.; Auger, F.A.; Juhasz, J.; Xu, W.; Li, H.; Germain, L. Reconstructed human skin produced in vitro and grafted on athymic mice. *Transplantation* **2002**, *73*, 1751–1757. [CrossRef] [PubMed]

129. Zhang, Y.; Iwata, T.; Nam, K.; Kimura, T.; Wu, P.; Nakamura, N.; Hashimoto, Y.; Kishida, A. Water absorption by decellularized dermis. *Heliyon* **2018**, *4*, e00600. [CrossRef] [PubMed]

130. Wolf, M.T.; Daly, K.A.; Brennan-Pierce, E.P.; Johnson, S.A.; Carruthers, C.A.; D'Amore, A.; Nagarkar, S.P.; Velankar, S.S.; Badylak, S.F. A hydrogel derived from decellularized dermal extracellular matrix. *Biomaterials* **2012**, *33*, 7028–7038. [CrossRef] [PubMed]

131. Lau, C.S.; Hassanbhai, A.; Wen, F.; Wang, D.; Chanchareonsook, N.; Goh, B.T.; Yu, N.; Teoh, S.H. Evaluation of decellularized tilapia skin as a tissue engineering scaffold. *J. Tissue Eng. Regen. Med.* **2019**, *13*, 1779–1791. [CrossRef]

132. Farrokhi, A.; Pakyari, M.; Nabai, L.; Pourghadiri, A.; Hartwell, R.; Jalili, R.; Ghahary, A. Evaluation of Detergent-Free and Detergent-Based Methods for Decellularization of Murine Skin. *Tissue Eng. Part A* **2018**, *24*, 955–967. [CrossRef]

133. Wallis, J.M.; Borg, Z.D.; Daly, A.B.; Deng, B.; Ballif, B.A.; Allen, G.B.; Jaworski, D.M.; Weiss, D.J. Comparative assessment of detergent-based protocols for mouse lung de-cellularization and re-cellularization. *Tissue Eng. Part C Methods* **2012**, *18*, 420–432. [CrossRef]

134. Hung, S.H.; Su, C.H.; Lin, S.E.; Tseng, H. Preliminary experiences in trachea scaffold tissue engineering with segmental organ decellularization. *Laryngoscope* **2016**, *126*, 2520–2527. [CrossRef]

135. Cozad, M.J.; Bachman, S.L.; Grant, S.A. Assessment of decellularized porcine diaphragm conjugated with gold nanomaterials as a tissue scaffold for wound healing. *J. Biomed. Mater. Res. Part A* **2011**, *99*, 426–434. [CrossRef] [PubMed]

136. Zang, M.; Zhang, Q.; Chang, E.I.; Mathur, A.B.; Yu, P. Decellularized tracheal matrix scaffold for tracheal tissue engineering: In vivo host response. *Plast. Reconstr. Surg.* **2013**, *132*, 549–559. [CrossRef]

137. Philipp, T.; Mgerliuclacuk, E.; Gillian, R.; Jonathan, M. Creation of Laryngeal Grafts from Primary Human Cells and Decellularized Laryngeal Scaffolds. *Tissue Eng.* **2019**, *26*, 543–555. [CrossRef]

138. Zang, M.; Zhang, Q.; Chang, E.I.; Mathur, A.B.; Yu, P. Decellularized tracheal matrix scaffold for tissue engineering. *Plast. Reconstr. Surg.* **2012**, *130*, 532–540. [CrossRef]

139. Remlinger, N.T.; Czajka, C.A.; Juhas, M.E.; Vorp, D.A.; Stolz, D.B.; Badylak, S.F.; Gilbert, S.; Gilbert, T.W. Hydrated xenogeneic decellularized tracheal matrix as a scaffold for tracheal reconstruction. *Biomaterials* **2010**, *31*, 3520–3526. [CrossRef]

140. Weiss, D.J. Concise review: Current status of stem cells and regenerative medicine in lung biology and diseases. *Stem Cells* **2014**, *32*, 16–25. [CrossRef] [PubMed]

141. O'Neill, J.D.; Anfang, R.; Anandappa, A.; Costa, J.; Javidfar, J.; Wobma, H.M.; Singh, G.; Freytes, D.O.; Bacchetta, M.D.; Sonett, J.R.; et al. Decellularization of human and porcine lung tissues for pulmonary tissue engineering. *Ann. Thorac. Surg.* **2013**, *96*, 1046–1056. [CrossRef] [PubMed]

142. Gilpin, S.E.; Guyette, J.P.; Gonzalez, G.; Ren, X.; Asara, J.M.; Mathisen, D.J.; Vacanti, J.P.; Ott, H.C. Perfusion decellularization of human and porcine lungs: Bringing the matrix to clinical scale. *J. Hear. Lung Transplant.* **2014**, *33*, 298–308. [CrossRef]

143. Giraldo-Gomez, D.M.; Leon-Mancilla, B.; Del Prado-Audelo, M.L.; Sotres-Vega, A.; Villalba-Caloca, J. Trypsin as enhancement in cyclical tracheal decellularization: Morphological and biophysical characterization. *Mater. Sci. Eng. C* **2016**, *59*, 930–937. [CrossRef]

144. Ershadi, R.; Rahim, M.; Jahany, S. Transplantation of the decellularized tracheal allograft in animal model (rabbit). *Asian J. Surg.* **2018**, *41*, 328–332. [CrossRef]

145. Orlando, G.; García-Arrarás, J.E.; Soker, T.; Booth, C.; Sanders, B.; Ross, C.L.; De Coppi, P.; Farney, A.C.; Rogers, J.; Stratta, R.J. Regeneration and bioengineering of the gastrointestinal tract: Current status and future perspectives. *Dig. Liver Dis.* **2012**, *44*, 714–720. [CrossRef] [PubMed]

146. Hussey, G.S.; Cramer, M.C.; Badylak, S.F. Extracellular Matrix Bioscaffolds for Building Gastrointestinal Tissue. *Cmgh* **2018**, *5*, 1–13. [CrossRef] [PubMed]

147. Hussey, G.S.; Keane, T.J.; Badylak, S.F. The extracellular matrix of the gastrointestinal tract: A regenerative medicine platform. *Nat. Rev. Gastroenterol. Hepatol.* **2017**, *14*, 540–552. [CrossRef] [PubMed]

148. Urbani, L.; Camilli, C.; Phylactopoulos, D.E.; Crowley, C.; Natarajan, D.; Scottoni, F.; Maghsoudlou, P.; McCann, C.J.; Pellegata, A.F.; Urciuolo, A.; et al. Multi-stage bioengineering of a layered oesophagus with in vitro expanded muscle and epithelial adult progenitors. *Nat. Commun.* **2018**, *9*, 4286. [CrossRef]

149. Meran, L.; Baulies, A.; Li, V.S.W. Intestinal Stem Cell Niche: The Extracellular Matrix and Cellular Components. *Stem Cells Int.* **2017**, *2017*, 7970385. [CrossRef]
150. Wang, Y.; Kim, R.; Hinman, S.S.; Zwarycz, B.; Magness, S.T.; Allbritton, N.L. Bioengineered Systems and Designer Matrices That Recapitulate the Intestinal Stem Cell Niche. *Cell. Mol. Gastroenterol. Hepatol.* **2018**, *5*, 440–453. [CrossRef]
151. Maghsoudlou, P.; Totonelli, G.; Loukogeorgakis, S.P.; Eaton, S.; De Coppi, P. A decellularization methodology for the production of a natural acellular intestinal matrix. *J. Vis. Exp.* **2013**. [CrossRef]
152. Kajbafzadeh, A.M.; Khorramirouz, R.; Masoumi, A.; Keihani, S.; Nabavizadeh, B. Decellularized human fetal intestine as a bioscaffold for regeneration of the rabbit bladder submucosa. *J. Pediatr. Surg.* **2018**, *53*, 1781–1788. [CrossRef]
153. Ashton, G.H.; Morton, J.P.; Myant, K.; Phesse, T.J.; Ridgway, R.A.; Marsh, V.; Wilkins, J.A.; Athineos, D.; Muncan, V.; Kemp, R.; et al. Focal Adhesion Kinase is required for intestinal regeneration and tumorigenesis downstream of Wnt/c-Myc signaling. *Dev. Cell* **2010**, *19*, 259–269. [CrossRef]
154. Giobbe, G.G.; Crowley, C.; Luni, C.; Campinoti, S.; Khedr, M.; Kretzschmar, K.; De Santis, M.M.; Zambaiti, E.; Michielin, F.; Meran, L.; et al. Extracellular matrix hydrogel derived from decellularized tissues enables endodermal organoid culture. *Nat. Commun.* **2019**, *10*, 5658. [CrossRef]
155. Badylak, S.F.; Lantz, G.C.; Coffey, A.; Geddes, L.A. Small intestinal submucosa as a large diameter vascular graft in the dog. *J. Surg. Res.* **1989**, *47*, 74–80. [CrossRef]
156. Daley, M.C.; Fenn, S.L.; Iii, L.D.B. *Cardiac Extracellular Matrix*; Springer International Publishing: New York City, NY, USA, 2018; Volume 1098, ISBN 978-3-319-97420-0.
157. Cobb, M.A.; Badylak, S.F.; Janas, W.; Simmons-Byrd, A.; Boop, F.A. Porcine small intestinal submucosa as a dural substitute. *Surg. Neurol.* **1999**, *51*, 99–104. [CrossRef]
158. Clarke, K.M.; Lantz, G.C.; Salisbury, S.K.; Badylak, S.F.; Hiles, M.C.; Voytik, S.L. Intestine submucosa and polypropylene mesh for abdominal wall repair in dogs. *J. Surg. Res.* **1996**, *60*, 107–114. [CrossRef] [PubMed]
159. Wallis, M.C.; Yeger, H.; Cartwright, L.; Shou, Z.; Radisic, M.; Haig, J.; Suoub, M.; Antoon, R.; Farhat, W.A. Feasibility study of a novel urinary bladder bioreactor. *Tissue Eng. Part A* **2008**, *14*, 339–348. [CrossRef]
160. Chen, M.K.; Badylak, S.F. Small bowel tissue engineering using small intestinal submucosa as a scaffold. *J. Surg. Res.* **2001**, *99*, 352–358. [CrossRef]
161. Oliveira, A.C.; Garzón, I.; Ionescu, A.M.; Carriel, V.; de la Cruz Cardona, J.; González-Andrades, M.; del Mar Perez, M.; Alaminos, M.; Campos, A. Evaluation of Small Intestine Grafts Decellularization Methods for Corneal Tissue Engineering. *PLoS ONE* **2013**, *8*, e66538. [CrossRef]
162. Syed, O.; Walters, N.J.; Day, R.M.; Kim, H.W.; Knowles, J.C. Evaluation of decellularization protocols for production of tubular small intestine submucosa scaffolds for use in oesophageal tissue engineering. *Acta Biomater.* **2014**, *10*, 5043–5054. [CrossRef]
163. Hodde, J.P.; Badylak, S.F.; Donald Shelbourne, K. The effect of range of motion on remodeling of small intestinal submucosa (SIS) when used as an achilles tendon repair material in the rabbit. *Tissue Eng.* **1997**, *3*, 27–37. [CrossRef]
164. Cook, J.L.; Tomlinson, J.L.; Arnoczky, S.P.; Fox, D.B.; Cook, C.R.; Kreeger, J.M. Kinetic study of the replacement of porcine small intestinal submucosa grafts and the regeneration of meniscal-like tissue in large avascular meniscal defects in dogs. *Tissue Eng.* **2001**, *7*, 321–334. [CrossRef]
165. Prevel, C.D.; Eppley, B.L.; Summerlin, D.J.; Sidner, R.; Jackson, J.R.; McCarty, M.; Badylak, S.F. Small intestinal submucosa: Utilization as a wound dressing in full-thickness rodent wounds. *Ann. Plast. Surg.* **1995**, *35*, 381–388. [CrossRef]
166. Moore, D.C.; Pedrozo, H.A.; Crisco, J.J.; Ehrlich, M.G. Preformed grafts of porcine small intestine submucosa (SIS) for bridging segmental bone defects. *J. Biomed. Mater. Res. Part A* **2004**, *69*, 259–266. [CrossRef] [PubMed]
167. Baptista, P.M.; Vyas, D.; Moran, E.; Wang, Z.; Soker, S. Human liver bioengineering using a whole liver decellularized bioscaffold. *Methods Mol. Biol.* **2013**, *1001*, 289–298. [CrossRef] [PubMed]
168. Mazza, G.; Rombouts, K.; Rennie Hall, A.; Urbani, L.; Vinh Luong, T.; Al-Akkad, W.; Longato, L.; Brown, D.; Maghsoudlou, P.; Dhillon, A.P.; et al. Decellularized human liver as a natural 3D-scaffold for liver bioengineering and transplantation. *Sci. Rep.* **2015**, *5*, 1–15. [CrossRef] [PubMed]

169. Mazza, G.; Al-Akkad, W.; Telese, A.; Longato, L.; Urbani, L.; Robinson, B.; Hall, A.; Kong, K.; Frenguelli, L.; Marrone, G.; et al. Rapid production of human liver scaffolds for functional tissue engineering by high shear stress oscillation-decellularization. *Sci. Rep.* **2017**, *7*, 1–14. [CrossRef]

170. Gonzalez-Perez, F.; Udina, E.; Navarro, X. *Extracellular Matrix Components in Peripheral Nerve Regeneration*, 1st ed.; Elsevier Inc.: Amsterdam, The Netherlands, 2013; Volume 108, ISBN 9780124104990.

171. Terenghi, G. Peripheral nerve injury and regeneration. *Histol. Histopathol.* **1995**, *10*, 709–718. [PubMed]

172. Baiguera, S.; Del Gaudio, C.; Lucatelli, E.; Kuevda, E.; Boieri, M.; Mazzanti, B.; Bianco, A.; Macchiarini, P. Electrospun gelatin scaffolds incorporating rat decellularized brain extracellular matrix for neural tissue engineering. *Biomaterials* **2014**, *35*, 1205–1214. [CrossRef]

173. Ghasemi-Mobarakeh, L.; Prabhakaran, M.P.; Morshed, M.; Nasr-Esfahani, M.H.; Baharvand, H.; Kiani, S.; Al-Deyab, S.S.; Ramakrishna, S. Application of conductive polymers, scaffolds and electrical stimulation for nerve tissue engineering. *J. Tissue Eng. Regen. Med.* **2011**, *5*, e17–e35. [CrossRef]

174. Heikkinen, A.; Pihlajaniemi, T.; Faissner, A.; Yuzaki, M. Neural ECM and Synaptogenesis. In *Progress in Brain Research*; Elsevier B.V.: Amsteram, The Netherlands, 2014; Volume 214, pp. 29–51.

175. Arslan, Y.E.; Efe, B.; Sezgin Arslan, T. A novel method for constructing an acellular 3D biomatrix from bovine spinal cord for neural tissue engineering applications. *Biotechnol. Prog.* **2019**, *35*, e2814. [CrossRef]

176. DelMonte, D.W.; Kim, T. Anatomy and physiology of the cornea. *J. Cataract Refract. Surg.* **2011**, *37*, 588–598. [CrossRef]

177. Eghrari, A.O.; Riazuddin, S.A.; Gottsch, J.D. *Overview of the Cornea: Structure, Function, and Development*, 1st ed.; Elsevier Inc.: Baltimore, MD, USA, 2015; Volume 134, ISBN 9780128010594.

178. Ghezzi, C.E.; Rnjak-Kovacina, J.; Kaplan, D.L. Corneal Tissue Engineering: Recent Advances and Future Perspectives. *Tissue Eng. Part B Rev.* **2015**, *21*, 278–287. [CrossRef]

179. Matthyssen, S.; Van den Bogerd, B.; Dhubhghaill, S.N.; Koppen, C.; Zakaria, N. Corneal regeneration: A review of stromal replacements. *Acta Biomater.* **2018**, *69*, 31–41. [CrossRef] [PubMed]

180. Mahdavi, S.S.; Abdekhodaie, M.J. Bioengineering Approaches for Corneal Regenerative Medicine. *Tissue Eng. Regen. Med.* **2020**. [CrossRef] [PubMed]

181. Chen, Z.; You, J.; Liu, X.; Cooper, S.; Hodge, C.; Sutton, G.; Crook, J.M.; Wallace, G.G. Biomaterials for corneal bioengineering. *Biomed. Mater.* **2018**, *13*, 32002. [CrossRef]

182. Zhang, B.; Xue, Q.; Li, J.; Ma, L.; Yao, Y.; Ye, H.; Cui, Z. 3D bioprinting for artificial cornea: Challenges and perspectives. *Med. Eng. Phys.* **2019**, *71*, 68–78. [CrossRef]

183. Wilson, S.L.; Sidney, L.E.; Dunphy, S.E.; Dua, H.S.; Hopkinson, A. Corneal Decellularization: A Method of Recycling Unsuitable Donor Tissue for Clinical Translation? *Curr. Eye Res.* **2016**, *41*, 769–782. [CrossRef] [PubMed]

184. Fernández-pérez, J.; Ahearne, M. The impact of decellularization methods on extracellular matrix derived hydrogels. *Nature* **2019**, *9*, 1–12. [CrossRef]

185. Alio del Barrio, J.L.; Chiesa, M.; Garagorri, N.; Garcia-Urquia, N.; Fernandez-Delgado, J.; Bataille, L.; Rodriguez, A.; Arnalich-Montiel, F.; Zarnowski, T.; Álvarez de Toledo, J.P.; et al. Acellular human corneal matrix sheets seeded with human adipose-derived mesenchymal stem cells integrate functionally in an experimental animal model. *Exp. Eye Res.* **2015**, *132*, 91–100. [CrossRef]

186. Liu, J.; Li, Z.; Li, J.; Liu, Z. Application of benzonase in preparation of decellularized lamellar porcine corneal stroma for lamellar keratoplasty. *J. Biomed. Mater. Res. Part A* **2019**, *107*, 2547–2555. [CrossRef]

187. Piccinini, E.; Bonfanti, P. Disassembling and Reaggregating the Thymus: The Pros and Cons of Current Assays. In *Immunological Tolerance*; Humana Press: New York, NY, USA, 2019; Volume 1899, ISBN 9781493989386.

188. Tajima, A.; Pradhan, I.; Trucco, M.; Fan, Y. Restoration of Thymus Function with Bioengineered Thymus Organoids. *Curr. Stem Cell Rep.* **2016**, *2*, 128–139. [CrossRef]

189. Tajima, A.; Pradhan, I.; Geng, X.; Trucco, M.; Fan, Y. Construction of Thymus Organoids from Decellularized Thymus Scaffolds. *Methods Mol. Biol.* **2019**, *1576*, 33–42.

190. Fan, Y.; Tajima, A.; Goh, S.K.; Geng, X.; Gualtierotti, G.; Grupillo, M.; Coppola, A.; Bertera, S.; Rudert, W.A.; Banerjee, I.; et al. Bioengineering Thymus Organoids to Restore Thymic Function and Induce Donor-Specific Immune Tolerance to Allografts. *Mol. Ther.* **2015**, *23*, 1262–1277. [CrossRef] [PubMed]

191. Badylak, S.F.; Freytes, D.O.; Gilbert, T.W. Extracellular matrix as a biological scaffold material: Structure and function. *Acta Biomater.* **2009**, *5*, 1–13. [CrossRef] [PubMed]

192. Brown, B.N.; Badylak, S.F. Extracellular matrix as an inductive scaffold for functional tissue reconstruction. *Transl. Res.* **2014**, *163*, 268–285. [CrossRef] [PubMed]

193. Folli, S.; Curcio, A.; Melandri, D.; Bondioli, E.; Rocco, N.; Catanuto, G.; Falcini, F.; Purpura, V.; Mingozzi, M.; Buggi, F.; et al. A New Human-Derived Acellular Dermal Matrix for Breast Reconstruction Available for the European Market: Preliminary Results. *Aesthetic Plast. Surg.* **2018**, *42*, 434–441. [CrossRef]

194. Neumann, A.; Sarikouch, S.; Breymann, T.; Cebotari, S.; Boethig, D.; Horke, A.; Beerbaum, P.; Westhoff-Bleck, M.; Bertram, H.; Ono, M.; et al. Early systemic cellular immune response in children and young adults receiving decellularized fresh allografts for pulmonary valve replacement. *Tissue Eng. Part A* **2014**, *20*, 1003–1011. [CrossRef]

195. Elkins, R.C.; Dawson, P.E.; Goldstein, S.; Walsh, S.P.; Black, K.S. Decellularized human valve allografts. *Ann. Thorac. Surg.* **2001**, *71*, S428–S432. [CrossRef]

196. Cheng, A.; Saint-Cyr, M. Comparison of Different ADM Materials in Breast Surgery. *Clin. Plast. Surg.* **2012**, *39*, 167–175. [CrossRef]

197. Zenn, M.; Venturi, M.; Pittman, T.; Spear, S.; Gurtner, G.; Robb, G.; Mesbahi, A.; Dayan, J. Optimizing Outcomes of Postmastectomy Breast Reconstruction with Acellular Dermal Matrix: A Review of Recent Clinical Data. *Eplasty* **2017**, *17*, e18. [PubMed]

198. Brooke, S.; Mesa, J.; Uluer, M.; Michelotti, B.; Moyer, K.; Neves, R.I.; MacKay, D.; Potochny, J. Complications in tissue expander breast reconstruction: A comparison of AlloDerm, DermaMatrix, and FlexHD acellular inferior pole dermal slings. *Ann. Plast. Surg.* **2012**, *69*, 347–349. [CrossRef]

199. Tal, H.; Moses, O.; Zohar, R.; Meir, H.; Nemcovsky, C. Root Coverage of Advanced Gingival Recession: A Comparative Study Between Acellular Dermal Matrix Allograft and Subepithelial Connective Tissue Grafts. *J. Periodontol.* **2002**, *73*, 1405–1411. [CrossRef]

200. Bruyneel, A.A.N.; Carr, C.A. Ambiguity in the Presentation of Decellularized Tissue Composition: The Need for Standardized Approaches. *Artif. Organs* **2017**, *41*, 778–784. [CrossRef] [PubMed]

201. Hoshiba, T.; Yamaoka, T. (Eds.) *Decellularized Extracellular Matrix: Characterization, Fabrication and Applications*; Royal Society of Chemistry: London, UK, 2019.

5

Biomaterials Loaded with Growth Factors/Cytokines and Stem Cells for Cardiac Tissue Regeneration

Saltanat Smagul [†], **Yevgeniy Kim** [†], **Aiganym Smagulova** [†], **Kamila Raziyeva, Ayan Nurkesh** and **Arman Saparov** *

Department of Medicine, School of Medicine, Nazarbayev University, Nur-Sultan 010000, Kazakhstan; ssmagul@nu.edu.kz (S.S.); Yevgeniy.Kim@nu.edu.kz (Y.K.); Aiganym.Smagulova@nu.edu.kz (A.S.); kamila.raziyeva@nu.edu.kz (K.R.); ayan.nurkesh@nu.edu.kz (A.N.)
* Correspondence: asaparov@nu.edu.kz
† These authors contributed equally to this work.

Abstract: Myocardial infarction causes cardiac tissue damage and the release of damage-associated molecular patterns leads to activation of the immune system, production of inflammatory mediators, and migration of various cells to the site of infarction. This complex response further aggravates tissue damage by generating oxidative stress, but it eventually heals the infarction site with the formation of fibrotic tissue and left ventricle remodeling. However, the limited self-renewal capability of cardiomyocytes cannot support sufficient cardiac tissue regeneration after extensive myocardial injury, thus, leading to an irreversible decline in heart function. Approaches to improve cardiac tissue regeneration include transplantation of stem cells and delivery of inflammation modulatory and wound healing factors. Nevertheless, the harsh environment at the site of infarction, which consists of, but is not limited to, oxidative stress, hypoxia, and deficiency of nutrients, is detrimental to stem cell survival and the bioactivity of the delivered factors. The use of biomaterials represents a unique and innovative approach for protecting the loaded factors from degradation, decreasing side effects by reducing the used dosage, and increasing the retention and survival rate of the loaded cells. Biomaterials with loaded stem cells and immunomodulating and tissue-regenerating factors can be used to ameliorate inflammation, improve angiogenesis, reduce fibrosis, and generate functional cardiac tissue. In this review, we discuss recent findings in the utilization of biomaterials to enhance cytokine/growth factor and stem cell therapy for cardiac tissue regeneration in small animals with myocardial infarction.

Keywords: biomaterials; stem cells; cytokines; growth factors; cardiac tissue regeneration; regenerative medicine

1. Introduction

Cardiovascular diseases (CVD) are the leading cause of mortality worldwide [1,2]. In 2017, about 17.8 million deaths globally were attributed to CVD and in the U.S. alone, CVD, which include heart disease and stroke, were among the top ten causes of death, accounting for 74% of total deaths [3]. Coronary heart disease causes the majority of deaths in CVD, with myocardial infarction (MI) often leading to heart failure. Tissue damage at the site of infarction triggers local inflammation that attracts neutrophils and monocytes to clear the area of cell debris and produce reactive oxygen species. Migration of monocytes with reparative functions induces the formation of new vasculature and collagen production and eventually, leads to tissue repair and fibrotic tissue formation [4–6]. One biomedical approach for improving cardiac tissue regeneration is the delivery of therapeutic growth factors and cytokines [7]. Growth factors and cytokines have attracted the attention of researchers and clinicians due to their angiogenic and antiapoptotic properties, as well as their ability to increase cell proliferation and mobilize endogenous cell migration [8]. Various factors and cytokines, including but not limited to, tumor necrosis factor-α (TNF-α) and interleukin-8 (IL-8), are also upregulated in MI and participate in triggering inflammatory cascade. Therefore, regulation of pro- and anti-inflammatory mediator functions can be used to ameliorate inflammation and to facilitate cardiac tissue regeneration [9]. However, there are some challenges associated with growth factors/cytokines. For example, the systemic administration of growth factors/cytokines is not efficient due to a short in vivo half-life and poor bioavailability at the target sites. This, in turn, requires repeated injections, resulting in more side effects and greater treatment costs [10,11]. Moreover, simultaneous and rapid diffusion can lead to formation of immature and unstable blood vessels in the case of therapy with angiogenic growth factors [12].

Biomaterials offer a controlled and sustained release of bound growth factors and cytokines, which makes them a promising tool for overcoming the aforementioned challenges [13,14]. Biomaterials of natural, synthetic or hybrid origins were developed. They demonstrated therapeutic benefits when used either alone or when loaded with agents such as growth factors, cytokines or stem cells [15]. The use of biomaterials alone exerts positive effects on cardiac tissue regeneration, possibly via mimicking the extracellular matrix (ECM) and providing direct mechanical support. Some biomaterials also help to increase electrical conductance in a fibrotic scar region, which is important for normal functioning of the heart [16,17].

The endogenous regenerative capacity of cardiac tissue is limited: adult cardiomyocyte proliferation, cardiac stem cell activation, and bone marrow progenitor cell migration are not efficient enough to regenerate fully functional cardiac tissue. Post-MI repair often involves tissue replacement with non-functional fibrotic scarring, which can later lead to heart failure. For these reasons, stem cell therapy is considered a promising approach in MI treatment, being particularly beneficial for reducing the infarcted area and promoting cardiac function recovery [18]. Different stem cell sources such as mesenchymal stem cells (MSCs), cardiac stem cells (CSCs), induced pluripotent stem cells (iPSCs), and others are now recognized for their potential use in cardiac tissue regeneration [19]. Stem cell benefits in MI treatment include differentiation capacity, stimulation of resident CSCs, reduction in inflammation, and ability to provide structural support by connective tissue formation and fibroblast differentiation [20]. Release of cytokines and growth factors by stem cells allows for immunomodulation, angiogenesis, and stimulation of adjacent cells via paracrine mechanisms [21,22]. However, harsh conditions at the infarction site present a significant burden for stem cell survival. These conditions include, but are not limited to, hypoxia, fibrogenesis, low blood supply, and inflammation [23]. Therefore, biomaterials can serve as a stem cell delivery system that increases the living potency of the cells after transplantation and enhance the exerted effects. This review will focus on recent findings on the use of biomaterials as drug delivery systems for growth factors, cytokines, and stem cells for improving cardiac tissue regeneration in small animal models of MI.

2. Biomaterials Loaded with Growth Factors and Cytokines for Cardiac Tissue Regeneration

The use of biomaterials is now rapidly evolving as a new approach for MI treatment [24,25]. They are composed of a plethora of various polymers and can be used as a drug delivery system in the field of regenerative medicine [26]. The most common types are polymeric micro and nanospheres, nanoparticles (NPs), nanofibrous structures, coacervates, hydrogels, cryogels, and scaffolds. They differ in their size and assembling materials, as well as in their morphology, i.e., sheet versus vesicle-like structures [27–33]. Hydrogels, in particular, are widely investigated in the area of CVD. Hydrogel is largely composed of water and a cross-linked polymer and physically resembles tissue [34]. Hydrogels made of cardiac ECM, alginate, hyaluronic acid (HA), natural biomaterials (collagen, fibrin, and heparin), synthetic polymers, and microparticles have been studied pre-clinically for cardiac repair [35]. The effects of hydrogel administration include direct mechanical strengthening [36], enhanced angiogenesis and regeneration of myocardial tissue, reduced apoptosis and scar size, and improved cardiac function recovery [37]. Moreover, multiple studies showed that the use of biomaterials alone favorably affects various cells in the post-MI environment such as macrophages, cardiomyocytes, fibroblasts, and endothelial cells [38]. Recently, hydrogels made of ECM-based biomaterials have drawn attention because of their ability to mimic native ECM and minimize immunogenicity [39]. McLaughlin and colleagues treated mice at the end of the proliferative phase of wound healing with the injectable biomaterial, which contained human recombinant collagen I and III, one of the main proteins in the ECM of heart tissue. The treatment reduced inflammation, polarized macrophages towards M2 phenotype, increased capillary density at the border zone, and improved cardiac function [40]. The application of the self-assembling peptide (SAP) cell-free hydrogel also significantly improved the functionality of the heart post-MI through increased angiogenesis and reduced scar formation [41]. The beneficial effects of biomaterials are shown to depend on the time of therapy administration. In the study by Blackburn and colleagues, 3h post-MI application of collagen-based hydrogel in a murine model reduced cell apoptosis as well as increased capillary density and as a result, improved left ventricular ejection fraction. The authors also reported that biomaterial therapy is ineffective after 14 days post-MI [37]. The mechanisms of the exerted effects of biomaterials are possibly mediated by modifying the inflammatory immune response. It was demonstrated that hydrogel treatment also reduced the number of macrophages and TNF-α production in cardiac tissue. The in vitro culture of macrophages on biomaterials demonstrated a decrease in pro-inflammatory cytokines and an increase in anti-inflammatory cytokines [37].

Fibrosis, and its consequent non-functional scar formation, is considered to be a major problem following MI, leading to left ventricle remodeling and heart failure. Several biomaterials were designed to improve conduction of electrical signals in the scar region. For example, pyrrole was grafted onto a chitosan biomaterial to produce a conductive polypyrrole (PPy)-chitosan hydrogel. In vivo experiments used a coronary artery ligation rat model of acute MI to show reduced QRS complex on an electrocardiogram and improved transverse conduction velocity in PPy-chitosan group. It was demonstrated that both chitosan alone and PPy-chitosan were effective in preserving heart function, but PPy-chitosan further improved the indices, suggesting better maintenance of heart function as compared to a non-conductive biomaterial [16]. Cui and colleagues tested PPy-chitosan in a cryoablation injury rat model and reported a significant improvement in longitudinal conduction velocity in comparison to the chitosan only group. Electromyography was used to assess the conductivity of scar tissue ex vivo, which showed a significant 300–350% increase in electrical signals in the myocardial scar tissue in the group treated with PPy-chitosan [42].

Extensive research has been performed to study the importance of growth factors, cytokines, and different components of ECM in the treatment of MI [43,44]. It was shown that transforming growth factor-β (TGF-β) stimulates both Smad3-dependent and independent activation of macrophages, with the involvement of Smad3 in phagocytosis activation, secretion of vascular endothelial growth factor (VEGF) and TGF-β1, and protection against adverse cardiac tissue remodeling [45]. IL-10 is also important because its deficiency increases necrosis and neutrophil migration, with an enlargement in infarct size. Moreover, IL-10 deficiency impairs the ability of endothelial progenitor cells to suppress cell apoptosis, reduce scar size, increase neovascularization, and improve left ventricle remodeling, which is mediated by upregulation of integrin-linked kinase [46]. In contrast, treatment with IL-10 suppresses inflammation, polarizes macrophages towards M2 phenotype, activates fibroblasts, and improves left ventricle remodeling [47]. Another important growth factor is VEGF, which can be released from cardiac macrophages to simulate angiogenesis and heart muscle repair by regulating endothelial cell proliferation, migration, and apoptosis [43,48]. Furthermore, VEGF-A, fibroblast growth factor (FGF), and stromal cell-derived factor-1 (SDF-1) can stimulate neovascularization [49]. IL-4 is also a key cytokine because IL-4 administration differentiates macrophages, which are derived from Ly6C[high] monocytes, into a M2 phenotype [50]. However, application of growth factors and cytokines in clinical practice is hindered by their short half-lives, decreased stability, and deactivation by enzymes [27]. For example, the half-life of VEGF is approximately thirty-four minutes in plasma [51]. Therefore, biomaterials can serve as promising tools for the protection, delivery, and sustained release of growth factors and cytokines [52]. Table 1 summarizes the use of biomaterials loaded with growth factors and cytokines for cardiac tissue regeneration.

The incorporation of growth factors and cytokines into engineered biomaterials, such as hydrogels and NPs, offers even more opportunities for MI therapy (Figure 1). As an example, the injection of heparan sulfate proteoglycans (HSPG), which is a major component of ECM, with basic FGF (bFGF), extended the bioavailability of the growth factor by protecting it from degradation, and improved angiogenesis and cardiac function in animals with MI [53]. Another group also used bFGF that was fused with glutathione-S-transferase (GST) and matrix metalloproteinase (MMP)-2/9 cleavable peptide TIMP, and then, incorporated the complex into a glutathione-modified collagen hydrogel. This approach allowed for the controlled release of bFGF after TIMP was cleaved by the secreted MMP-2/9 at the site of tissue infarction. The use of this type of hydrogel decreased collagen deposition, increased vascularization, and improved heart function in rats with MI [54]. The mechanism of bFGF, which is a paracrine signaling protein, is mediated through binding to FGF receptor-heparan sulfate complex and further activation of tyrosine kinase. Downstream signaling proceeds via RAS-mitogen-activated protein kinase RAS-(MAPK) and phosphatidylinositide 3-kinase (PI3K) pathways [55]. In a separate study, sustained and targeted delivery of neuregulin-1β (NRG), which is a member of epidermal growth factor that regulates cardiomyocyte development and proliferation, by a hydroxyethyl methacrylate hyaluronic acid (HEMA-HA) hydrogel, demonstrated a cardioprotective effect and significantly improved ventricular function and structure [38]. The cardioprotective effect was assessed by the amount of caspase-3 in murine hearts post-MI, which was significantly reduced in the NRG-hydrogel group in comparison to the control groups treated with phosphate-buffered saline, NRG, or hydrogel alone. Caspase-3 is a key mediator of the terminal apoptotic pathway and its downregulation is associated with reduced infarct size, decreased apoptotic index of myocytes, and enhanced heart function in an experimental model of MI [56]. Awada and colleagues demonstrated that sequential delivery of VEGF followed by platelet-derived growth factor (PDGF) using a fibrin gel/heparin coacervate delivery system improves angiogenesis and cardiac function and reduces scar formation and inflammation at two and four weeks after MI in a rat model [57]. Mechanistically, VEGF promotes angiogenesis by activating or affecting different pathways and proteins, including PI3K, VRAP, Src tyrosine kinase, MAPK, and phospholipase C [58]. Recent reports show the critical role of multiple types of tyrosine and serine/threonine phosphatases, such as Shp2 and low molecular weight protein tyrosine phosphatase, in negative/positive regulation of VEGFR-2 signaling [59]. Interestingly, although VEGF demonstrated positive effects on MI in the experimental animal models, the results were not very promising according to several clinical trials [49]. One possible reason is the short period of protein bioactivity in vivo [60].

Although natural hydrogels are widely used in experiments [61], synthetic and hybrid hydrogels are also broadly investigated [62]. Synthetic glycosaminoglycan mimetic peptide nanofiber developed by Rufaihah and colleagues promoted the formation of new blood vessels and the differentiation of cardiomyocytes in rats [63]. Carlini and colleagues designed synthetic cyclic SAPs that were delivered to the heart through a catheter and rapidly formed a hydrogel after cleavage by enzymes MMP-2/9 and elastase, which are endogenous to the site of infarction in a rat model of MI. In addition to their low viscosity and ability to form a gel-like structure, the novel SAPs showed hemocompatibility, biocompatibility, and non-thrombogenicity that open up the possibility for implementation in drug delivery for the treatment of MI [35]. A novel hybrid temperature-responsive poly(N-isopropylacrylamide) gelatin-based injectable hydrogel was developed for cardiac tissue engineering and it exhibited a high level of cardiomyocyte and cardiac fibroblast survival and enhanced cytoskeletal organization [64]. Moreover, myeloid-derived growth factor (Mydgf) was incorporated into an injectable citrate-based polyester hydrogel to investigate its effects on improving cardiac tissue repair following MI. The combination of the released Mydgf and citrate, which is an important substrate in cellular energy metabolism, reduced cell apoptosis and scar formation as well as improved angiogenesis and cardiac function [65]. In the study by Waters and colleagues, therapeutic biomolecules, such as growth factors and cytokines, secreted by human adipose-derived stem cells (ADSCs), were loaded into laponite/gelatin hydrogel and injected into the peri-infarct region in an acute MI rat model, which resulted in increased angiogenesis and reduced fibrosis as well as a significant improvement in ejection fraction and cardiac output [66]. The hydrogel could accommodate growth factors due to laponite, which is a synthetic nanoclay composed of discoid NPs that can bind growth factors through electrostatic forces.

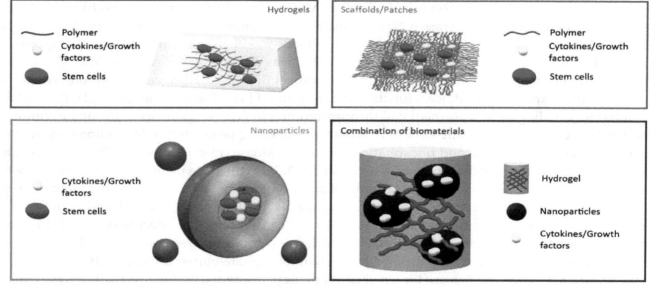

Figure 1. Representative images of biomaterials. Hydrogels, scaffolds/patches, and nanoparticles loaded with growth factors/cytokines and stem cells, and their combination are shown.

Along with hydrogels, nanoscale carriers (Table 1) are extensively studied for cardiac tissue repair following MI [67]. Targeted delivery, maintenance of protein stability, presence in blood circulation for an extended time, and controlled release of loaded agents make NPs attractive carriers for cardiac tissue therapy. For the purpose of targeting MI, Nguyen and colleagues developed NPs that respond to a specific enzymatic stimulus of MMP-9 and MMP-2 enzymes, which are upregulated upon infarction. This method allows for better accumulation at the MI site and longer clearance from the system [68]. Moreover, DNA enzymes conjugated to gold NPs have been demonstrated to produce an anti-inflammatory effect and improve cardiac function in a rat model of acute MI via silencing TNF-α and downregulating pro-inflammatory mediators, such as IL-12β, IL-1β, IL-6, as well as inducible nitric oxide synthase [69]. Another group loaded liraglutide in poly(lactic-co-glycolic acid)-poly(ethylene glycol) nanoparticles (NP-liraglutide) and delivered it to the infarcted rats via intramyocardial injection to overcome the challenges posed by its short half-life [70]. As a result, the NP-liraglutide system is retained in the myocardium over four weeks, thus, enhancing heart function, attenuating adverse cardiac remodeling, stimulating angiogenesis, and suppressing cardiomyocyte apoptosis. Although NPs appear to be a promising drug delivery system, the main concerns are their toxicity and tendency to aggregate, which lead to changes in physical and chemical properties and the formation of protein corona on the surface of NPs that prevents specific targeting [67].

Hydrogels and NPs can be used separately, as previously mentioned, or in combination. For example, a sulfonated hydrogel incorporated with VEGF and IL-10 and combined with PDGF-loaded micelle NPs showed a sequential and sustained release of all three factors for 28 days in vitro and a significant increase in the formation of mature vessels in vivo on a subcutaneous injection murine model [12]. As a result, this novel system significantly promoted angiogenesis and demonstrated the potential to ameliorate inflammation for improving cardiac repair post-MI. Another study used a novel, shear-thinning biocompatible and catheter-deliverable HA-based hydrogel loaded with dimeric fragment of hepatocyte growth factor (HGFdf) and a variant of stromal cell-derived factor 1α (ESA) to demonstrate a dual stage release that decreased infarct size and improved angiogenesis and heart function following MI [71]. ESA is a potent chemokine that attracts endothelial progenitor cells to infarcted areas and displays significant pro-angiogenic and wound healing effects. Moreover, hepatocyte growth factor prevents tissue fibrosis by inhibiting TGF-β production and stimulating MMP-1 to increase collagen degradation, as well as possessing pro-angiogenic and cardiomyogenic properties [72].

Another type of biomaterial is a cardiac patch that is directly applied to the myocardium. An acellular epicardial patch, developed from hydrogel, was also shown to prevent left ventricle remodeling and improve cardiac function in acute and subacute MI models in rats [73]. Wan and colleagues developed a novel cardiac patch derived from human heart valves. It is thought that the use of a human heart valve-derived scaffold (hHVS) may be superior to other approaches in cardiac repair by providing a native myocardial ECM. An in vitro study showed increased cellular proliferation and induction of cardiomyogenic differentiation of cells attached to a hHVS. An in vivo experiment demonstrated that patch application of hHVS alone reduced infarct size in a murine MI model. However, c-kit+ stem cell-seeded hHVS was more effective [74]. Cardiac patches have also been used for growth factor delivery (Table 1). Rodness and colleagues demonstrated that VEGF-containing calcium-alginate microsphere patches increased capillary density and improved tissue regeneration and cardiac function [75]. Transplanted human cardiomyocyte patches, which contained cardiomyocytes derived from human iPSCs and NPs loaded with FGF1 and CHIR99021, an inhibitor of the enzyme glycogen synthase kinase-3, reduced infarction size and improved angiogenesis and cardiac function. The combination of these factors reduced apoptosis and increased proliferation of transplanted cardiomyocytes [76].

In summary, biomaterials including micro and nanospheres, lipid NPs, nanofibrous structures, coacervate, hydrogels, and scaffolds appear to be a promising drug delivery system for cardiac tissue repair following MI. They can be administered alone or loaded with powerful therapeutic agents, such as growth factors and cytokines, that regulate cardiac tissue regeneration following MI. Biomaterials loaded with growth factors/cytokines have been shown to enhance angiogenesis and tissue regeneration, reduce cardiac cell death and scar size, ameliorate inflammation, and improve cardiac function (Table 1).

Table 1. Biomaterials loaded with growth factors and cytokines for cardiac tissue regeneration.

Biomaterial	Growth Factors/Cytokine	Effect	References
Heparan sulfate proteoglycans	bFGF	Extended bioavailability of the growth factor by protecting it from degradation, and improved angiogenesis and cardiac function	[53]
Glutathione-modified collagen hydrogel	bFGF fused with glutathione-S-transferase and MMP-2/9 cleavable peptide TIMP	Decreased collagen deposition, increased vascularization, and improved heart function	[54]
Hydroxyethyl methacrylate hyaluronic acid hydrogel	Neuregulin-1β	Improved ventricular function and structure	[38]
Fibrin gel/heparine coacervate	VEGF and PDGF	Improved angiogenesis and cardiac function, and reduced scar formation and inflammation	[57]
Citrate-based polyester hydrogel	Mydgf	Reduced cell apoptosis and scar formation, and improved angiogenesis and cardiac function	[65]

Table 1. *Cont.*

Biomaterial	Growth Factors/Cytokine	Effect	References
Laponite/gelatin hydrogel	ADSC secretome	Improved angiogenesis, ejection fraction, and cardiac output, and reduced fibrosis	[66]
Poly(lactic-co-glycolic acid)–poly(ethylene glycol) nanoparticles	Liraglutide	Improved heart function, attenuated adverse cardiac remodeling, stimulated angiogenesis, and suppressed cardiomyocyte apoptosis	[70]
A sulfonated hydrogel and poly(ethylene glycol)-blockpoly(serinol hexamethylene urea)-block-poly(ethylene glycol) micelle nanoparticles	VEGF, IL-10, and PDGF	Improved angiogenesis and demonstrated potential amelioration of inflammation to optimize cardiac repair post-MI	[12]
Hyaluronic acid-based hydrogel	HGFdf and ESA	Decreased infarct size, and improved angiogenesis and heart function	[71]
Calcium-alginate microsphere patch	VEGF	Improved tissue regeneration and cardiac function, and increased capillary density	[75]
Human cardiomyocyte patch with polylactic-co-glycolic acid nanoparticles	FGF1 and CHIR99021	Reduced infarction size and improved angiogenesis and cardiac function. The combination of factors reduced apoptosis and increased proliferation of transplanted cardiomyocytes	[76]

3. Biomaterials Loaded with Stem Cells for Cardiac Tissue Regeneration

Stem cells possess self-regenerating, differentiating, and immunomodulating properties, as well as release trophic factors. Therefore, they have been considered to be promising tools for cardiac tissue regeneration [20,77,78]. Many reports have demonstrated the therapeutic potential of various stem cell types, such as bone marrow-derived stem cells (BMSCs), ADSCs, cardiac-derived stem cells/cardiac progenitor cells (CPCs), and others, on myocardial tissue regeneration [79–83]. Moreover, stem cells have shown their therapeutic efficiency in several clinical trials [84]. Treatment with MSCs can improve left ventricle remodeling and function through decreasing scar size, promoting angiogenesis, and improving contractility [85,86]. Stem cells mediate cardioprotection by lowering the number of apoptotic myocytes at the site of injection. The mechanism responsible for protection includes insulin-like growth factor 1 (IGF-1)-mediated activation of stress-signaling and inflammatory response pathways and the suppression of cardiac transcription factor, nuclear factor kappa B [20]. Stem cells also support neoangiogenesis in post-MI tissue through positive regulation of VEGF, angiopoietin-1 (Ang-1), epidermal growth factor (EGF), and PDGF. Cell survival and proliferation is regulated by the AKT signaling pathway [87]. Despite the beneficial effects of stem cells on post-MI tissue regeneration, limitations such as low engraftment and survival rates in a harsh microenvironment compromise the clinical translatability of this approach [88]. Poor engraftment of transplanted cells is linked to mechanical loss during injection, loss of viability during long-lasting pre-conditioning, hypoxia, nutritional deficiencies, and low cell proliferation rate in vivo [88]. Therefore, various approaches are now being examined to increase engraftment and enhance the survival and stability of stem cells. One such approach is the use of biomaterials. Table 2 summarizes the use of biomaterials loaded with stem cells for cardiac tissue regeneration.

Several stem cell delivery systems are now utilized, including direct needle injection, nanogels, polymers, and inorganic nanomaterials [89–91]. Needle injection is the preferred method in clinics as it is less invasive. However, it has low cell retention, with less than 5% of transplanted cells reaching and remaining in cardiac tissue [92]. Recently, Park and colleagues proposed a new efficient direct MSC injection method to treat MI. MSCs were used in favor of other stem cell types based on their efficiency in reducing apoptosis and inflammation, as well as their ability to enhance vascularization and cardioprotection. In their study, they applied electrostatic interactions between bioengineered cationic mussel adhesive protein (MAP) and anionic HA. The resulting MAP/HA coacervate, named the adhesive protein-based immiscible condensed liquid system (APICLS), was successfully loaded with MSCs. APICLS was shown to be an innovative platform to treat MI, where stem cells demonstrated higher viability and retention and therefore, recovered infarcted tissue more effectively [92]. Another promising biomaterial for in vivo stem cell delivery is a collagen-based hydrogel transglutaminase cross-linked gelatin (Col-Tgel). The Col-Tgel-ADSCs system was shown to greatly improve MI treatment by enhancing engraftment of stem cells. ADSCs, which are the MSCs derived from adipose tissue, are shown to have several advantages over the bone marrow-derived MSCs. These include a more attractive cost and yield, a less invasive method for isolation, and a higher rate of cell growth [93]. In the study by Blocki and colleagues, injectable microcapsules made of agarose and ECM components were developed to enhance the survival of bone marrow-derived MSCs after their transplantation to rats with acute MI. The design was safe and efficient as evidenced by the absence of fibrotic response and persistence of the cells in the infarcted myocardium for four weeks after injection. In contrast, when these cells were injected without microcapsules, i.e., as cell suspension, they were detectable in post-MI hearts for only two days following transplantation [94]. Gallagher and colleagues showed that delivering MSCs using an arginylglycylaspartic acid (RGD)-modified HA hydrogel improves MSCs survival in the ischemic area. This effect was achieved due to HA being a natural ECM component and RGD being a tripeptide sequence that promoted MSC attachment to the hydrogel [95]. Another study successfully improved post-MI heart recovery in rats by delivering iPSCs in erythropoietin-linked hydrogel. The hydrogel was administered by injection into the myocardium [96]. Moreover, Cai and colleagues developed a novel designer self-assembling peptide (DSAP) consisting of the existing synthetic SAP and angiopoetin-1-derived pro-survival peptide QHREDGS in order to improve engraftment and retention of MSCs. This system significantly improved the survival of rat MSCs when they were injected into rats with MI [97]. Enhanced cell survival could be attributed to the presence of the QHREDGS peptide in the SAP. This peptide is an integrin-binding motif of Ang-1, a growth factor that stimulates endothelial cell survival, migration, and differentiation [98]. It was shown that QHREDGS peptide could mediate the same effects on its own, without being a part of Ang-1, when it is incorporated into various biomaterials such as hydrogels, for example [97]. However, the exact mechanism by which it promotes cell survival is still to be elucidated.

Pro-survival peptides were also used in the study by Lee and colleagues. In particular, they utilized collagen–dendrimer biomaterial crosslinked with pro-survival peptide analogues, namely, bone morphogenetic protein-2 peptide analogue, erythropoietin peptide analogue, and FGF2 peptide analogue, to augment the survival of CPCs in the MI model of mice [99]. CPCs that were transplanted with pro-survival factors enriched the collagen matrix and showed significantly greater long-term survival and engraftment compared to cells without the matrix. The authors described the molecular mechanism of enhanced cellular survival. Thus, the pro-survival matrix caused an increase in the expression of genes involved in the MAPK and phosphatidylinositol-3-OH kinase-protein kinase B (PI3K-AKT) pathways, while inhibiting pro-apoptotic pathways. In another study, silica-coated magnetic nanoparticles (MNPs) and an external magnet were utilized to enhance the survival of transplanted cells [100]. Embryonic cardiomyocytes, embryonic stem cell-derived cardiomyocytes, and BMSCs were incorporated into MNPs. Afterwards, the cell-MNP delivery system was intramyocardially injected into a murine model of MI, and a magnet was placed close to the chest of the animals to force the cells into the infarcted tissue. The treatment had drastically enhanced cell engraftment by 7-fold and 3.4-fold, two and eight weeks after application, respectively. The increased engraftment of the transplanted cells was due to a decrease in the loss of cells via the injection channel, which increased their proliferation and reduced apoptosis. The graphene oxide/alginate microgels constructed for cell delivery also demonstrated a favorable approach to promote MI recovery of the left ventricle during transplantation of MSCs [101].

Cardiac cell patches (Table 2) can be constructed from natural or synthetic materials, albeit natural materials are more favorable due to their biocompatibility and comparatively low cost [102]. Studies show that cardiac patches loaded with stem cells, where MSCs are preferable compared to CPCs, embryonic or iPSCs, facilitate a higher engraftment rate of transplanted cells. Moreover, cell patches also provide a positive impact on cardiomyogenesis and angiogenesis [102,103]. Wang and colleagues transplanted poly(ε-caprolactone)/gelatin patch loaded with MSCs into the epicardium of the murine model of MI. The patch reduced MI-induced damage by promoting angiogenesis, lymphangiogenesis, and cardiomyogenesis, decreasing scar size and enhancing the release of paracrine factors from stem cells. They also showed an increase in the expression of hypoxia-inducible factor 1α, TGF-β, VEGF, and SDF1 factors and a negative regulation of CXCL14. Cytokine release enhanced the recruitment of endogenous c-kit+ cells and activated the epicardium [103]. Chen and colleagues designed a novel chitosan and silk fibroin microfibrous cardiac patch that significantly improved the survival of murine adipose tissue-derived MSCs in infarcted hearts of a rat model [104]. This was achieved due to the structural resemblance of the patch to the native ECM of the heart. Thus, the patch provided a suitable environment for the retention and survival of the transplanted cells. Nevertheless, the detailed mechanism of this process is yet to be identified. Similarly, in the study by Gaetani and colleagues, it was shown that a 3D-printed HA/gelatin cardiac patch could support long-term survival and differentiation of the CPCs when they were tested on the mouse model of MI [105]. Su and colleagues used a cardiac patch not only to provide an adhesion and retention framework for stem cells, but to also nutritionally support them [106]. Specifically, they developed a vascularized fibrin gel that could accommodate CSCs. Such a construct would help stem cells receive nutrients through biomimetic blood vessels (BMV) within the hydrogel and consequently, enrich their survival. In addition, the BMV were made of fibronectin, a constituent of the natural ECM, and hence, provided the appropriate environment for the transplanted cells. Moreover, Dong and colleagues constructed a patch made of gold NPs coated with a combination of ECM and silk proteins [107]. The patch was loaded with rat bone marrow-derived MSCs and tested in a cryoinjury model of MI in rats. The construct was found to greatly improve stem cell survival and retention as well as significantly decrease the infarct size 28 days post-infarction. The authors proposed several mechanisms to achieve beneficial effects of the patch on cell viability. Namely, the construct possesses antioxidant properties and acts as a mechanical scaffold, thus, protecting the transplanted cells from the harsh environment in the infarcted region. Gao and colleagues used an ECM scaffold to deliver human iPSC-derived cardiomyocytes, smooth muscle cells, and endothelial cells to mice with MI [108]. This treatment significantly reduced infarct size and improved cell proliferation, cardiac function, and angiogenesis. Furthermore, Tang and colleagues developed new microneedle patches loaded with cardiac stromal cells (CSCs) for post-MI tissue regeneration. Poly(vinyl alcohol)-made microneedles served as channels between myocardial tissue and regenerative factors released from CSCs. In vivo studies on a rat MI model showed that microneedle patches could promote angiogenesis, reduce fibrosis, and repair the left ventricular wall [109]. Combinatorial dual stem cell delivery is another approach to enhance the survival of transplanted stem cells. Park and colleagues used MSCs seeded on polycaprolactone patch and iPSC-derived cardiomyocytes for in vivo treatment of the rat MI model. Analysis with immunohistochemistry, gene expression, and echocardiography demonstrated significant enhancement in MI recovery. Cardiomyocytes contributed to myocardium regeneration, while growth-promoting paracrine factors from MSCs accelerated angiogenesis as well as caused iPSC-cardiomyocytes to resemble adult-like cardiomyocyte morphology [110]. Interestingly, the cardiac patch can be 3D printed for iPSC-derived cell delivery to effectively enhance post-MI treatment [111].

Another positive effect of biomaterials on stem cell therapy is the enhanced release of paracrine factors produced by the cells. Melhem and colleagues developed a microchanneled hydrogel patch that can sustain a continuous release of stem cell synthesized factors [112]. The patch was loaded with human bone marrow-derived MSCs and tested in vitro and in the murine model of MI. Patch-protected MSCs released a variety of angiogenic, anti-inflammatory, cardioprotective, antifibrotic, and antiapoptotic factors in vitro. Furthermore, the sustainable release of paracrine factors by the system was confirmed by the assessment of the VEGF release profile for one week. Over this period of time, the amount of VEGF linearly increased. The microchanneled hydrogel patch loaded with MSCs showed other benefits as well. Namely, mice treated with the patch showed significant improvement in cardiac function, which was established by echocardiographic examinations of ejection fraction and stroke volume five weeks after infarction. Importantly, the therapeutic effects of the treatment were significantly greater with MSCs, the patch without MSCs, or MSCs alone as compared to the patch without microchannels. Moreover, the effects of the patch did not depend on the number of transplanted cells, implicating that the construct could reduce the number of stem cells required for treatment. Similarly, Mayfield and colleagues showed that single cell hydrogel microencapsulation of human CSCs significantly improves the production of pro-angiogenic/cardioprotective cytokines, angiogenesis, and angiogenic cells recruitment after direct intramyocardial injection into mice with MI [113]. Less is known about biomaterial distribution after injection in vivo; Ahmadi and colleagues reported that a collagen matrix is retained mostly in the injected area with minimal distribution to non-target areas [114]. Han and colleagues utilized iron NPs that were co-cultured with rat cardiomyoblasts to boost the therapeutic efficiency of human bone marrow-derived MSCs [115]. The modified MSCs showed increased expression of various paracrine factors, namely, bFGF, HGF, VEGF, Ang-1, urokinase type plasminogen activator, placental growth factor, and monocyte chemoattractant protein-1. Moreover, pre-treated MSCs reduced infarct size, prevented fibrosis, decreased apoptosis of myocardial cells, increased angiogenesis, and improved cardiac function and the survival of rats with acute MI overall. The authors stated that improvements in the therapeutic potential of MSCs should be attributed to the increased expression of connexin 43 gap junction protein by cardiomyoblasts, which was stimulated by iron NPs. The greater expression of connexin, in turn, leads to a more efficient electrophysiologic and paracrine crosstalk between MSCs and cardiomyoblasts [116–118].

Yet another advantageous effect of biomaterials on stem cell treatment is their ability to accommodate factors that could act synergistically with stem cells, thus, enhancing their therapeutic actions [119,120]. For instance, Yokoyama and colleagues tested the efficiency of statins and human ADSCs combinations incorporated into NPs [121]. The treatment was injected into the tail vein of mice with MI, and its therapeutic effects were assessed for four weeks after infarction. The statin-ADSCs encapsulating NPs significantly increased the ejection fraction and several other parameters, which reflect left ventricular function. This positive effect of statin-ADSCs combination was superior compared to the use of statins or ADSCs alone. The mechanism by which the treatment brought about these improvements is likely through stimulation of the sustained and localized release of the statins by ADSCs. This, in turn, resulted in the inhibition of local inflammation, promotion of circulating stem cell recruitment, and stimulation of their differentiation to cardiomyocytes and angiogenesis [121]. Importantly, in this study, treatment efficiency was achieved with a smaller cell number of ADSCs than has ever been reported. A conductive hydrogel was also used to deliver plasmid DNA encoding endothelial nitric oxide synthase and ADSCs by injection into the infarcted myocardium. The results again demonstrated improved cardiac function with the conductive hydrogel [17]. Yao and colleagues also combined adipose-derived MSCs with a nitric oxide (NO) releasing system [122]. They utilized a naphthalene hydrogel that could maintain a controllable release of NO. In addition to demonstrating an excellent cell survival rate, the hydrogel stimulated the synthesis of angiogenic factors VEGF and SDF-1α in the MI model of mice. In yet another study, the therapeutic benefits of ADSCs were enhanced by NRG1 growth factor [123]. The ADSCs-NRG1 mixture was encapsulated into microparticles and injected into rats with MI. This combination improved cell survival as demonstrated by the persistence of the transplanted cells at three months after injection. Furthermore, ADSCs induced the shift of macrophages found in the infarcted myocardium from pro-inflammatory M1 to regenerative M2 phenotype. At the same time, NRG1 reduced the infarct size and stimulated cardiomyocyte proliferation. Compared to a separate administration, the combined treatment with ADSCs-NRG1 microparticles resulted in a more pronounced regeneration of the damaged myocardium. Chung and colleagues showed that cardiac patch-supported co-transplantation of CSCs and VEGF had a synergistic effect on angiogenesis, cell proliferation, and the recruitment of stem cells [124]. In particular, they developed a poly(l-lactic acid) mat and loaded it with rat CSCs and VEGF. When the system was tested in the rat MI models, it had greater angiogenic and cardiomyogenic effects compared to either VEGF with the patch or CSCs with the patch. Thus, numerous developments have been achieved in recent years in the usage of biomaterials to deliver stem cells to the infarction site. These include coacervates, various modifications of the hydrogels, NPs, and cardiac patches. Moreover, the combination of NPs and hydrogels also promoted transplanted cell survival. Stem cells derived from different sources were loaded into biomaterials alone, preconditioned or loaded in combination with bioactive molecules. Transplanted or tail vein injected stem cells delivering biomaterials have significantly enhanced recovery of the MI in small animal models and show promising results for their therapeutic applications. Small animals are commonly used in cardiovascular research due to their small size, low cost, short gestation time, and ease in maintenance and genetic manipulations [125]. However, there are limitations to their use that are responsible for their high failure rates in human clinical trials. These include a small heart size and anatomical differences in the coronary artery and conduction system. [126–128].

Table 2. Biomaterials loaded with stem cells for cardiac tissue regeneration.

Biomaterial	Stem Cells	Effect	References
Mussel adhesive protein/HA coacervate	MSCs	Increased MSCs survival and retention	[92]
Collagen-based hydrogel	ADSCs	Improved engraftment of stem cells	[93]
Microcapsules made of agarose and ECM components	MSCs	Increased MSCs survival	[94]
Arginylglycylaspartic acid (RGD) modified HA hydrogel	MSCs	Increased MSCs survival	[95]
Erythropoietin linked hydrogel	iPSCs	Improved the post-MI heart recovery	[96]
Synthetic SAP and angiopoetin-1-derived pro-survival peptide QHREDGS	MSCs	Increased cells survival and cardiac function	[97]
Collagen–dendrimer	CPCs	Increased long-term survival	[99]
Silica-coated SOMag5 magnetic nanoparticles	Embryonic cardiomyocytes, embryonic stem cell-derived cardiomyocytes and BMSCs	Improved cell engraftment	[100]
Graphene oxide/alginate microgel	MSCs	Improved MI recovery	[101]
Poly(ε-caprolactone)/gelatin patch	MSCs	Increased angiogenesis, lymphangiogenesis, cardiomyogenesis, and paracrine factors released by stem cells and reduced scar size	[103]

Table 2. *Cont.*

Biomaterial	Stem Cells	Effect	References
Chitosan and silk fibroin microfibrous cardiac patch	MSCs	Increased MSC survival	[104]
Hyaluronic acid/gelatin cardiac patch	CPCs	Increased long-term CPCs survival and differentiation	[105]
Vascularized fibrin hydrogel patch	CSCs	Increased cell survival	[106]
Gold nanoparticles coated with a combination of ECM and silk proteins	MSCs	Increased cell survival and retention and decreased infarct size	[107]
ECM scaffold with the usage of methacrylated gelatin	iPSC-derived cardiomyocytes, smooth muscle cells and endothelial cells	Reduced infarct size and improved cell proliferation, cardiac function, and angiogenesis	[108]
Poly(vinyl alcohol) microneedle patch	CSCs	Improved angiogenesis, reduced fibrosis, and repaired left ventricular wall	[109]
Polycaprolactone patch	MSCs and iPSC-derived cardiomyocytes	Improved MI recovery and angiogenesis	[110]
Microchanneled poly(ethylene glycol) dimethacrylate hydrogel patch	MSCs	Improved cardiac function	[112]
Agarose hydrogel microcapsules supplemented with fibronectin and fibrinogen	CSCs	Improved production of pro-angiogenic/cardioprotective cytokines, angiogenesis, and angiogenic cells recruitment after direct intramyocardial injection	[113]
Iron nanoparticles	MSCs	Reduced the infarct size, prevented fibrosis, decreased apoptosis of myocardial cells, increased angiogenesis, and improved cardiac function	[115]
Statin-conjugated poly(lactic-co-glycolic acid) nanoparticles	ADSCs	Increased the ejection fraction and several other parameters which reflect the left ventricular function. Inhibited local inflammation, promoted recruitment of circulating stem cells, and stimulated their differentiation to cardiomyocytes and angiogenesis	[121]
Tetraaniline-polyethylene glycol diacrylate and thiolated hyaluronic acid conductive hydrogel	ADSCs	Improved neovascularization, regeneration of the damaged myocardium, and post-infarction cardiac function	[17]
Naphthalene hydrogel	MSCs	Increased cells survival, and stimulated the synthesis of angiogenic factors VEGF and SDF-1α	[122]

Table 2. *Cont.*

Biomaterial	Stem Cells	Effect	References
Poly(lactic-co-glycolic acid) microparticles	ADSCs	Improved cells survival, and induced the shift of macrophage found in the infarcted myocardium from pro-inflammatory M1 to regenerative M2 phenotype	[123]
Poly(l-lactic acid) mat	CSCs	Improved angiogenic and cardiomyogenic effects	[124]

4. Conclusions

Biomaterials are being actively investigated for their use in tissue engineering and regenerative medicine due to their biodegradability and biocompatibility properties. Another important property of biomaterials is their ability to incorporate various growth factors and cytokines and to spatially and temporally control their release. Thus, biomaterials can serve as a good platform for the controlled and sustained delivery of growth factors and cytokines to ameliorate inflammation, improve angiogenesis, reduce fibrosis, and generate functional cardiac tissue. Moreover, biomaterials can be used to address some of the challenges associated with stem cell therapy of cardiovascular diseases. Specifically, they can improve stem cell survival and retention, enhance the delivery of the factors produced by the cells, support differentiation, and boost their therapeutic efficacy overall. However, despite the promising results of biomaterials in MI treatment, additional studies should be performed to improve their biocompatibility and biodegradability. Furthermore, the best source of transplanted stem cells and optimal doses of various growth factors and cytokines should be determined in order to create functional cardiac tissue and improve heart function. However, small animals do not fully recapitulate all the aspects of disease phenotypes, although they do replicate some of features. Therefore, translational aspects should be carefully interpreted with respect to these issues.

Author Contributions: Conceptualization and writing-review and editing, A.S. (Arman Saparov); writing-original draft preparation, S.S., Y.K., A.S. (Aiganym Smagulova), K.R. and A.N. All authors have read and agreed to the published version of the manuscript.

References

1. McClellan, M.; Brown, N.; Califf, R.M.; Warner, J.J. Call to action, urgent challenges in cardiovascular disease, a presidential advisory from the American Heart Association. *Circulation* **2019**, *139*, e44–e54. [CrossRef] [PubMed]

2. Roth, G.A.; Johnson, C.; Abajobir, A.; Abd-Allah, F.; Abera, S.F.; Abyu, G.; Ahmed, M.; Aksut, B.; Alam, T.; Alam, K.; et al. Global, regional, and national burden of cardiovascular diseases for 10 causes, 1990 to 2015. *J. Am. Coll. Cardiol.* **2017**, *70*, 1–25. [CrossRef] [PubMed]

3. Virani, S.S.; Alonso, A.; Benjamin, E.J.; Bittencourt, M.S.; Callaway, C.W.; Carson, A.P.; Chamberlain, A.M.; Chang, A.R.; Cheng, S.; Delling, F.N.; et al. Heart disease and stroke statistics—2020 update, a report from the American Heart Association. *Circulation* **2020**, *141*, E139–E596. [CrossRef]

4. Swirski, F.K.; Matthias, N. Cardioimmunology: The immune system in cardiac homeostasis and disease. *Nat. Rev. Immunol.* **2018**, *18*, 733–744. [CrossRef] [PubMed]

5. Saparov, A.; Ogay, V.; Nurgozhin, T.; Chen, W.C.; Mansurov, N.; Issabekova, A.; Zhakupova. Role of the immune system in cardiac tissue damage and repair following myocardial infarction. *Inflamm. Res.* **2017**, *66*, 739–751. [CrossRef] [PubMed]

6.   Andreadou, I.; Cabrera-Fuentes, H.A.; Devaux, Y.; Frangogiannis, N.G.; Frantz, S.; Guzik, T.; Liehn, E.A.; Gomes, C.P.; Schulz, R.; Hausenloy, D.J. Immune cells as targets for cardioprotection, new players and novel therapeutic opportunities. *Cardiovasc. Res.* **2019**, *115*, 1117–1130. [CrossRef]

7.   Hashimoto, H.; Eric, N.O.; Rhonda, B.D. Therapeutic approaches for cardiac regeneration and repair. *Nat. Rev. Cardiol.* **2018**, *15*, 585–600. [CrossRef]

8.   Rebouças, J.D.S.; Santos-Magalhães, N.S.; Formiga, F.R. Cardiac regeneration using growth factors, advances and challenges. *Arq. Bras. Cardiol.* **2016**, *107*, 271–275. [CrossRef]

9.   Zarrouk-Mahjoub, S.; Zaghdoudi, M.; Amira, Z.; Chebi, H.; Khabouchi, N.; Finsterer, J.; Mechmeche, R.; Ghazouani, E. Pro-and anti-inflammatory cytokines in post-infarction left ventricular remodeling. *Int. J. Cardiol.* **2016**, *221*, 632–636. [CrossRef]

10.   Ferrini, A.; Stevens, M.M.; Sattler, S.; Rosenthal, N. Toward regeneration of the heart, bioengineering strategies for immunomodulation. *Front. Cardiovasc. Med.* **2019**, *6*, 26. [CrossRef]

11.   Pascual-Gil, S.; Garbayo, E.; Díaz-Herráez, P.; Prosper, F.; Blanco-Prieto, M.J. Heart regeneration after myocardial infarction using synthetic biomaterials. *J. Control. Release* **2015**, *203*, 23–38. [CrossRef] [PubMed]

12.   Rocker, A.J.; Lee, D.J.; Shandas, R.; Park, D. Injectable Polymeric Delivery System for Spatiotemporal and Sequential Release of Therapeutic Proteins To Promote Therapeutic Angiogenesis and Reduce Inflammation. *ACS Biomater. Sci. Eng.* **2020**, *6*, 1217–1227. [CrossRef]

13.   Dormont, F.; Varna, M.; Couvreur, P. Nanoplumbers, biomaterials to fight cardiovascular diseases. *Mater. Today* **2018**, *21*, 122–143. [CrossRef]

14.   Nurkesh, A.; Jaguparov, A.; Jimi, S.; Saparov, A. Recent Advances in the Controlled Release of Growth Factors and Cytokines for Improving Cutaneous Wound Healing. *Front. Cell Dev. Biol.* **2020**, *8*, 638. [CrossRef]

15.   Saludas, L.; Pascual-Gil, S.; Prósper, F.; Garbayo, E.; Blanco-Prieto, M. Hydrogel based approaches for cardiac tissue engineering. *Int. J. Pharm.* **2017**, *523*, 454–475. [CrossRef] [PubMed]

16.   Mihic, A.; Cui, Z.; Wu, J.; Vlacic, G.; Miyagi, Y.; Li, S.H.; Lu, S.; Sung, H.W.; Weisel, R.D.; Li, R.K. A Conductive Polymer Hydrogel Supports Cell Electrical Signaling and Improves Cardiac Function after Implantation into Myocardial Infarct. *Circulation* **2015**, *132*, 772–784. [CrossRef]

17.   Wang, W.; Tan, B.; Chen, J.; Bao, R.; Zhang, X.; Liang, S.; Shang, Y.; Liang, W.; Cui, Y.; Fan, G.; et al. An injectable conductive hydrogel encapsulating plasmid DNA-eNOs and ADSCs for treating myocardial infarction. *Biomaterials* **2018**, *160*, 69–81. [CrossRef]

18.   Maghin, E.; Garbati, P.; Quarto, R.; Piccoli, M.; Bollini, S. Young at Heart, Combining Strategies to Rejuvenate Endogenous Mechanisms of Cardiac Repair. *Front. Bioeng. Biotechnol.* **2020**, *8*, 447. [CrossRef]

19.   Carvalho, E.; Verma, P.; Hourigan, K.; Banerjee, R. Myocardial infarction, stem cell transplantation for cardiac regeneration. *Regen. Med.* **2015**, *10*, 1025–1043. [CrossRef]

20.   Karantalis, V.; Hare, J.M. Use of mesenchymal stem cells for therapy of cardiac disease. *Circ. Res.* **2015**, *116*, 1413–1430. [CrossRef]

21.   Shafei, A.E.S.; Ali, M.A.; Ghanem, H.G.; Shehata, A.I.; Abdelgawad, A.A.; Handal, H.R.; Talaat, K.A.; Ashaal, A.E.; El-Shal, A.S. Mesenchymal stem cell therapy, A promising cellbased therapy for treatment of myocardial infarction. *J. Gene Med.* **2017**, *19*, e2995. [CrossRef] [PubMed]

22.   Katarzyna, R. Adult stem cell therapy for cardiac repair in patients after acute myocardial infarction leading to ischemic heart failure, an overview of evidence from the recent clinical trials. *Curr. Cardiol. Rev.* **2017**, *13*, 223–231. [CrossRef] [PubMed]

23.   Bar, A.; Cohen, S. Inducing Endogenous Cardiac Regeneration, Can *Biomaterials* Connect the Dots? *Front. Bioeng. Biotechnol.* **2020**, *8*, 126. [CrossRef] [PubMed]

24.   Domenech, M.; Polo-Corrales, L.; Ramirez-Vick, J.E.; Freytes, D.O. Tissue engineering strategies for myocardial regeneration, acellular versus cellular scaffolds? *Tissue Eng. Part B Rev.* **2016**, *22*, 438–458. [CrossRef] [PubMed]

25.   Ashtari, K.; Nazari, H.; Ko, H.; Tebon, P.; Akhshik, M.; Akbari, M.; Alhosseini, S.N.; Mozafari, M.; Mehravi, B.; Soleimani, M. Electrically conductive nanomaterials for cardiac tissue engineering. *Adv. Drug Deliv. Rev.* **2019**, *144*, 162–179. [CrossRef]

26.   Saghazadeh, S.; Rinoldi, C.; Schot, M.; Kashaf, S.S.; Sharifi, F.; Jalilian, E.; Nuutila, K.; Giatsidis, G.; Mostafalu, P.; Derakhshandeh, H. Drug delivery systems and materials for wound healing applications. *Adv. Drug Deliv. Rev.* **2018**, *127*, 138–166. [CrossRef]

27. Wang, Z.; Wang, Z.; Lu, W.W.; Zhen, W.; Yang, D.; Peng, S. Novel biomaterial strategies for controlled growth factor delivery for biomedical applications. *NPG Asia Mater.* **2017**, *9*, e435. [CrossRef]

28. Kakkar, A.; Traverso, G.; Farokhzad, O.C.; Weissleder, R.; Langer, R. Evolution of macromolecular complexity in drug delivery systems. *Nat. Rev. Chem.* **2017**, *1*, 1–17.

29. Mansurov, N.; Chen, W.C.; Awada, H.; Huard, J.; Wang, Y.; Saparov, A. A controlled release system for simultaneous delivery of three human perivascular stem cell-derived factors for tissue repair and regeneration. *J. Tissue Eng. Regen. Med.* **2018**, *12*, e1164–e1172. [CrossRef]

30. Jimi, S.; Jaguparov, A.; Nurkesh, A.; Sultankulov, B.; Saparov, A. Sequential Delivery of Cryogel Released Growth Factors and Cytokines Accelerates Wound Healing and Improves Tissue Regeneration. *Front. Bioeng. Biotechnol.* **2020**, *8*, 345. [CrossRef]

31. Sultankulov, B.; Berillo, D.; Kauanova, S.; Mikhalovsky, S.; Mikhalovska, L.; Saparov, A. Composite Cryogel with Polyelectrolyte Complexes for Growth Factor Delivery. *Pharmaceutics* **2019**, *11*, 650. [CrossRef] [PubMed]

32. Sultankulov, B.; Berillo, D.; Sultankulova, K.; Tokay, T.; Saparov, A. Progress in the Development of Chitosan-Based *Biomaterials* for Tissue Engineering and Regenerative Medicine. *Biomolecules* **2019**, *9*, 470. [CrossRef] [PubMed]

33. Jo, H.; Gajendiran, M.; Kim, K. Influence of PEG chain length on colloidal stability of mPEGylated polycation based coacersomes for therapeutic protein delivery. *J. Ind. Eng. Chem.* **2020**, *82*, 234–242. [CrossRef]

34. Li, J.; Mooney, D.J. Designing hydrogels for controlled drug delivery. *Nat. Rev. Mater.* **2016**, *1*, 1–17. [CrossRef]

35. Carlini, A.S.; Gaetani, R.; Braden, R.L.; Luo, C.; Christman, K.L.; Gianneschi, N.C. Enzyme-responsive progelator cyclic peptides for minimally invasive delivery to the heart post-myocardial infarction. *Nat. Commun.* **2019**, *10*, 1–14. [CrossRef]

36. Matsumura, Y.; Zhu, Y.; Jiang, H.; D'Amore, A.; Luketich, S.K.; Charwat, V.; Yoshizumi, T.; Sato, H.; Yang, B.; Uchibori, T. Intramyocardial injection of a fully synthetic hydrogel attenuates left ventricular remodeling post myocardial infarction. *Biomaterials* **2019**, *217*, 119289. [CrossRef] [PubMed]

37. Blackburn, N.J.; Sofrenovic, T.; Kuraitis, D.; Ahmadi, A.; McNeill, B.; Deng, C.; Rayner, K.J.; Zhong, Z.; Ruel, M.; Suuronen, E.J. Timing underpins the benefits associated with injectable collagen biomaterial therapy for the treatment of myocardial infarction. *Biomaterials* **2015**, *39*, 182–192. [CrossRef] [PubMed]

38. Lister, Z.; Rayner, K.J.; Suuronen, E.J. How *Biomaterials* Can Influence Various Cell Types in the Repair and Regeneration of the Heart after Myocardial Infarction. *Front. Bioeng. Biotechnol.* **2016**, *4*, 62. [CrossRef] [PubMed]

39. Li, H.; Bao, M.; Nie, Y. Extracellular matrix-based biomaterials for cardiac regeneration and repair. *Heart Fail. Rev.* **2020**. [CrossRef]

40. McLaughlin, S.; McNeill, B.; Podrebarac, J.; Hosoyama, K.; Sedlakova, V.; Cron, G.; Smyth, D.; Seymour, R.; Goel, K.; Liang, W.; et al. Injectable human recombinant collagen matrices limit adverse remodeling and improve cardiac function after myocardial infarction. *Nat. Commun.* **2019**, *10*, 1–14. [CrossRef]

41. Firoozi, S.; Pahlavan, S.; Ghanian, M.H.; Rabbani, S.; Tavakol, S.; Barekat, M.; Yakhkeshi, S.; Mahmoudi, E.; Soleymani, M.; Baharvand, H. A Cell-Free SDKP-Conjugated Self-Assembling Peptide Hydrogel Sufficient for Improvement of Myocardial Infarction. *Biomolecules* **2020**, *10*, 205. [CrossRef]

42. Cui, Z.; Ni, N.C.; Wu, J.; Du, G.Q.; He, S.; Yau, T.M.; Weisel, R.D.; Sung, H.W.; Li, R.K. Polypyrrole-chitosan conductive biomaterial synchronizes cardiomyocyte contraction and improves myocardial electrical impulse propagation. *Theranostics* **2018**, *8*, 2752–2764. [CrossRef]

43. Thiagarajan, H.; Thiyagamoorthy, U.; Shanmugham, I.; Nandagopal, G.D.; Kaliyaperumal, A. Angiogenic growth factors in myocardial infarction, a critical appraisal. *Heart Fail. Rev.* **2017**, *22*, 665–683. [CrossRef] [PubMed]

44. Frangogiannis, N.G. The extracellular matrix in myocardial injury, repair, and remodeling. *J. Clin. Investig.* **2017**, *127*, 1600–1612. [CrossRef] [PubMed]

45. Chen, B.; Huang, S.; Su, Y.; Wu, Y.J.; Hanna, A.; Brickshawana, A.; Graff, J.; Frangogiannis, N.G. Macrophage Smad3 protects the infarcted heart, stimulating phagocytosis and regulating inflammation. *Circ. Res.* **2019**, *125*, 55–70. [CrossRef] [PubMed]

46. Yue, Y.; Wang, C.; Benedict, C.; Huang, G.; Truongcao, M.; Roy, R.; Cimini, M.; Garikipati, V.N.S.; Cheng, Z.; Koch, W.J. Interleukin-10 Deficiency Alters Endothelial Progenitor Cell–Derived Exosome Reparative Effect on Myocardial Repair via Integrin-Linked Kinase Enrichment. *Circ. Res.* **2020**, *126*, 315–329. [CrossRef]

47. Jung, M.; Ma, Y.; Iyer, R.P.; DeLeon-Pennell, K.Y.; Yabluchanskiy, A.; Garrett, M.R.; Lindsey, M.L. IL-10 improves cardiac remodeling after myocardial infarction by stimulating M2 macrophage polarization and fibroblast activation. *Basic Res. Cardiol.* **2017**, *112*, 33. [CrossRef]

48. Ferraro, B.; Leoni, G.; Hinkel, R.; Ormanns, S.; Paulin, N.; Ortega-Gomez, A.; Viola, J.R.; de Jong, R.; Bongiovanni, D.; Bozoglu, T.; et al. Pro-angiogenic macrophage phenotype to promote myocardial repair. *J. Am. Coll. Cardiol.* **2019**, *73*, 2990–3002. [CrossRef]

49. Cahill, T.J.; Choudhury, R.P.; Riley, P.R. Heart regeneration and repair after myocardial infarction, translational opportunities for novel therapeutics. *Nat. Rev. Drug Discov.* **2017**, *16*, 699. [CrossRef]

50. Shiraishi, M.; Shintani, Y.; Shintani, Y.; Ishida, H.; Saba, R.; Yamaguchi, A.; Adachi, H.; Yashiro, K.; Suzuki, K. Alternatively activated macrophages determine repair of the infarcted adult murine heart. *J. Clin. Investig.* **2016**, *126*, 2151–2166. [CrossRef]

51. Oduk, Y.; Zhu, W.; Kannappan, R.; Zhao, M.; Borovjagin, A.V.; Oparil, S.; Zhang, J. VEGF nanoparticles repair the heart after myocardial infarction. *Am. J. Physiol.-Heart C* **2018**, *314*, H278–H284. [CrossRef] [PubMed]

52. Hachim, D.; Whittaker, T.E.; Kim, H.; Stevens, M.M. Glycosaminoglycan-based biomaterials for growth factor and cytokine delivery: Making the right choices. *J. Control. Release* **2019**, *313*, 131–147. [CrossRef] [PubMed]

53. Shi, J.; Fan, C.; Zhuang, Y.; Sun, J.; Hou, X.; Chen, B.; Xiao, Z.; Chen, Y.; Zhan, Z.; Zhao, Y. Heparan sulfate proteoglycan promotes fibroblast growth factor-2 function for ischemic heart repair. *Biomater. Sci.* **2019**, *7*, 5438–5450. [CrossRef] [PubMed]

54. Fan, C.; Shi, J.; Zhuang, Y.; Zhang, L.; Huang, L.; Yang, W.; Chen, B.; Chen, Y.; Xiao, Z.; Shen, H. Myocardial-Infarction-Responsive Smart Hydrogels Targeting Matrix Metalloproteinase for On-Demand Growth Factor Delivery. *Adv. Mater.* **2019**, *31*, 1902900. [CrossRef] [PubMed]

55. Itoh, N.; Ohta, H.; Nakayama, Y.; Konishi. Roles of FGF signals in heart development, health, and disease. *Front. Cell Dev. Biol.* **2016**, *4*, 110. [CrossRef]

56. Teringova, E.; Tousek, P. Apoptosis in ischemic heart disease. *J. Transl. Med.* **2017**, *15*, 87. [CrossRef]

57. Awada, H.K.; Johnson, N.R.; Wang, Y. Sequential delivery of angiogenic growth factors improves revascularization and heart function after myocardial infarction. *J. Control. Release* **2015**, *207*, 7–17. [CrossRef]

58. Taimeh, Z.; Loughran, J.; Birks, E.J.; Bolli, R. Vascular endothelial growth factor in heart failure. *Nat. Rev. Cardiol.* **2013**, *10*, 519. [CrossRef]

59. Corti, F.; Simons, M. Modulation of VEGF receptor 2 signaling by protein phosphatases. *Pharmacol. Res.* **2017**, *115*, 107–123. [CrossRef]

60. Cassani, M.; Fernandes, S.; Vrbsky, J.; Ergir, E.; Cavalieri, F.; Forte, G. Combining Nanomaterials and Developmental Pathways to Design New Treatments for Cardiac Regeneration, the Pulsing Heart of Advanced Therapies. *Front. Bioeng. Biotechnol.* **2020**, *8*, 323. [CrossRef]

61. Catoira, M.C.; Fusaro, L.; Di Francesco, D.; Ramella, M.; Boccafoschi, F. Overview of natural hydrogels for regenerative medicine applications. *J. Mater. Sci. Mater. Med.* **2019**, *30*, 115. [CrossRef] [PubMed]

62. Gyles, D.A.; Castro, L.D.; Silva, J.O.C., Jr.; Ribeiro-Costa, R.M. A review of the designs and prominent biomedical advances of natural and synthetic hydrogel formulations. *Eur. Polym. J.* **2017**, *88*, 373–392. [CrossRef]

63. Rufaihah, A.J.; Yasa, I.C.; Ramanujam, V.S.; Arularasu, S.C.; Kofidis, T.; Guler, M.O.; Tekinay, A.B. Angiogenic peptide nanofibers repair cardiac tissue defect after myocardial infarction. *Acta Biomater.* **2017**, *58*, 102–112. [CrossRef] [PubMed]

64. Navaei, A.; Truong, D.; Heffernan, J.; Cutts, J.; Brafman, D.; Sirianni, R.W.; Vernon, B.; Nikkhah, M. PNIPAAm-based biohybrid injectable hydrogel for cardiac tissue engineering. *Acta Biomater.* **2016**, *32*, 10–23. [CrossRef]

65. Yuan, Z.; Tsou, Y.; Zhang, X.; Huang, S.; Yang, Y.; Gao, M.; Ho, W.; Zhao, Q.; Ye, X.; Xu, X. Injectable citrate-based hydrogel as an angiogenic biomaterial improves cardiac repair after myocardial infarction. *ACS Appl. Mater. Interfaces* **2019**, *11*, 38429–38439. [CrossRef]

66. Waters, R.; Alam, P.; Pacelli, S.; Chakravarti, A.; Ahmed, R.; Paul, A. Stem cell-inspired secretome-rich injectable hydrogel to repair injured cardiac tissue. *Acta Biomater.* **2018**, *69*, 95–106. [CrossRef]

67. Ho, Y.T.; Poinard, B.; Kah, J.C.Y. Nanoparticle drug delivery systems and their use in cardiac tissue therapy. *Nanomedicine* **2016**, *11*, 693–714. [CrossRef]

68.   Nguyen, M.M.; Carlini, A.; Chien, M.; Sonnenberg, S.; Luo, C.; Braden, R.; Osborn, K.; Li, Y.; Gianneschi, N.; Christman, K. Enzyme-Responsive Nanoparticles for Targeted Accumulation and Prolonged Retention in Heart Tissue after Myocardial Infarction. *Adv. Mater.* **2015**, *27*, 5547–5552. [CrossRef]

69.   Somasuntharam, I.; Yehl, K.; Carroll, S.; Maxwell, J.; Martinez, M.; Che, P.; Brown, M.; Salaita, K.; Davis, M. Knockdown of TNF-α by DNAzyme gold nanoparticles as an anti-inflammatory therapy for myocardial infarction. *Biomaterials* **2016**, *83*, 12–22. [CrossRef]

70.   Qi, Q.; Lu, L.; Li, H.; Yuan, Z.; Chen, G.; Lin, M.; Ruan, Z.; Ye, X.; Xiao, Z.; Zhao, Q. Spatiotemporal delivery of nanoformulated liraglutide for cardiac regeneration after myocardial infarction. *Int. J. Nanomed.* **2017**, *12*, 4835–4848. [CrossRef]

71.   Steele, A.N.; Paulsen, M.; Wang, H.; Stapleton, L.; Lucian, H.; Eskandari, A.; Hironaka, C.; Farry, J.; Baker, S.; Thakore, A.; et al. Multi-phase catheter-injectable hydrogel enables dual-stage protein-engineered cytokine release to mitigate adverse left ventricular remodeling following myocardial infarction in a small animal model and a large animal model. *Cytokine* **2020**, *127*, 154974. [CrossRef] [PubMed]

72.   Park, S.; Nguyen, N.; Pezhouman, A.; Ardehali, R. Cardiac fibrosis, potential therapeutic targets. *Transl. Res.* **2019**, *209*, 121–137. [CrossRef] [PubMed]

73.   Lin, X.; Liu, Y.; Bai, A.; Cai, H.; Bai, Y.; Jiang, W.; Yang, H.; Wang, X.; Yang, L.; Sun, N.; et al. A viscoelastic adhesive epicardial patch for treating myocardial infarction. *Nat. Biomed. Eng.* **2019**, *3*, 632–643. [CrossRef] [PubMed]

74.   Wan, L.; Chen, Y.; Wang, Z.; Wang, W.; Schmull, S.; Dong, J.; Xue, S.; Imboden, H.; Li, J. Human heart valve-derived scaffold improves cardiac repair in a murine model of myocardial infarction. *Sci. Rep.* **2017**, *7*, 1–11.

75.   Rodness, J.; Mihic, A.; Miyagi, Y.; Wu, J.; Weisel, R.; Li, R. VEGF-loaded microsphere patch for local protein delivery to the ischemic heart. *Acta Biomater.* **2016**, *45*, 169–181. [CrossRef]

76.   Fan, C.; Tang, Y.; Zhao, M.; Lou, X.; Pretorius, D.; Menasche, P.; Zhu, W.; Zhang, J. CHIR99021 and fibroblast growth factor 1 enhance the regenerative potency of human cardiac muscle patch after myocardial infarction in mice. *J. Mol. Cell. Cardiol.* **2020**, *141*, 1–10. [CrossRef]

77.   Donndorf, P.; Strauer, B.E.; Haverich, A.; Steinhoff, G. Stem cell therapy for the treatment of acute myocardial infarction and chronic ischemic heart disease. *Curr. Pharm. Biotechnol.* **2013**, *14*, 12–19.

78.   Saparov, A.; Chen, C.W.; Beckman, S.A.; Wang, Y.; Huard, J. The role of antioxidation and immunomodulation in postnatal multipotent stem cell-mediated cardiac repair. *Int. J. Mol. Sci.* **2013**, *14*, 16258–16279. [CrossRef]

79.   Luo, L.; Tang, J.; Nishi, K.; Yan, C.; Dinh, P.; Cores, J.; Kudo, T.; Zhang, J.; Li, T.; Cheng, K. Fabrication of synthetic mesenchymal stem cells for the treatment of acute myocardial infarction in mice. *Circ. Res.* **2017**, *120*, 1768–1775. [CrossRef]

80.   Chang, M.L.; Chiu, Y.J.; Li, J.S.; Cheah, K.P.; Lin, H.H. Analyzing Impetus of Regenerative Cellular Therapeutics in Myocardial Infarction. *J. Clin. Med.* **2020**, *9*, 1277. [CrossRef]

81.   Chen, C.W.; Okada, M.; Proto, J.; Gao, X.; Sekiya, N.; Beckman, S.; Corselli, M.; Crisan, M.; Saparov, A.; Tobita, K.; et al. Human pericytes for ischemic heart repair. *Stem cells* **2013**, *31*, 305–316. [CrossRef] [PubMed]

82.   Jiang, Y.; Lian, X.L. Heart regeneration with human pluripotent stem cells, Prospects and challenges. *Bioact. Mater.* **2020**, *5*, 74–81. [CrossRef] [PubMed]

83.   Parizadeh, S.M.; Jafarzadeh-Esfehani, R.; Ghandehari, M.; Parizadeh, M.R.; Ferns, G.A.; Avan, A.; Hassanian, S.M. Stem cell therapy, A novel approach for myocardial infarction. *J. Cell. Physiol.* **2019**, *234*, 16904–16912. [CrossRef] [PubMed]

84.   Higuchi, A.; Ku, N.; Tseng, Y.; Pan, C.; Li, H.; Kumar, S.; Ling, Q.; Chang, Y.; Alarfaj, A.; Munusamy, M.; et al. Stem cell therapies for myocardial infarction in clinical trials, bioengineering and biomaterial aspects. *Lab. Investig.* **2017**, *97*, 1167–1179. [CrossRef] [PubMed]

85.   Karantalis, V.; DiFede, D.; Gerstenblith, G.; Pham, S.; Symes, J.; Zambrano, J.; Fishman, J.; Pattany, P.; McNiece, I.; Conte, J.; et al. Autologous mesenchymal stem cells produce concordant improvements in regional function, tissue perfusion, and fibrotic burden when administered to patients undergoing coronary artery bypass grafting, the Prospective Randomized Study of Mesenchymal Stem Cell Therapy in Patients Undergoing Cardiac Surgery (PROMETHEUS) trial. *Circ. Res.* **2014**, *114*, 1302–1310. [PubMed]

86. Luger, D.; Lipinski, M.; Westman, P.; Glover, D.; Dimastromatteo, J.; Frias, J.; Albelda, M.; Sikora, S.; Kharazi, A.; Vertelov, G.; et al. Intravenously delivered mesenchymal stem cells, systemic anti-inflammatory effects improve left ventricular dysfunction in acute myocardial infarction and ischemic cardiomyopathy. *Circ. Res.* **2017**, *120*, 1598–1613. [CrossRef]

87. Bao, L.; Meng, Q.; Li, Y.; Deng, S.; Yu, Z.; Liu, Z.; Zhang, L.; Fan, H. C-Kit Positive cardiac stem cells and bone marrow–derived mesenchymal stem cells synergistically enhance angiogenesis and improve cardiac function after myocardial infarction in a paracrine manner. *J. Card. Fail.* **2017**, *23*, 403–415. [CrossRef]

88. Kanda, P.; Davis, D.R. Cellular mechanisms underlying cardiac engraftment of stem cells. *Expert Opin. Biol. Ther.* **2017**, *17*, 1127–1143. [CrossRef]

89. Yao, Y.; Liao, W.; Yu, R.; Du, Y.; Zhang, T.; Peng, Q. Potentials of combining nanomaterials and stem cell therapy in myocardial repair. *Nanomedicine (Lond)* **2018**, *13*, 1623–1638. [CrossRef]

90. Amer, M.H.; Rose, F.; White, L.; Shakesheff, K. A detailed assessment of varying ejection rate on delivery efficiency of mesenchymal stem cells using narrow-bore needles. *Stem cells Transl. Med.* **2016**, *5*, 366–378. [CrossRef]

91. Aguado, B.A.; Mulyasasmita, W.; Su, J.; Lampe, K.; Heilshorn, S. Improving viability of stem cells during syringe needle flow through the design of hydrogel cell carriers. *Tissue Eng. Part A* **2012**, *18*, 806–815. [CrossRef] [PubMed]

92. Park, T.Y.; Oh, J.; Cho, J.; Sim, S.; Lee, J.; Cha, H. Stem cell-loaded adhesive immiscible liquid for regeneration of myocardial infarction. *J. Control. Release* **2020**, *321*, 602–615. [CrossRef] [PubMed]

93. Chen, Y.; Li, C.; Li, C.; Chen, J.; Li, Y.; Xie, H.; Lin, C.; Fan, M.; Guo, Y.; Gao, E.; et al. Tailorable Hydrogel Improves Retention and Cardioprotection of Intramyocardial Transplanted Mesenchymal Stem Cells for the Treatment of Acute Myocardial Infarction in Mice. *J. Am. Heart Assoc.* **2020**, *9*, e013784. [CrossRef] [PubMed]

94. Blocki, A.; Beyer, S.; Dewavrin, J.; Goralczyk, A.; Wang, Y.; Peh, P.; Ng, M.; Moonshi, S.; Vuddagiri, S.; Raghunath, M.; et al. Microcapsules engineered to support mesenchymal stem cell (MSC) survival and proliferation enable long-term retention of MSCs in infarcted myocardium. *Biomaterials* **2015**, *53*, 12–24. [CrossRef]

95. Gallagher, L.B.; Dolan, E.; O'Sullivan, J.; Levey, R.; Cavanagh, B.; Kovarova, L.; Pravda, M.; Velebny, V.; Farrell, T.; O'Brien, F.; et al. Pre-culture of mesenchymal stem cells within RGD-modified hyaluronic acid hydrogel improves their resilience to ischaemic conditions. *Acta Biomater.* **2020**, *107*, 78–90. [CrossRef]

96. Chow, A.; Stuckey, D.; Kidher, E.; Rocco, M.; Jabbour, R.; Mansfield, C.; Darzi, A.; Harding, S.; Stevens, M.; Athanasiou, T. Human Induced Pluripotent Stem Cell-Derived Cardiomyocyte Encapsulating Bioactive Hydrogels Improve Rat Heart Function Post Myocardial Infarction. *Stem Cell Rep.* **2017**, *9*, 1415–1422. [CrossRef]

97. Cai, H.; Wu, F.; Wang, Q.; Xu, P.; Mou, F.; Shao, S.; Luo, Z.; Zhu, J.; Xuan, S.; Lu, R.; et al. Self-assembling peptide modified with QHREDGS as a novel delivery system for mesenchymal stem cell transplantation after myocardial infarction. *Faseb J.* **2019**, *33*, 8306–8320. [CrossRef]

98. Harel, S.; Mayaki, D.; Sanchez, V.; Hussain, S. NOX2, NOX4, and mitochondrial-derived reactive oxygen species contribute to angiopoietin-1 signaling and angiogenic responses in endothelial cells. *Vasc. Pharm.* **2017**, *92*, 22–32. [CrossRef]

99. Lee, A.S.; Inayathullah, M.; Lijkwan, M.; Zhao, X.; Sun, W.; Park, S.; Hong, W.; Parekh, M.; Malkovskiy, A.; Lau, E.; et al. Prolonged survival of transplanted stem cells after ischaemic injury via the slow release of pro-survival peptides from a collagen matrix. *Nat. Biomed. Eng.* **2018**, *2*, 104–113. [CrossRef]

100. Ottersbach, A.; Mykhaylyk, O.; Heidsieck, A.; Eberbeck, D.; Rieck, S.; Zimmermann, K.; Breitbach, M.; Engelbrecht, B.; Brügmann, T.; Hesse, M.; et al. Improved heart repair upon myocardial infarction, Combination of magnetic nanoparticles and tailored magnets strongly increases engraftment of myocytes. *Biomaterials* **2018**, *155*, 176–190. [CrossRef]

101. Choe, G.; Kim, S.; Park, J.; Park, J.; Kim, S.; Kim, Y.; Ahn, Y.; Jung, D.; Williams, D.; Lee, J. Anti-oxidant activity reinforced reduced graphene oxide/alginate microgels, Mesenchymal stem cell encapsulation and regeneration of infarcted hearts. *Biomaterials* **2019**, *225*, 119513. [CrossRef] [PubMed]

102. Kai, D.; Wang, Q.; Wang, H.; Prabhakaran, M.; Zhang, Y.; Tan, Y.; Ramakrishna, S. Stem cell-loaded nanofibrous patch promotes the regeneration of infarcted myocardium with functional improvement in rat model. *Acta. Biomater.* **2014**, *10*, 2727–2738. [CrossRef] [PubMed]

103. Wang, Q.L.; Wang, H.; Li, Z.; Wang, Y.; Wu, X.; Tan, Y. Mesenchymal stem cell-loaded cardiac patch promotes epicardial activation and repair of the infarcted myocardium. *J. Cell Mol. Med.* **2017**, *21*, 1751–1766. [CrossRef]

104. Chen, J.; Zhan, Y.; Wang, Y.; Han, D.; Tao, B.; Luo, Z.; Ma, S.; Wang, Q.; Li, X.; Fan, L.; et al. Chitosan/silk fibroin modified nanofibrous patches with mesenchymal stem cells prevent heart remodeling post-myocardial infarction in rats. *Acta. Biomater.* **2018**, *80*, 154–168. [CrossRef] [PubMed]

105. Gaetani, R.; Feyen, D.; Verhage, V.; Slaats, R.; Messina, E.; Christman, K.; Giacomello, A.; Doevendans, P.; Sluijter, J. Epicardial application of cardiac progenitor cells in a 3D-printed gelatin/hyaluronic acid patch preserves cardiac function after myocardial infarction. *Biomaterials* **2015**, *61*, 339–348. [CrossRef]

106. Su, T.; Huang, K.; Daniele, M.; Hensley, M.; Young, A.; Tang, J.; Allen, T.; Vandergriff, A.; Erb, P.; Ligler, F.; et al. Cardiac Stem Cell Patch Integrated with Microengineered Blood Vessels Promotes Cardiomyocyte Proliferation and Neovascularization after Acute Myocardial Infarction. *ACS Appl. Mater. Interfaces* **2018**, *10*, 33088–33096. [CrossRef]

107. Dong, Y.; Hong, M.; Dai, R.; Wu, H.; Zhu, P. Engineered bioactive nanoparticles incorporated biofunctionalized ECM/silk proteins based cardiac patches combined with MSCs for the repair of myocardial infarction, In vitro and in vivo evaluations. *Sci. Total Environ.* **2020**, *707*, 135976. [CrossRef]

108. Gao, L.; Kupfer, M.; Jung, J.; Yang, L.; Zhang, P.; Da Sie, Y.; Tran, Q.; Ajeti, V.; Freeman, B.; Fast, V.; et al. Myocardial Tissue Engineering With Cells Derived From Human-Induced Pluripotent Stem Cells and a Native-Like, High-Resolution, 3-Dimensionally Printed Scaffold. *Circ. Res.* **2017**, *120*, 1318–1325. [CrossRef]

109. Tang, J.; Wang, J.; Huang, K.; Ye, Y.; Su, T.; Qiao, L.; Hensley, M.; Caranasos, T.; Zhang, J.; Gu, Z.; et al. Cardiac cell-integrated microneedle patch for treating myocardial infarction. *Sci. Adv.* **2018**, *4*, eaat9365. [CrossRef]

110. Park, S.J.; Kim, R.; Park, B.; Lee, S.; Choi, S.; Park, J.; Choi, J.; Kim, S.; Jang, J.; Cho, D.; et al. Dual stem cell therapy synergistically improves cardiac function and vascular regeneration following myocardial infarction. *Nat. Commun.* **2019**, *10*, 3123. [CrossRef]

111. Yeung, E.; Fukunishi, T.; Bai, Y.; Bedja, D.; Pitaktong, I.; Mattson, G.; Jeyaram, A.; Lui, C.; Ong, C.; Inoue, T.; et al. Cardiac regeneration using human-induced pluripotent stem cell-derived biomaterial-free 3D-bioprinted cardiac patch in vivo. *J. Tissue Eng. Regen. Med.* **2019**, *13*, 2031–2039. [CrossRef]

112. Melhem, M.R.; Park, J.; Knapp, L.; Reinkensmeyer, L.; Cvetkovic, C.; Flewellyn, J.; Lee, M.; Jensen, T.; Bashir, R.; Kong, H.; et al. 3D Printed Stem-Cell-Laden, Microchanneled Hydrogel Patch for the Enhanced Release of Cell-Secreting Factors and Treatment of Myocardial Infarctions. *ACS Biomater. Sci. Eng.* **2017**, *3*, 1980–1987. [CrossRef]

113. Mayfield, A.E.; Tilokee, E.; Latham, N.; McNeill, B.; Lam, B.; Ruel, M.; Suuronen, E.; Courtman, D.; Stewart, D.; Davis, D. The effect of encapsulation of cardiac stem cells within matrix-enriched hydrogel capsules on cell survival, post-ischemic cell retention and cardiac function. *Biomaterials* **2014**, *35*, 133–142. [CrossRef] [PubMed]

114. Ahmadi, A.L.; Thorn, S.; Alarcon, E.; Kordos, M.; Padavan, D.; Hadizad, T.; Cron, G.; Beanlands, R.; DaSilva, J.; Ruel, M.; et al. PET imaging of a collagen matrix reveals its effective injection and targeted retention in a mouse model of myocardial infarction. *Biomaterials* **2015**, *49*, 18–26. [CrossRef] [PubMed]

115. Han, J.; Kim, B.; Shin, J.; Ryu, S.; Noh, M.; Woo, J.; Park, J.; Lee, Y.; Lee, N.; Hyeon, T.; et al. Iron oxide nanoparticle-mediated development of cellular gap junction crosstalk to improve mesenchymal stem cells' therapeutic efficacy for myocardial infarction. *ACS Nano* **2015**, *9*, 2805–2819. [CrossRef] [PubMed]

116. Lemcke, H.; Gaebel, R.; Skorska, A.; Voronina, N.; Lux, C.; Petters, J.; Sasse, S.; Zarniko, N.; Steinhoff, G.; David, R. Mechanisms of stem cell based cardiac repair-gap junctional signaling promotes the cardiac lineage specification of mesenchymal stem cells. *Sci. Rep.* **2017**, *7*, 9755. [CrossRef]

117. Marcu, I.C.; Illaste, A.; Heuking, P.; Jaconi, M.; Ullrich, N. Functional Characterization and Comparison of Intercellular Communication in Stem Cell-Derived Cardiomyocytes. *Stem Cells* **2015**, *33*, 2208–2218. [CrossRef]

118. Sottas, V.; Wahl, C.; Trache, M.; Bartolf-Kopp, M.; Cambridge, S.; Hecker, M.; Ullrich, N. Improving electrical properties of iPSC-cardiomyocytes by enhancing Cx43 expression. *J. Mol. Cell Cardiol.* **2018**, *120*, 31–41. [CrossRef]

119. Liang, W.; Chen, J.; Li, L.; Li, M.; Wei, X.; Tan, B.; Shang, Y.; Fan, G.; Wang, W.; Liu, W. Conductive Hydrogen Sulfide-Releasing Hydrogel Encapsulating ADSCs for Myocardial Infarction Treatment. *ACS Appl. Mater. Interfaces* **2019**, *11*, 14619–14629. [CrossRef]

120. Tang, J.; Su, T.; Huang, K.; Dinh, P.; Wang, Z.; Vandergriff, A.; Hensley, M.; Cores, J.; Allen, T.; Li, T.; et al. Targeted repair of heart injury by stem cells fused with platelet nanovesicles. *Nat. Biomed. Eng.* **2018**, *2*, 17–26. [CrossRef]

121. Yokoyama, R.; Ii, M.; Tabata, Y.; Hoshiga, M.; Ishizaka, N.; Asahi, M. Cardiac Regeneration by Statin-Polymer Nanoparticle-Loaded Adipose-Derived Stem Cell Therapy in Myocardial Infarction. *Stem Cells Transl. Med.* **2019**, *8*, 1055–1067. [CrossRef]

122. Yao, X.; Liu, Y.; Gao, J.; Yang, L.; Mao, D.; Stefanitsch, C.; Li, Y.; Zhang, J.; Ou, L.; Kong, D.; et al. Nitric oxide releasing hydrogel enhances the therapeutic efficacy of mesenchymal stem cells for myocardial infarction. *Biomaterials* **2015**, *60*, 130–140. [CrossRef] [PubMed]

123. Díaz-Herráez, P.; Saludas, L.; Pascual-Gil, S.; Simón-Yarza, T.; Abizanda, G.; Prósper, F.; Garbayo, E.; Blanco-Prieto, M. Transplantation of adipose-derived stem cells combined with neuregulin-microparticles promotes efficient cardiac repair in a rat myocardial infarction model. *J. Control. Release* **2017**, *249*, 23–31. [CrossRef]

124. Chung, H.J.; Kim, J.; Kim, H.; Kyung, H.; Katila, P.; Lee, J.; Yang, T.; Yang, Y.; Lee, S. Epicardial delivery of VEGF and cardiac stem cells guided by 3-dimensional PLLA mat enhancing cardiac regeneration and angiogenesis in acute myocardial infarction. *J. Control. Release* **2015**, *205*, 218–230. [CrossRef] [PubMed]

125. Kumar, M.; Kasala, E.; Bodduluru, L.; Dahiya, V.; Sharma, D.; Kumar, V.; Lahkar, M. Animal models of myocardial infarction, mainstay in clinical translation. *Regul. Toxicol. Pharmacol.* **2016**, *76*, 221–230. [CrossRef] [PubMed]

126. Zaragoza, C.; Gomez-Guerrero, C.; Martin-Ventura, J.; Blanco-Colio, L.; Lavin, B.; Mallavia, B.; Tarin, C.; Mas, S.; Ortiz, A.; Egido, J. Animal models of cardiovascular diseases. *J. Biomed. Biotechnol.* **2011**, *2011*, 497841. [CrossRef]

127. Seyhan, A.A. Lost in translation, the valley of death across preclinical and clinical divide–identification of problems and overcoming obstacles. *Transl. Med. Commun.* **2019**, *4*, 1–19. [CrossRef]

128. Camacho, P.; Fan, H.; Liu, Z.; He, J.Q. Small mammalian animal models of heart disease. *Am. J. Cardiovasc. Dis.* **2016**, *6*, 70.

6

The Specific Molecular Composition and Structural Arrangement of *Eleutherodactylus Coqui* Gular Skin Tissue Provide Its High Mechanical Compliance

Justin Hui [1], Shivang Sharma [2], Sarah Rajani [1] and Anirudha Singh [2,3,*]

[1] Department of Biomedical Engineering, Johns Hopkins University, Baltimore, MD 21218, USA;
    jhui6@jhu.edu (J.H.); srajani2@jhu.edu (S.R.)

[2] Department of Chemical & Biomolecular Engineering, Johns Hopkins University, Baltimore, MD 21218, USA;
    ssharm55@jhu.edu

[3] Department of Urology, The James Buchanan Brady Urological Institute,
    The Johns Hopkins School of Medicine, Baltimore, MD 21287, USA

* Correspondence: asingh29@jhu.edu

Abstract: A male *Eleutherodactylus Coqui* (*EC*, a frog) expands and contracts its gular skin to a great extent during mating calls, displaying its extraordinarily compliant organ. There are striking similarities between frog gular skin and the human bladder as both organs expand and contract significantly. While the high extensibility of the urinary bladder is attributed to the unique helical ultrastructure of collagen type III, the mechanism behind the gular skin of *EC* is unknown. We therefore aim to understand the structure–property relationship of gular skin tissues of *EC*. Our findings demonstrate that the male EC gular tissue can elongate up to 400%, with an ultimate tensile strength (UTS) of 1.7 MPa. Species without vocal sacs, *Xenopus Laevis* (*XL*) and *Xenopus Muelleri* (*XM*), elongate only up to 80% and 350% with UTS~6.3 MPa and ~4.5 MPa, respectively. Transmission electron microscopy (TEM) and histological staining further show that *EC* tissues' collagen fibers exhibit a layer-by-layer arrangement with an uninterrupted, knot-free, and continuous structure. The collagen bundles alternate between a circular and longitudinal shape, suggesting an out-of-plane zig-zag structure, which likely provides the tissue with greater extensibility. In contrast, control species contain a nearly linear collagen structure interrupted by thicker muscle bundles and mucous glands. Meanwhile, in the rat bladder, the collagen is arranged in a helical structure. The bladder-like high extensibility of *EC* gular skin tissue arises despite it having eight-fold lesser elastin and five times more collagen than the rat bladder. To our knowledge, this is the first study to report the structural and molecular mechanisms behind the high compliance of *EC* gular skin. We believe that these findings can lead us to develop more compliant biomaterials for applications in regenerative medicine.

Keywords: collagen; elastin; bladder; compliance; microarchitecture; biomimicry

1. Introduction

For millions of years, animals have developed many unique organs for the sole purpose of attracting a suitable mate. These organs take advantage of phenomena such as color [1], scent [2], sound [3], and visual size [4]. Many of these phenomena are exemplified and exploited by many male frogs (*anuran*) to attract a suitable female [5]. Most commonly known is the sound of a frog and the iconic inflation of its gular skin (vocal sac). The characteristic inflation of the gular skin serves as a visual stimulus during mating season and facilitates the energetic mechanical effectiveness of air movement during calls [5]. Investigating the extraordinary inflating action of the gular skin can uncover novel mechanisms behind their ability to stretch and guide the development of artificial compliant biomaterials for regenerative medicine.

Most families in the order *anuran* have inflatable gular skin, such as *Hylidae*, *Eleutherodactylidae*, and *Leptopelis*; however, the size, shape, and color can vary significantly between species [6–8]. Commonly associated with frogs is the single inflated gular skin located under the floor of the gular; however, some species can have two external vocal sacs or no external vocal sac at all [5]. The vocal sac is imperative to the success of the male frog in attracting its female counterpart. *Anuran* vocal sacs play the same role in enhancing the calling effectiveness of the male frog to penetrate their often heavily forested habitats. The calling process begins with the frog filling its lungs with air and subsequently passing it over its vocal cords to produce a call. To reduce the time required to inhale after every call, the vocal sac stores and pushes air back into the lungs, effectively removing the need to inhale [5]. The vocal sac is able to increase the energetic mechanical efficiency of this process because of its elastic property. It stores the energy while the muscles push the air into the sac and release the energy to push the air back into the lungs, much like a helical spring or rubber balloon. The evolution of this elastic organ allows frogs to be very efficient in their calling and produce thousands of calls per night [5]. Doing so enables females to pinpoint the male's location for mating.

In this study, we focused on the male *Eleutherodactylidae Coqui* (*EC*) for its single inflatable gular skin. The highly stretchable nature of the gular skin in frogs is reminiscent of organs in several other species, such as the body of *Lagocephalus Gloveri* [9] and *Diodon Holocanthus* (pufferfish) [10], *Nerodia Sipedon* (snake) gular [11], the gular skin of *Fregata Magnificens* (frigate bird) [12], and the urinary bladder [13]. These organs possess the material phenomena known as elasticity and compliance that allow them to expand to such great volumes. Compliant materials can stretch easily under low forces, as opposed to stiff materials, which require significant force to deform the material slightly. These mechanical features originate from the composition of structural proteins and their architectures.

Elastin is a protein often associated with higher elasticity and compliance in tissues [14,15]. Several groups have incorporated elastin to develop highly elastic biomaterials, with the caveat that most methods include crosslinkers that are inherently cytotoxic (i.e., glutaraldehyde, hexamethylene diisocyanate) [16]. Other strategies used either peptide materials [17–19] or micro/nanofiber reinforced materials [20,21] to enhance elasticity and compliance; however, they are still unable to match the high extensibility of the urinary bladder. The inability of these materials to recapitulate the biomechanics of the bladder warrants further investigation into other mechanisms of elasticity and compliance, such as tissue ultrastructure. Using scanning electron microscopy (SEM), Murakumo et al. identified a unique helical architecture of collagen type III within the bladder, suggesting a role of collagen architecture in its biomechanics [13]. Although some groups have incorporated micro/nanofibers [20,21] into the material, the fibers are deposited in its fully elongated state and simply act to increase tensile strength instead of compliance. While it is possible to electrospin intertwined nanofibers [22], the authors of the study did not apply their technique in scaffold construction. We believe that the configuration or microarchitecture of collagen in tissues likely plays a significant role. Thus, the characterization of collagen microarchitectures in several compliant tissues in nature could reveal alternative ultrastructures that may be translated into biomaterial design.

Here, we characterize the commonly observed biomechanics and biocomposition of the gular skin, shown in Figure 1A. To fully understand the mechanism behind the highly inflatable gular skin, we performed experiments to characterize the tensile strength, tissue morphology, ultrastructure, and collagen/elastin content. We further compared the male EC gular skin tissue with frogs without a visually inflatable gular skin, namely *XL*, *XM*, female *EC*, and the male rat urinary bladder. In addition, leg tissues (Figure 1B) were dissected from the frogs and underwent the same analyses to identify the key features that allow the *EC*'s gular skin to be functionally unique. We performed TEM, tensile tests, histology, and biochemical assays. We demonstrate that the gular skin of *EC* can achieve comparable elongation to the rat bladder and is more compliant than *Xenopus Laevis* (*XL*) and *Xenopus Muelleri* (*XM*).

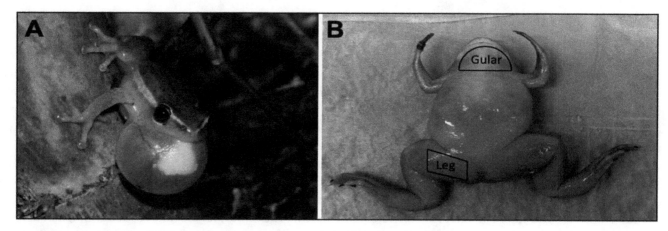

Figure 1. (**A**) A male *Hyperolius Cinnamomeoventris,* with inflated gular skin tissue to resonate its mating call. Reproduced from [5]. (**B**) Gross visual of tissue dissection areas. Shown here is the Xenopus Laevis specimen.

Furthermore, we identified a unique and more sophisticated collagen ultrastructure in *EC* gular skin, different from the helical structure observed in the urinary bladder. *EC* gular skin has a combination of several structures, such as layering, crimping, and twisting. The significance of these features is further emphasized by the statistically insignificant amount of elastin in most of the test samples. We believe that biomimicry of the collagen microstructure present in the *EC*'s gular skin may provide researchers with an alternative solution to reconstruct a more mechanically significant scaffold for regenerative medicine.

2. Results

2.1. Mechanical Properties

The uniaxial tensile tests illustrate marked differences between the frogs of genus *Xenopus* and those of *Eleutherodactylidae,* in addition to location-dependent properties. Figure 2 shows representative stress vs. strain and membrane tension characteristics of the species' gular skin tissue and leg skin tissue, with rat bladder as a comparison. Mean and standard deviation (SD) values are summarized in Table 1. It is evident in both Figure 2A,B that *Xenopus* tissues were stiffer relative to those of *Eleutherodactylidae,* as observed by the steep rise in the stress vs. strain curve before ultimate tensile strength (UTS) was reached. As shown in Table 1, there is a general trend of leg tissue being stiffer at 20% elongation, indicated by a higher secant modulus.

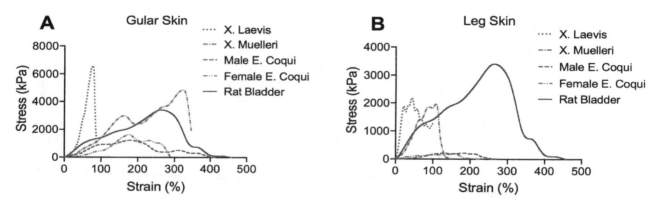

Figure 2. Representative uniaxial stress–strain curves of (**A**) gular skin tissue and (**B**) leg skin of different anuran species and their comparison to the rat bladder.

Furthermore, male *EC* tissues had lower secant modulus values than all samples tested, highlighting their uniqueness. Interestingly, gular skin tissue from *XM* displayed similar elongation at failure when compared to male *EC*, which were, on average, 350% and 398%, respectively. However, *XM* required much higher stress to elongate the same amount as *EC*, shown in Figure 2A. *XL* behaved as expected, with a very steep slope (indicative of a high secant modulus) followed by quick failure at 104% and 108% for gular and leg tissue, respectively, in stark contrast to samples from *EC* (gular: 398%, leg: 348%). Gular tissue of the male *EC* exhibited an average ultimate tensile strength (UTS) of 1263 ± 134 kPa, half that of its female counterpart (2142 ± 1789 kPa) and a quarter that of the *XL* (4461 ± 2215 kPa) and *XM* (4156 ± 1973 kPa). Comparisons between male and female *EC* show that male gular tissue exhibited higher elongation at break, of ~400% and ~340%, respectively. Figure 2B illustrates the stress vs. strain behavior of leg tissue, displaying similar trends as gular tissue. *Xenopus* samples had higher secant modulus values than *Eleutherodactylidae* in addition to a lower elongation at failure. The most striking observation is the clear difference between the final elongations between the tissue types. In all cases, the gular tissue of a single specimen had a greater elongation than the leg tissue.

Further comparisons with male rat bladder illustrated that there is a closer biomechanical resemblance of male *EC* gular tissue to the urinary bladder. The rat bladder had an average elongation of ~410%, similar to that of male and female *EC* tissues. Additionally, it had a mean peak stress of ~3000 kPa, highlighting its ability to retain urine and prevent failure.

We determined the secant modulus for assessing the stiffness of the sample. From Table 1, it is evident that leg tissues were generally stiffer than the same species' gular tissue, indicated by a higher mean value of the secant modulus. Male EC exhibited slightly higher mean secant modulus values of 373 kPa and 318 kPa for gular and leg tissue, respectively. As expected, the male EC had lower secant modulus values compared to the other species, and XL leg tissue exhibited the highest mean secant modulus of 4863 kPa. In general, leg tissues had higher secant modulus values than gular tissues. However, these trends did not show any statistical significance when compared to the rat bladder (Table 1).

Table 1. Summary of the mechanical properties.

Specimen	Tissue	UTS (kPa)	Strain at Peak Stress (%)	Breaking Strain (%)	Secant Modulus (20% strain) (kPa)
		Mean ± SD (*p*)	Mean ± SD (*p*)	Mean ± SD (*p*)	Mean ± SD (*p*)
Coqui (M)	Gular	1263 ± 134 (0.188)	187 ± 22 (0.999)	398 ± 86 (0.999)	373 ± 194 (0.808)
	Leg	193 ± 40 (0.006)	83 ± 53 (0.999)	348 ± 88 (0.999)	318 ± 333 (0.762)
Coqui (F)	Gular	2142 ± 1789 (0.861)	226 ± 43 (0.999)	337 ± 83 (0.999)	786 ± 1052 (0.991)
	Leg	936 ± 1265 (0.078)	101 ± 59 (0.999)	252 ± 39 (0.999)	1475 ± 1823 (0.999)
XL	Gular	4461 ± 2215 (0.330)	78 ± 12.5 (0.999)	104 ± 17 (0.999)	3117 ± 1400 (0.143)
	Leg	2553 ± 1775 (0.996)	61 ± 23 (0.999)	108 ± 12 (0.999)	4863 ± 2118 (0.003)
XM	Gular	4156 ± 1973 (0.581)	271 ± 70 (0.999)	350 ± 9 (0.999)	807 ± 668 (0.993)
	Leg	2524 ± 560 (0.932)	126 ± 16 (0.999)	175 ± 28 (0.999)	2877 ± 2265 (0.258)
Rat	Bladder	2985 ± 853	233 ± 28	412 ± 163	1288 ± 850

2.2. Tissue Morphology

H&E and trichrome-stained gular tissue cross-sections are shown in Figure 3. Gular tissue dissected from frogs belonging to the genus *Xenopus* are substantially thicker and contain a higher density of muscle bundles and mucous glands. Observations of the collagen structure between the families showed marked contrasts. The collagen structure in the male *EC* had a continuous and crimp structure, in stark contrast to the female *EC*, *XL*, and *XM*, which have more spread out crimp structures. Additionally, the tissue sample of the genus *Xenopus* has a nearly linear collagen structure. Interestingly, *XL* and *XM* have distinct regions where the collagen is perpendicularly aligned. These areas seemingly discretize the tissue into collagen segments.

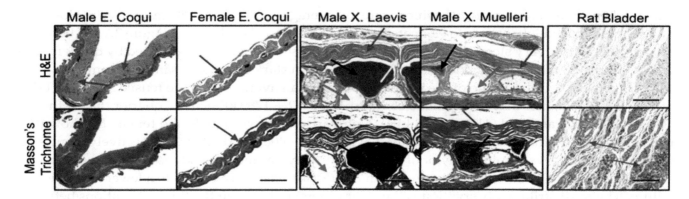

Figure 3. Tissue morphology by histology of gular tissue dissected from various species of frogs. Scale bar 100 μm. Muscle bundles (black arrows), mucous glands (red arrows), collagen structure (green arrows), perpendicularly aligned collagen (yellow arrows), urothelium (orange), lamina propria (purple), and detrusor muscle (grey).

Histological cross-sections of leg tissue show similar trends with the collagen structures between the frog species. *EC* tissue has a more pronounced crimp structure compared to *XL* and *XM*. Once again, the muscle bundles and mucous glands constitute a more substantial portion of the tissue in *Xenopus* samples. Collagen in the male *EC* leg has a less compact crimp structure than that found in the gular tissue. Similar to the gular tissue of the *Xenopus* samples, the collagen structure is more linear and has minimal crimping. Leg tissues do not exhibit the same inflatable function as the gular tissue; therefore, it does not require the crimped collagen structure. Lastly, sectioning of fresh rat bladder showed an abundance of muscle bundles with intermittent collagen fibers, dissimilar to the frog tissues. Collagen ultrastructures found in gular tissues were not visually identified in the rat bladder seen in Figures 3 and 4.

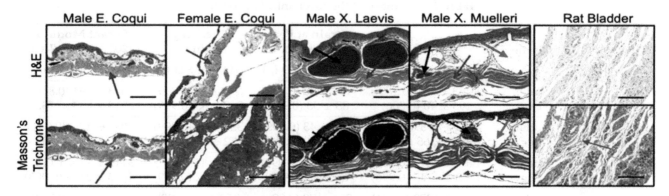

Figure 4. Hematoxylin and eosin (H&E) staining of leg tissue dissected from various species of frogs. Scale bar 100 μm. Muscle bundles (black arrows), mucous glands (red arrows), collagen structure (green arrows), perpendicularly aligned collagen (yellow arrows), urothelium (orange), lamina propria (purple), and detrusor muscle (grey).

2.3. Tissue Ultrastructure

TEM was used in this study to properly observe the collagen microstructure and orientation, shown in Figure 5. Images at 3400× magnification reinforce the observations from histology and show unique patterns in the *EC* gular tissue compared to *XL* and *XM*. Higher magnification images (13,500×) were obtained from the areas highlighted with the red box in the first row of Figure 5A–D and shown in the second row (Figure 5E–H). At higher magnification, the intricate orientation and design of the collagen layers can be identified. Figure 5E,F show a tri-dimensional collagen arrangement defined by the waviness, transitions from circular to rectangular cross-section within a layer, and contrast differences showing axial twisting. Although the layering pattern is present in the *Xenopus* tissues, other patterns are not observed. In contrast, there are discontinuities within the collagen layers of the *Xenopus* tissues shown in Figure 5C,D.

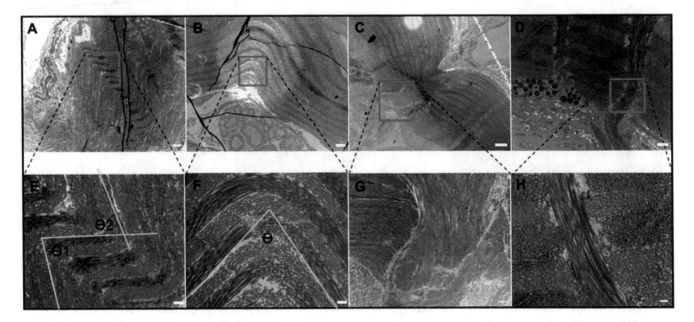

Figure 5. TEM images of gular tissue for (**A**) male EC, 3400×. Scale bar 2 μm; (**B**) female EC, 3400×. Scale bar 2 μm; (**C**) XL, 1000×. Scale bar 10 μm; (**D**) XM, 4200×. Scale bar 10 μm; (**E**) male EC, 13,500×. Scale bar 500 nm. Green lines represent the crimp angle measurements ($\theta 1 = 80°$, $\theta 2 = 70°$). (**F**) Female EC, 13,500×. Scale bar 500 nm. Green lines represent the crimp angle measurements ($\theta = 77°$). (**G**) XL, 3400×. Scale bar 2 μm. (**H**) XM, 13,500×. Scale bar 500 nm. Red boxes indicate the area magnified.

Furthermore, Figure 5E,F can be used to identify key physical characteristics of the crimped collagen structure. Firstly, there is a clear "layering" structure within the bulk tissue, which indicates an alternating collagen bundle orientation across the tissue. In the male *EC*, a collagen layer (Figure 5E) is approximately 500 nm thick, whereas the female *EC* is 667 nm thick. The two figures also show very clearly that the male *EC* gular tissue has more layers within the same field of view. Secondly, calculations were conducted to identify the crimp angle of the collagen bundles. As shown in Figure 5, the angles between the two folded layers of the male *EC* tissues are 80° and 70°, while there is a single fold of 77° for the female *EC* tissues. Those of the *Xenopus* genus were virtually 0°, meaning that they were nearly straight throughout the tissue.

Some similar features were observed in the rat bladder and are shown in Figure 6. Rat bladder and *EC* gular skin had several features in common, such as out-of-plane orientation and crimping (Figure 6A,B). As shown in Figure 6D, collagen strands change from longitudinal to circular cross-sectional orientation, similar to those seen in Figure 5E,F. However, the rat bladder lacks the layered structure found in *EC* tissues. Rat bladder had a thick collagen bundle (2.5 μm), shifting its orientation all at once, whereas *EC* tissues had smaller bundles stacked on top of each other and alternating their orientation. Figure 6B,E resemble a crimped collagen bundle. This was a short bundle and was seldom identified across the tissue. As expected, a helical structure was observed in the rat bladder (Figure 6C,F). The twisted collagen strands identified in Figure 6F show very clearly the formation of spring-like structures, reminiscent of those found in the human bladder. These observations were frequently found throughout the tissue cross-section.

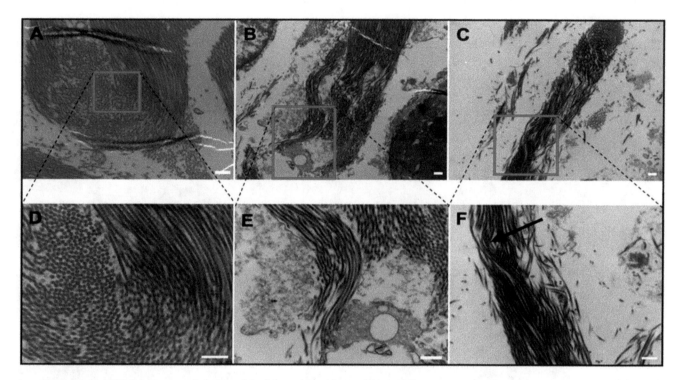

Figure 6. TEM images of the rat bladder. (**A**) Alternating collagen orientation. 17,500×. Scale bar 500 nm. (**B**) Crimp structure. 13,500×. Scale bar 500 nm. (**C**) Helical structure. 9700×. Scale bar 500 nm. (**D**) Magnified alternating collagen orientation. 33,000×. Scale bar 500 nm. (**E**) Magnified crimp structure. 24,500×. Scale bar 500 nm. (**F**) Magnified helical structure. 17,500×. Scale bar 500 nm.

2.4. Biochemical Analysis

As shown in Figure 7, *EC* tissues had, on average, much higher elastin concentrations than the *Xenopus* samples. On the other hand, there are minimal differences between the elastin content in male and female *EC* tissues. As summarized in Table 2, there was an average of 58.5 ± 16.9 and 64.5 ± 9.8 μg/mg of wet tissue in male and female gular skin, respectively. The elastin content found in *EC* gular tissues was higher than the *Xenopus* gular tissues, with 22.1 ± 5.2 and 42.1 ± 11.9 μg/mg of wet tissue for *XL* and *XM*, respectively. In general, leg tissues contained less elastin than the gular tissues, and, once again, *EC* samples contained more elastin than *Xenopus* samples (Figure 7A). Collagen assays show the opposite trend, with *EC* tissues containing less collagen than *Xenopus*. Male *EC* gular tissue contained an average of 202.0 ± 41.3 μg/mg of wet tissue and female *EC* had 234.5 ± 86.7 μg/mg of wet tissue. These values were much lower than those of *XL* and *XM*, with 377.8 ± 86.7 μg/mg and 400.8 ± 36.5 μg/mg of wet tissue, respectively (Figure 7B). As expected, male *EC* leg tissue contained more collagen than the gular tissues; however, this trend was not found in the other samples. Although *EC* tissues had more elastin than *Xenopus* tissues, the similar concentrations of elastin between male and female *EC* further imply the importance of the collagen structure in the extensibility of the tissue. Lastly, the average elastin to collagen ratio was calculated to show that the *EC* gular tissue had higher elastin to collagen ratios of 0.29 and 0.27 for males and females, respectively. In stark contrast, *Xenopus* gular tissues were 0.06 and 0.1 for *XL* and *XM*, respectively. As a comparison, the male rat bladder had significantly more elastin and less collagen, with averages of 484.8 ± 121.9 and 96.1 ± 19.4 μg/mg of wet tissue, respectively. Rat bladder has approximately eight times more elastin than the male EC. As expected, the rat bladder had five times more elastin than collagen, unlike frog tissues, where the opposite is observed.

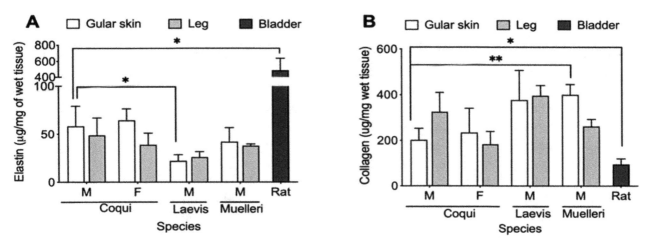

Figure 7. (A) Elastin content and (B) collagen content for the respective tissue sample. M: male, F: female. $p < 0.05$ (*) and $p < 0.01$ (**).

Table 2. Summary of elastin and collagen content in tissue samples.

		Elastin (µg/mg Wet Tissue)	Collagen (µg/mg Wet Tissue)	Elastin/Collagen Ratio
Specimen	Tissue	Mean ± SD	Mean ± SD	
Female Coqui	Gular	64.5 ± 9.8	234.5 ± 86.7	0.27
	Leg	39.0 ± 9.8	183.1 ± 45.9	0.21
Male Coqui	Gular	58.5 ± 16.9	202.0 ± 41.3	0.29
	Leg	48.8 ± 14.8	325.0 ± 69.5	0.15
XL	Gular	22.1 ± 5.2	377.8 ± 105.3	0.06
	Leg	26.0 ± 4.6	396.7 ± 36.3	0.07
XM	Gular	42.1 ± 11.9	400.8 ± 36.5	0.10
	Leg	37.9 ± 1.4	262.1 ± 24.3	0.14
Rat	Bladder	484.8 ± 121.9	96.1 ± 19.4	5.05

3. Discussion

To understand the fundamental mechanisms behind the iconic inflatable gular tissue seen in select frog species, in the present study, we analyzed multiple frog tissues. Among all tested frog tissues, *EC* tissue displayed lower secant moduli in uniaxial tensile tests, more sophisticated collagen microarchitectures, and higher elastin to collagen ratios. These mechanical characteristics allow the tissue to meet the functional requirements of the gular tissue to move air between the vocal sac and the lungs mechanically. The low UTS (1263 ± 134 kPa) of the male *EC* gular tissue reflects the easily extensible tissue, allowing the frog to inflate its gular skin without excessive force. Interestingly, the gular tissue in all species had a higher UTS when compared to its leg tissue. This is likely indicative of the need for the tissue to withstand repeated stress from fast air movement. Furthermore, gular tissue from male *EC* had the highest average elongation of nearly 400% compared to the other species. In contrast to expectations, the gular tissue of *XM* was able to elongate, on average, 350% ± 9% of its original length, surpassing the female *EC* but not achieving the same elongation of the male *EC*. On the other hand, *XL* elongated substantially less, with only an average of 104% ± 17%. The leg tissues of all species had lower peak stress and strain values compared to their gular tissue counterparts. In all species, leg skin tissue displayed half the peak stress of the gular skin except for male *EC*. The male *EC* gular skin had a UTS of 1263 kPa while the leg skin had 193 kPa. This trend shows that the gular skin is a mechanically more durable material than that of the leg. These UTS values also occurred at smaller strain values than those of the gular skin. Male *EC* gular skins' UTS occurred at 187%, compared to its leg tissue at 83%, as shown in Table 1. The high elongation and moderate UTS of

EC gular tissue reflect its role as a mechanically dynamic tissue. Its frequent expansion and loading require it to sustain higher stress while still elongating, much like the rat urinary bladder. Gular tissue of the male EC and XM had comparable elongations to the male rat bladder (~412%). The rat bladder also had UTS averaging ~3000 kPa, higher than tissue from EC but lower than XM and XL. The bladder can withstand higher stresses without compensating for elongation, which is essential to retain urine pressure at higher volumes. We believe that the high variability in the data, as indicated by Table 1, is due to the limited number of EC frogs and inherent biological- and age-related variations.

TEM and histological staining illustrated the relationship between microarchitecture and the mechanical characteristics of the tissues. The typical crimp pattern observed in H&E and trichrome stains were also found in other elastic tissues [9]. Evident in Figure 3 is the lack of large muscle bundles in EC tissues. Therefore, the large muscle bundles likely contribute to the tissues' higher UTS. Furthermore, the more crimped structure found in the EC suggests a greater elongation when stretched. The nearly linear collagen structure found in Xenopus tissues (Figures 3 and 4) has clear linear collagen bundles, which restrict the tissues' ability to elongate. In combination with the stress vs. strain behaviors, the crimped collagen structure is critical to the extensibility of the tissue. However, collagen fibers from the skin of the leg were less compact, which explains the lower stress values and lower strain at failure.

Further investigations with high magnification TEM show a sophisticated collagen microstructure in the male EC gular tissue. Firstly, it displayed increased folding of the crimp structure, with two folds of crimp angles of 80° and 70°, compared to the female, with a single fold of crimp angle of 77°. The compact collagen allows the tissue to stretch with less force and to a higher degree. Secondly, the collagen bundles can be observed to alternate between a circular and longitudinal shape, suggesting an out-of-plane zig-zag structure. This structure likely provides the tissue with greater extensibility because the collagen bundles have additional degrees of freedom to realign in space.

Additionally, each collagen strand twists about its central axis, allowing it to unwind during elongation for higher extension (Figure 5). Lastly, the layering structure throughout the cross-section of the tissue has an alternating round and longitudinal shape, indicating a mesh-like structure. This technique is documented to aid in dissipated energy during loading [23]. Taken altogether, the gular tissue of the male EC has a compact three-dimensional structure that allows collagen bundles to extend significantly. Its female counterpart also displayed these features but to a lesser extent. The crimp structure was less compact, indicated by the larger curvature seen in Figure 5B. The unique features found in EC were not present in the Xenopus frogs but, instead, many locations were identified to be potential areas of stress concentration and elongation inhibition. Figure 5C,D show apparent discontinuities in the overall collagen structure in the Xenopus gular tissues. The discontinuities show clear breaks where the collagen aligned parallel to the tissue is impeded by collagen aligned perpendicularly. These areas are a likely cause of the higher secant modulus as they are the site of stress concentration during loading. Architecturally, EC tissue is unique because of its continuous hierarchical collagen structure. We believe that the multi-dimensional crimp structure allows for easy elongation of the tissue in all directions, which is key to mate calling. The difference between male and female gular tissue lies in the compactness of the structure, where a compact crimp allows greater elongation under lower forces. This property is also known as the "crimp angle" and is often used to describe the collagen structure in human tendons [24–26]. As mentioned in previous studies, the crimped collagen structure was determined to act as a recoiling system during muscle relaxation. This structure in the gular tissue likely serves a similar purpose in recoiling the tissue post-inflation and working in tandem with elastin.

The lamina propria in the rat bladder consisted of collagen interspersed between muscle bundles (Figures 3 and 4). This feature is in clear contrast to the frog tissues, where EC tissues consisted of no muscle, and XM/XL had muscle bundles set apart from a collagen layer. Both TEM and histology showed that collagen in the rat bladder was less compact and less regularly structured compared to frog tissues.

As expected, the biochemical analysis showed a higher concentration of elastin in the *EC* tissues than in *Xenopus* tissues (see Figure 7). A high elastin content is one of the major contributing factors to elastic tissues and plays a key role in elongation. Male and female *EC* gular tissue had 58 ± 17 and 64 ± 10 µg/mg of wet tissue, respectively, whereas *XL* and *XM* only had 22 ± 5 and 42 ± 12 µg/mg of wet tissue, respectively. The higher concentration of elastin present in *EC* gular tissue sets it apart, bio-compositionally, from *XL* and *XM*. The biocomposition translates to a higher elongation, evident in the mechanical characteristics shown in Figure 2. Furthermore, male and female *EC* gular tissues have similar elastin concentrations, but the male *EC* has larger elongation. This emphasizes the benefits of the collagen ultrastructure for the mechanics of the tissue.

Although very dissimilar under histology stains, TEM was able to uncover similar collagen ultrastructure with the rat bladder. As shown in Figure 6A,B, the rat bladder and *EC* gular skin share ultrastructural features such as out-of-plane zig-zag and crimp structures. Additionally, *EC* gular skin and rat bladder had similar average elongations, of 398% and 412%, respectively. This clearly shows the importance of the collagen ultrastructure on the elongation of the tissue. Furthermore, rat bladder contained 485 ± 122 µg elastin/mg of wet tissue, eight-fold higher than *EC* gular skin. The significantly higher amount of elastin in the rat bladder did not translate to higher elongation. However, an abundance of elastin quickly recoils collagen bundles during contraction. [13] These results show that this unique collagen ultrastructure or high amounts of elastin can lead to higher elongation of a material. Further investigations are required in order to elucidate the individual impact of high elastin content and various collagen ultrastructures (i.e., crimp or helical) on compliance.

Tensile tests on these tissues show the much higher UTS of the gular tissue compared to that of the leg. This is counter-intuitive because of the gular tissue's lower collagen content; however, the increase in mechanical strength can be attributed to the hierarchical microstructure seen in the male *EC*. Shown in a previous study [27], the microstructure of the material has a significant role in the mechanical behavior of the material. Evidently shown here is that reducing performance compromises is one of the key reasons behind intricate microarchitectures. The layered structure of the collagen provides the tissue with much of its strength [27–30], while the waviness enhances tissue elongation. This hierarchical structure provides a balance between tensile strength and compliance, two mechanical properties that are not often positively correlated. As shown in a previous study [13], elastin is only regionally present in the bladder, and the helical shape of collagen provides much of the bladder's compliance. While collagen type I provides the bladder with its strength, collagen type III and elastin provide its compliance [13,31–35]. The unique conformation of collagen type III and elastin provide recoverable deformation due to hydrophobic interactions. This conformation may not be required to achieve the same biomechanical action, as gular skin did not contain coiled collagen or a high elastin content. The compliance of male *EC* tissue is a result of its tissue-specific collagen ultrastructure, thus reinforcing the importance of material architecture in bulk material properties. The functional similarities between the gular tissue and the bladder prove that the same fundamental mechanisms can be used in biomaterial design for bladder reconstruction.

Taken altogether, we believe that the collagen architecture shown here in EC gular skin tissue can be relevant in tissue engineering of large-deforming compliant tissues. The unique orthogonal arrangement of collagen layers can also be potentially considered for corneal tissue engineering [36]. Future work will involve recapitulating this structure using 3D bioprinting techniques to develop reinforced hydrogels [37–39]. Additionally, the area of decellularized tissues in tissue engineering has generated considerable attention in recent works [40,41]. Decellularized EC gular tissues have the potential to act as a temporary functional graft, although additional experiments will need to be conducted to validate its potential.

4. Materials and Methods

4.1. Specimen Collection

Euthanized male and female *E. Coqui* were obtained from Atlanta Botanical Garden (Atlanta, GA, USA). The females were tested as a species-specific control as females do not tend to inflate their gular tissue. The number of biological replicates was limited because these frogs were taken from live populations, and excessive sampling of frogs causes ecological instability. In total, four male ECs and two female ECs were collected for all the experiments. Although the age of the frogs may influence the mechanics of the tissue, *E. Coqui* was obtained from a population, and an exact age could not be determined. The estimated age was between 1 and 5 years. In total, four euthanized male *X. Laevis* and four male *X. Muelleri* were obtained from Xenopus Express Inc. These species were used as negative controls as they do not display any external tissue inflation during calls. All specimens were euthanized on-site before either fixation or freezing and were transported overnight. Samples used for tensile testing were received frozen in ice and immediately dissected for sample collection. Samples were then stored at −20 °C until testing was performed. Fixed samples were used for histology and microscopy, while frozen samples were used for tensile tests and biochemical assays. The skin tissue of the desired regions was dissected in the laboratory with surgical scissors and scalpels. The specimen's gular and leg skin tissue were of particular interest; areas examined can be seen in Figure 1b. Rat bladders were dissected in accordance with protocols approved by the Johns Hopkins University Animal Care and Use Committee (RA17M330, 7/23/2018).

4.2. Uniaxial Tensile Test

We determined the mechanical properties of the unfixed and hydrated skin tissues by conducting uniaxial tensile tests using the Instron Tensile testing equipment (MTS Criterion™ 40). Samples ($n = 3$) per group (male *EC:* three biological repeats; female *EC:* two biological repeats; male *XL:* three biological repeats; male *XM:* three biological repeats) for both gular skin and leg were cut into small rectangular pieces and measured for their thickness, width, and length. Cut samples ranged from 1 mm to 3 mm wide and 4 mm long. Average sample thickness (0.1–0.4 mm) was measured using an electronic caliper, which varied due to differences in species and the site of extraction (gular vs. leg). Generally, *EC* samples were thinner than those of *XL/XM*. The sample dimensions were limited to the size of the frogs, as EC is extremely small (~1.5 inches for male). Reference state dimensions were established after mounting the tissue onto the apparatus and tared before testing began. Preconditioning was not performed prior to tensile loading. We also assumed that the tissues would behave isotropically under tensile loading. We acknowledge that the biaxial test would be more predictive of tissue expansion; however, because of the small size of *EC*, we opted for comparative uniaxial analysis. Each side of the cut samples was then glued in between two pieces of thick paper to increase the surface area that is clamped onto the instrument, similar to a study conducted by Dahms et al. [42]. The strain was calculated using the distance between the clamps. The samples were extended with a strain rate of 0.5 mm/min, using a 5N load cell. Engineering stress, membrane tension, strain, and secant modulus (at 20% strain) were calculated and visualized with GraphPad Prism 8.0. Secant moduli were calculated using the stress and strain values before and after 20% strain.

4.3. Histology

Tissue samples were washed in Phosphate Buffer Saline (PBS), formalin-fixed, and paraffin-embedded for staining using hematoxylin and eosin (H&E) and Masson's trichrome, in accordance with our previous study [43]. Paraffin blocks were sectioned at 6 microns onto glass slides for brightfield microscopy imaging.

4.4. Transmission Electron Microscopy

Samples were fixed in 3% paraformaldehyde, 1.5% glutaraldehyde, 5 mM $CaCl_2$, 2.5% sucrose, and 0.1% tannic acid in 0.1 M sodium cacodylate buffer, pH 7.2. After buffer rinse, samples were fixed in 1.0% osmium tetroxide for 1 h on ice in the dark. Following a rinse with distilled water, the samples were stained with 2.0% aqueous uranyl acetate (0.22 μm filtered) for 1 h in the dark, dehydrated using a graded series of ethanol, and embedded in Eponate 12 resin (Ted Pella, Redding, CA, USA). Samples were polymerized for 2–3 days at 37 °C and were stored at 60 °C overnight. Thin sections, 60–90 nm, were cut at a depth of 50 μm from the surface with a diamond knife using a Reichert-Jung Ultracut E ultramicrotome and placed on naked copper grids and were stained with 2% uranyl acetate in 50% methanol and observed using a Philips/FEI BioTwin CM120 TEM.

4.5. Crimp Angle Measurement

The crimp angle analyzed through TEM imaging of unloaded samples was calculated by drawing a line along the straight edges of a collagen layer and measuring the angle between the two lines using ImageJ software (NIH). Measurement lines are shown in green in Figure 5E,F for reference.

4.6. Collagen and Elastin Content

The collagen and elastin concentrations in both gular and leg tissues ($n = 3$) were determined using Biocolor Sircol™ Insoluble Collagen and Biocolor Fastin™ Elastin assays (Accurate Chemical, Westbury, NY). Briefly, samples ranging from 0.1 to 4.0 mg wet weight were thawed and dissociated in the fragmentation reagent provided by the kit. We performed tissue fragmentation for 3 h at 65 °C under intermittent vortex mixing. Dissociated collagen was collected through centrifugation and dyed using the Sircol Dye Reagent for 30 min, followed by centrifugation. Excess liquid was drained and exposed to an acid-salt wash to remove the unbound dye and centrifuged to collect the collagen-dye pellet and remove excess dye. The alkali reagent was added and vortexed to remove the bound dye and immediately used for colorimetric absorbance measurements. Elastin was extracted according to the protocol described briefly here. Elastin was extracted using 0.25 M oxalic acid at 100 °C for 2 h and precipitated using the provided precipitating agent. The solution was centrifuged to form an elastin pellet, exposed to the dye reagent for 90 min, and vortexed intermittently. Following centrifugation, the elastin-bound dye was released using the provided dissociation reagent and used immediately for colorimetric absorbance reading. Results from both assays were expressed in micrograms of collagen or elastin per milligram of the wet tissue sample.

4.7. Statistical Methods

Statistical analysis was conducted on GraphPad Prism 8.0 using two-way ANOVA (Holm-Sidak) for biochemical assays and two-way ANOVA (Dunnett's) for tensile properties, with an α value of 0.05. All sample groups were tested against each other for biochemical assays, and tensile properties were compared against the rat bladder as a control.

5. Conclusions

In conclusion, our results emphasize the role of protein microarchitecture on the mechanical properties of a material. We first illustrated the unique mechanical properties of the male EC gular tissue, contrasted with other tissue types, other species of frogs, gender, and rat bladder. The high elongation and moderate UTS demonstrated the tissues' ability to reduce performance trade-offs. That is, the tissue was able to elongate while maintaining higher stresses. Importantly, the specific multi-dimensional hierarchical collagen structure allowed the tissue to elongate as well as maintain moderately higher stresses. Together, the two components allow the unique functionality of the gular tissue to inflate and withstand the internal air pressure during calling. The elongation similarities of EC gular skin and rat bladder are reinforced by the ultrastructural likeness of the two tissues. Taken together, these findings demonstrate the potential of the *EC* gular skin as a novel biomimetic example for developing materials for tissue engineering of large deforming tissues, such as the bladder. The characterization of this microarchitecture can provide a simple template for future artificial biomaterial scaffold designs for regenerative medicine applications such as biomimicry in 3D printed bladder tissues.

Author Contributions: Conceptualization, J.H. and A.S; methodology, J.H. and A.S.; investigation, J.H., S.S. and S.R., writing—original draft preparation, J.H.; writing—review and editing, S.R. and A.S. All authors have read and agreed to the published version of the manuscript.

Acknowledgments: We greatly thank Chelsea Thomas and Brad Wilson at the Atlanta Botanical Garden for their help in animal procurement and processing. We would also like to thank the Department of Materials Science and Engineering of Johns Hopkins University for allowing us to perform experiments on the MTS Criterion tensile testing equipment.

Abbreviations

UTS	Ultimate Tensile Strength
EC	Eleutherodactylus Coqui
XL	Xenopus Laevis (XL)
XM	Xenopus Muelleri
TEM	Transmission Electron Microscopy
SEM	Scanning Electron Microscopy
H&E	Haemotoxylin and Eosin

References

1. McCoy, D.E.; McCoy, V.E.; Mandsberg, N.K.; Shneidman, A.V.; Aizenberg, J.; Prum, R.O.; Haig, D. Structurally assisted super black in colourful peacock spiders. *Proc. R. Soc. B Biol. Sci.* **2019**, *286*, 20190589. [CrossRef]

2. Roberts, S.A.; Davidson, A.J.; Beynon, R.J.; Hurst, J.L. Female attraction to male scent and associative learning: The house mouse as a mammalian model. *Anim. Behav.* **2014**, *97*, 313–321. [CrossRef]

3. Halfwerk, W.; Jones, P.L.; Taylor, R.C.; Ryan, M.J.; Page, R.A. Risky Ripples Allow Bats and Frogs to Eavesdrop on a Multisensory Sexual Display. *Science* **2014**, *343*, 413–416. [CrossRef]

4. Solberg, E.J.; Saether, B.-E. Fluctuating asymmetry in the antlers of moose (Alces alces): Does it signal male quality? *Proc. R. Soc. B Biol. Sci.* **1993**, *254*, 251–255. [CrossRef]

5. Starnberger, I.; Preininger, D.; Hödl, W. The anuran vocal sac: A tool for multimodal signalling. *Anim. Behav.* **2014**, *97*, 281–288. [CrossRef] [PubMed]

6. Hayes, M.P.; Krempels, D.M. Vocal Sac Variation among Frogs of the Genus Rana from Western North America. *Copeia* **1986**, *1986*, 927–936. [CrossRef]

7. Elias-Costa, A.J.; Montesinos, R.; Grant, T.; Faivovich, J. The vocal sac of Hylodidae (Amphibia, Anura): Phylogenetic and functional implications of a unique morphology. *J. Morphol.* **2017**, *278*, 1506–1516. [CrossRef] [PubMed]

8. Kelley, D.B. Vocal communication in frogs. *Curr. Opin. Neurobiol.* **2004**, *14*, 751–757. [CrossRef] [PubMed]

9. Kirti; Khora, S.S. Mechanical properties of pufferfish (*Lagocephalus gloveri*) skin and its collagen arrangement. *Mar. Freshw. Behav. Physiol.* **2016**, *49*, 327–336. [CrossRef]

10. Brainerd, E.L. Pufferfish inflation: Functional morphology of postcranial structures inDiodon holocanthus (Tetraodontiformes). *J. Morphol.* **1994**, *220*, 243–261. [CrossRef]

11. Close, M.; Cundall, D. Snake lower jaw skin: Extension and recovery of a hyperextensible keratinized integument. *J. Exp. Zoöl. Part A Ecol. Genet. Physiol.* **2013**, *321*, 78–97. [CrossRef] [PubMed]

12. Madsen, V.; Balsby, T.J.S.; Dabelsteen, T.; Osorno, J.L. Bimodal Signaling of a Sexually Selected Trait: Gular Pouch Drumming in the Magnificent Frigatebird. *Condor* **2004**, *106*, 156–160. [CrossRef]

13. Murakumo, M.; Ushiki, T.; Abe, K.; Matsumura, K.; Shinno, Y.; Koyanagi, T. Three-Dimensional Arrangement of Collagen and Elastin Fibers in the Human Urinary Bladder: A Scanning Electron Microscopic Study. *J. Urol.* **1995**, *154*, 251–256. [CrossRef]

14. Viidik, A.; Danielson, C.C.; Oxlund, H.; Danielsen, C. On fundamental and phenomenological models, structure and mechanical properties of collagen, elastin and glycosaminoglycan complexes. *Biorheol* **1982**, *19*, 437–451. [CrossRef]

15. Carter, F.J.; Frank, T.; Davies, P.J.; Cuschieri, A. Puncture forces of solid organ surfaces. *Surg. Endosc.* **2000**, *14*, 783–786. [CrossRef]

16. Rodríguez-Cabello, J.C.; De Torre, I.G.; Ibañez-Fonseca, A.; Alonso, M.; Ibañez-Fonzeca, A. Bioactive scaffolds based on elastin-like materials for wound healing. *Adv. Drug Deliv. Rev.* **2018**, *129*, 118–133. [CrossRef]

17. Croisier, F.; Liang, S.; Schweizer, T.; Balog, S.; Mionić, M.; Snellings, R.; Cugnoni, J.; Michaud, V.; Frauenrath, H. A toolbox of oligopeptide-modified polymers for tailored elastomers. *Nat. Commun.* **2014**, *5*, 4728. [CrossRef]

18. Voorhaar, L.; Diaz, M.M.; Leroux, F.; Rogers, S.; Abakumov, A.M.; Van Tendeloo, G.; Van Assche, G.; Van Mele, B.; Hoogenboom, R. Supramolecular thermoplastics and thermoplastic elastomer materials with self-healing ability based on oligomeric charged triblock copolymers. *NPG Asia Mater.* **2017**, *9*, e385. [CrossRef]

19. Urry, D.W.; Pattanaik, A.; Xu, J.; Woods, T.C.; McPherson, D.T.; Parker, T.M. Elastic protein-based polymers in soft tissue augmentation and generation. *J. Biomater. Sci. Polym. Ed.* **1998**, *9*, 1015–1048. [CrossRef]

20. Ghafari, A.M.; Rajabi-Zeleti, S.; Naji, M.; Ghanian, M.H.; Baharvand, H. Mechanical reinforcement of urinary bladder matrix by electrospun polycaprolactone nanofibers. *Sci. Iran.* **2017**, *24*, 3476–3480. [CrossRef]

21. Feng, C.; Liu, C.; Liu, S.; Wang, Z.; Yu, K.; Zeng, X. Electrospun Nanofibers with Core–Shell Structure for Treatment of Bladder Regeneration. *Tissue Eng. Part A* **2019**, *25*, 1289–1299. [CrossRef] [PubMed]

22. Chang, G.; Shen, J. Fabrication of Microropes via Bi-electrospinning with a Rotating Needle Collector. *Macromol. Rapid Commun.* **2010**, *31*, 2151–2154. [CrossRef] [PubMed]

23. Tang, Z.; Wang, Y.; Podsiadlo, P.; Kotov, N.A. Biomedical Applications of Layer-by-Layer Assembly: From Biomimetics to Tissue Engineering. *Adv. Mater.* **2006**, *18*, 3203–3224. [CrossRef]

24. Franchi, M.; Ottani, V.; Stagni, R.; Ruggeri, A. Tendon and ligament fibrillar crimps give rise to left-handed helices of collagen fibrils in both planar and helical crimps. *J. Anat.* **2010**, *216*, 301–309. [CrossRef]

25. Shim, V.; Fernandez, J.; Besier, T.; Hunter, P. Investigation of the role of crimps in collagen fibers in tendon with a microstructually based finite element model. In Proceedings of the 2012 Annual International Conference of the IEEE Engineering in Medicine and Biology Society, San Diego, CA, USA, 28 August–1 September 2012; pp. 4871–4874. [CrossRef]

26. Benjamin, M.; Kaiser, E.; Milz, S. Structure-function relationships in tendons: A review. *J. Anat.* **2008**, *212*, 211–228. [CrossRef]

27. Jia, Z.; Yu, Y.; Wang, L. Learning from nature: Use material architecture to break the performance tradeoffs. *Mater. Des.* **2019**, *168*, 107650. [CrossRef]

28. Naleway, S.E.; Porter, M.M.; McKittrick, J.; Meyers, M.A. Structural Design Elements in Biological Materials: Application to Bioinspiration. *Adv. Mater.* **2015**, *27*, 5455–5476. [CrossRef]

29. Bouville, F.; Maire, E.; Meille, S.; Van De Moortèle, B.; Stevenson, A.; Deville, S. Strong, tough and stiff bioinspired ceramics from brittle constituents. *Nat. Mater.* **2014**, *13*, 508–514. [CrossRef]

30. Wegst, U.G.; Bai, H.; Saiz, E.; Tomsia, A.P.; Ritchie, R.O. Bioinspired structural materials. *Nat. Mater.* **2014**, *14*, 23–36. [CrossRef]

31. Eilber, K.S.; Sukotjo, C.; Raz, S.; Nishimura, I. Alteration of collagen three-dimensional architecture in noncompliant human urinary bladder. *Retinal Degener. Dis.* **2003**, *539*, 791–801. [CrossRef]

32. Chang, S.L.; Howard, P.S.; Koo, H.P.; Macarak, E.J. Role of type III collagen in bladder filling. *Neurourol. Urodyn. Off. J. Int. Cont. Soc.* **1998**, *17*, 135–145. [CrossRef]

33. Aitken, K.; Bägli, D.J. The bladder extracellular matrix. Part I: Architecture, development and disease. *Nat. Rev. Urol.* **2009**, *6*, 596–611. [CrossRef] [PubMed]

34. Fratzl, P.; Misof, K.; Zizak, I.; Rapp, G.; Amenitsch, H.; Bernstorff, S. Fibrillar Structure and Mechanical Properties of Collagen. *J. Struct. Biol.* **1998**, *122*, 119–122. [CrossRef] [PubMed]

35. Landau, E.H.; Jayanthi, V.R.; Churchill, B.M.; Shapiro, E.; Gilmour, R.F.; Khoury, A.E.; Macarak, E.J.; McLorie, G.A.; Steckler, R.E.; Kogan, B.A. Loss of Elasticity in Dysfunctional Bladders: Urodynamic and Histochemical Correlation. *J. Urol.* **1994**, *152*, 702–705. [CrossRef]

36. Majumdar, S.; Wang, X.; Sommerfeld, S.D.; Chae, J.J.; Athanasopoulou, E.; Shores, L.; Duan, X.; Amzel, L.M.; Stellacci, F.; Schein, O.; et al. Cyclodextrin Modulated Type I Collagen Self-Assembly to Engineer Biomimetic Cornea Implants. *Adv. Funct. Mater.* **2018**, *28*, 1804076. [CrossRef]

37. Agrawal, A.; Rahbar, N.; Calvert, P. Strong fiber-reinforced hydrogel. *Acta Biomater.* **2013**, *9*, 5313–5318. [CrossRef]

38. Eslami, M.; Vrana, N.E.; Zorlutuna, P.; Sant, S.; Jung, S.; Masoumi, N.; Khavari-Nejad, R.A.; Javadi, G.; Khademhosseini, A. Fiber-reinforced hydrogel scaffolds for heart valve tissue engineering. *J. Biomater. Appl.* **2014**, *29*, 399–410. [CrossRef]

39. Iviglia, G.; Cassinelli, C.; Torre, E.; Baino, F.; Morra, M.; Vitale-Brovarone, C. Novel bioceramic-reinforced hydrogel for alveolar bone regeneration. *Acta Biomater.* **2016**, *44*, 97–109. [CrossRef]

40. Nie, X.; Chuah, Y.J.; Zhu, W.; He, P.; Peck, Y.; Wang, D.-A. Decellularized tissue engineered hyaline cartilage graft for articular cartilage repair. *Biomaterials* **2020**, *235*, 119821. [CrossRef]

41. Bejleri, D.; Davis, M.E. Decellularized Extracellular Matrix Materials for Cardiac Repair and Regeneration. *Adv. Health Mater.* **2019**, *8*, e1801217. [CrossRef]

42. Dahms, S.E.; Piechota, H.J.; Dahiya, R.; Lue, T.F.; Tanagho, E.A. Composition and biomechanical properties of the bladder acellular matrix graft: Comparative analysis in rat, pig and human. *BJU Int.* **1998**, *82*, 411–419. [CrossRef] [PubMed]

43. Singh, A.; Lee, D.; Jeong, H.; Yu, C.; Li, J.; Fang, C.H.; Sabnekar, P.; Liu, X.; Yoshida, T.; Sopko, N.A.; et al. Tissue-Engineered Neo-Urinary Conduit from Decellularized Trachea. *Tissue Eng. Part A* **2018**, *24*, 1456–1467. [CrossRef] [PubMed]

7

Devitalizing Effect of High Hydrostatic Pressure on Human Cells—Influence on Cell Death in Osteoblasts and Chondrocytes

Janine Waletzko [1,*], Michael Dau [1], Anika Seyfarth [2], Armin Springer [3,4], Marcus Frank [3,4], Rainer Bader [2] and Anika Jonitz-Heincke [2]

[1] Department of Oral, Maxillofacial and Plastic Surgery, University Medical Center Rostock, 18057 Rostock, Germany; michael.dau@med.uni-rostock.de

[2] Department of Orthopedics, Biomechanics and Implant Technology Research Laboratory, University Medical Center Rostock, 18057 Rostock, Germany; anika.seyfarth@med.uni-rostock.de (A.S.); rainer.bader@med.uni-rostock.de (R.B.); anika.jonitz-heincke@med.uni-rostock.de (A.J.-H.)

[3] Medical Biology and Electron Microscopy Center, University Medical Center Rostock, 18057 Rostock, Germany; armin.springer@med.uni-rostock.de (A.S.); marcus.frank@med.uni-rostock.de (M.F.)

[4] Department Life, Light & Matter, University of Rostock, 18059 Rostock, Germany

*   Correspondence: janine.waletzko@med.uni-rostock.de

Abstract: Chemical and physical processing of allografts is associated with a significant reduction in biomechanics. Therefore, treatment of tissue with high hydrostatic pressure (HHP) offers the possibility to devitalize tissue gently without changing biomechanical properties. To obtain an initial assessment of the effectiveness of HHP treatment, human osteoblasts and chondrocytes were treated with different HHPs (100–150 MPa, 250–300 MPa, 450–500 MPa). Devitalization efficiency was determined by analyzing the metabolic activity via WST-1(water-soluble tetrazolium salt) assay. The type of cell death was detected with an apoptosis/necrosis ELISA (enzyme-linked immune sorbent assay) and flow cytometry. Field emission scanning electron microscopy (FESEM) and transmission electron microscopy (TEM) were carried out to detect the degree of cell destruction. After HHP treatment, the metabolic activities of both cell types decreased, whereas HHP of 250 MPa and higher resulted in metabolic inactivation. Further, the highest HHP range induced mostly necrosis while the lower HHP ranges induced apoptosis and necrosis equally. FESEM and TEM analyses of treated osteoblasts revealed pressure-dependent cell damage. In the present study, it could be proven that a pressure range of 250–300 MPa can be used for cell devitalization. However, in order to treat bone and cartilage tissue gently with HHP, the results of our cell experiments must be verified for tissue samples in future studies.

Keywords: high hydrostatic pressure; devitalization; decellularization; allografts; regenerative medicine; bone and cartilage regeneration

---

1. Introduction

Musculoskeletal disorders are the world's leading cause of chronic pain and impaired physical function, which result in loss of life quality [1]. In particular, the treatment of bone-cartilage defects is challenging.

Reasons for a defective or unstable bone structure are on the one hand systemic bone loss due to osteoporosis and on the other hand local trauma or cancerous diseases [2,3]. Often, bone can regenerate itself, but in the case of larger bone defects, where whole segments are missing, this process fails [4]. The gold standard to heal bone defects is still the transplantation of autologous material, which has the advantage that rejections can be avoided [5]. However, the material is naturally limited, and the transplantation of autologous tissue always requires a second surgical treatment which leads to the problem of donor site morbidity [5]. In contrast to autografts, the availability of allografts is less limited. However, the remodeling of allografts into the recipient tissue is critical. Numerous allografts are degraded by recipient's tissue instead of promoting bone remodeling [6].

Cartilage defects can lead to osteoarthritis when left untreated [7]. One of the main problems of cartilage regeneration is the avascularity of cartilage tissue and low cell turnover, which limit the ability of cartilage healing and self-regeneration [8]. Additionally, for the treatment of cartilage defects, no optimal healing method has been found since regeneration approaches resulted mostly in fibrous tissue instead of hyaline extracellular matrix [9]. The transplantation of scaffolds loaded with mesenchymal stem cells seems to be the most promising approach so far. Nevertheless, an immunological reaction of the recipient in response to the transferred stem cells cannot be completely ruled out, and previously used scaffolds can hardly imitate the complex physiology of osteochondral tissue [8,10]

A promising alternative for patients with musculoskeletal disorders could be the transplantation of allografts which morphologically correspond to the recipient tissue. Additionally, allogenic tissues have only a low potential of immunological response compared to xenogeneic grafts. For the preparation of allografts, strong chemical or physical methods are currently used, which are associated with a significant reduction in biomechanics and a change in the biological behavior regarding remodeling [11]. High hydrostatic pressure (HHP) technology which is widely used in the food industry for the decontamination of food with simultaneous retaining properties such as taste and vitamins, could be a way to devitalize tissues while maintaining their biomechanical properties [12]. If HHP is suitable for providing replacement materials, the kind of cell devitalization should be taken into account during tissue processing. Based on the level of the applied pressure, cells react in either an apoptotic or a necrotic manner [12]. Pressures around 200 MPa seem to induce apoptosis while pressures higher than 300 MPa are associated with necrotic-like pathways [12]. Induction of necrosis provokes a release of proinflammatory molecules and danger signals which can further trigger the immunological response [13]. With a guided initiation of apoptosis using HHP the immunological potential of allografts might be reduced without the need to use other tissue-destructive methods.

The aim of this study was the identification of necessary HHP ranges to initiate apoptosis in human osteoblasts and chondrocytes and establish cell-specific treatment protocols. To assess the potential of HHP treatment for allograft processing, the effects of different HHP ranges on cell survival and cell death were analyzed. Hence, human osteoblasts and chondrocytes were treated with three different HHPs, ranging from 100–150 MPa, 250–300 MPa, and 450–500 MPa, to determine cell viability and HHP-dependent cell death pathways. Additionally, different methodological approaches were carried out to further characterize cell apoptosis and necrosis. To avoid a loss of treated cells due to the experimental setting, the HHP treatment was performed on freshly pelleted cells. Moreover, field emission scanning microscopy (FESEM) and transmission electron microscopy (TEM) were used to provide a detailed insight into cell structures after HHP treatment. Finally, cell specific characterization of HHP treatment served as an indication to transfer cell-specific protocols to tissues in a next step to generate treatment protocols for allogenic tissue such as en bloc bone grafts.

2. Results

2.1. Cellular Activity

In both cell types, no significant difference in the metabolic activity between the control group and the group treated with 100–150 MPa was detectable (Figure 1). In contrast, HHPs ranging between 250–300 MPa and higher led to a significant decrease in the cell activity in comparison to the control group and the group treated with the lowest HHP range (osteoblasts: $p = 0.0429$ (control vs. 250–300 MPa); $p = 0.0161$ (control vs. 450–500 MPa); $p = 0.0285$ (100–150 MPa vs. 250–300 MPa); $p = 0.0107$ (100–150 MPa vs. 450–500 MPa); chondrocytes: $p = 0.0155$ (control vs. 250–300 MPa);

$p = 0.0150$ (control vs. 450–500 MPa)). However, a significant difference in the metabolic activity of cells treated with 250–300 MPa and 450–500 MPa was not detectable.

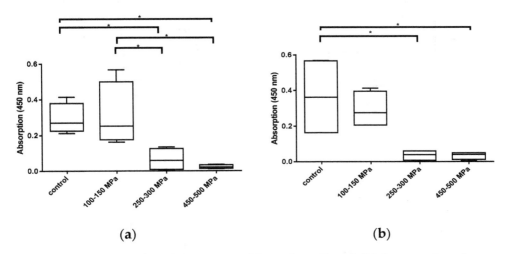

Figure 1. Metabolic activity of (**a**) human osteoblasts ($n = 4$) and (**b**) human chondrocytes ($n = 4$) following treatment with different HHPs ranging from 100–150 MPa, 250–300 MPa, and 450–500 MPa. After HHP exposure, cells were incubated at 37 °C and 5% CO_2 over a period of 24 h. Subsequently, metabolic activity was determined with water-soluble tetrazolium salt (WST-) 1 assay. Data are presented as boxplots. Significant differences between stimulation groups were determined via one-way ANOVA: * $p \leq 0.05$.

2.2. Induction of Cell Death Following Exposure to HHP

Cell death analysis after HHP treatment was carried out via Cell Death Detection ELISA. In human osteoblasts, HHP exposure at the lowest range (100–150 MPa) led to only a slight increase in necrosis and apoptosis without significant differences compared to untreated cells. In contrast, a significant increase in necrosis (Figure 2a; $p = 0.0206$) and apoptosis (Figure 2b; $p = 0.0022$) of human osteoblasts was shown after HHP treatment of 250–300 MPa. In addition, HHP treatment of 450–500 MPa also resulted in a significant increase in necrosis compared to the control group ($p = 0.0233$) while this HHP range apparently had no effect on osteoblasts regarding apoptosis (Figure 2a,b).

In Figure 2c,d the apoptosis and necrosis rates of human chondrocytes after HHP treatment are shown. A significant increase in necrosis took place in response to an HHP range of 250–300 MPa compared to untreated cells ($p = 0.0184$). In addition, the results of apoptosis (Figure 2d) revealed that the lowest HHP (100–150 MPa) induced mostly apoptosis while higher HHP ranges had no apoptotic effect. However, these data were not statistically significant.

2.3. Differentiation of Necrosis, Apoptosis, and Late Apoptosis Following HHP Treatment of Pelleted Cells

Since Cell Death ELISA data revealed that HHP treatment of 250–300 MPa had the strongest effect on the tested cell types, a more detailed cell death analysis took place via flow cytometry. In comparison to this pressure, pelleted cells were also treated with a pressure of 100–150 MPa again for analysis via flow cytometry. The used gating strategy is shown in Figure 3 on the basis of untreated human osteoblasts. After gating the cells from the cell debris with the side scatter height (SSC-H) and forward scatter height (FSC-H) channels, fluorescence (FL)4-H on the y-axis for apoptosis and FL2-H on the x-axis for necrosis were set. Then, four quadrants (Q)1 to Q4 were adjusted, which represent annexin single positive cells (Q1; apoptosis), annexin/propidium iodide (PI) double positive cells (Q2; late apoptotic and necrotic cells), PI single positive cells (Q3; necrosis), and annexin/PI double negative cells (Q4; vital cells), respectively.

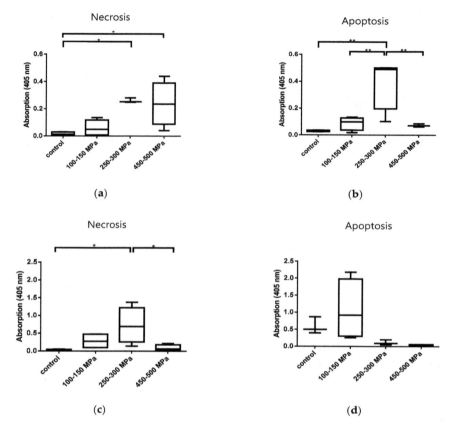

Figure 2. Analysis of cell death induced by HHP treatment of (**a**,**b**) human osteoblasts ($n = 4$), and (**c**,**d**) human chondrocytes ($n = 4$). Graphics (**a**,**c**) show necrosis and graphics (**b**,**d**) show apoptosis, both detected via Cell Death Detection ELISA. After HHP treatment, the supernatants were collected for necrosis detection. The residual cells were lysed for 30 min with 200 μL lysis buffer provided by the kit. After centrifugation at 118× g for 8 min, supernatants were collected for apoptosis detection. All data are presented as boxplots. Significant differences between groups were determined via one-way ANOVA. * $p \leq 0.05$; ** $p \leq 0.01$.

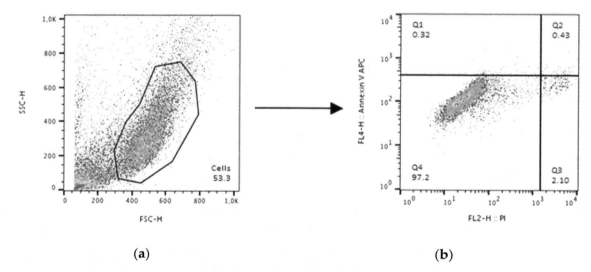

Figure 3. Gating strategy for cell death detection with annexin and propidium iodide (PI) exemplified using untreated human osteoblasts. Analysis took place with FloJo. (**a**) Gating of cells from the debris using the side scatter height (SSC-H) and forward scatter height (FSC-H) channel. This gate was used to detect the different stainings in (**b**) where quadrant (Q)1 represents annexin single positive cells, Q2 annexin/PI double positive cells, Q3 PI single positive cells and Q4 annexin/PI double negative cells.

The majority of untreated groups of both cell types could be detected as double negative (alive, around 80%). The remaining parts of the cells were distributed between apoptotic, necrotic and late apoptotic/necrotic detection for both osteoblasts and chondrocytes. A similar distribution could be observed for cells treated with 100–150 MPa for 10 min (Figure 4a,b). Human osteoblasts (Figure 4a) were significantly late apoptotic/necrotic following HHP treatment of 250–300 MPa in comparison to the control group ($p < 0.0001$), while the live, apoptotic and necrotic cells were less than 5%. In contrast to this, human chondrocytes were detected as necrotic and late apoptotic/necrotic in a wide range after treatment with 250–300 MPa (Figure 4b). However, a significant increase in late apoptotic/necrotic cells after treatment with 250–300 MPa in comparison to the control group could be detected ($p < 0.0001$), and an increase in necrotic cells after HHP treatment of 250–300 MPa in comparison to the control group ($p = 0.0041$) could be observed.

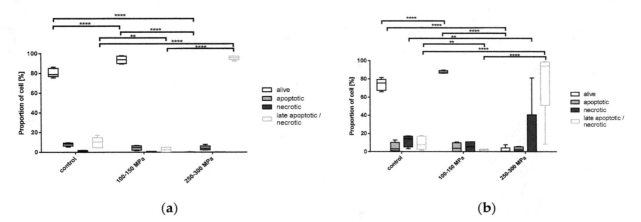

(a)   (b)

Figure 4. Analysis of cell death detection after HHP treatment (100–150 MPa and 250–300 MPa, 10 min) of (**a**) human osteoblast cell pellets ($n \geq 4$) and (**b**) human chondrocyte cell pellets ($n \geq 5$) with annexin and propidium iodide (PI). For each cell type, cell pellets were prepared for HHP treatment. Afterwards, treated cells were washed with autoMACS® Running Buffer, stained with annexin and PI, and analyzed with the FACS Calibur. Data are shown as box plots. Significance between groups was detected via two-way ANOVA. ** $p \leq 0{,}01$; **** $p < 0.0001$.

2.4. Evaluation of the Cellular Damage of Osteoblasts after HHP Treatment Based on Field Emission Electron Microscopy (FESEM) and Transmission Electron Microscopy (TEM)

Imaging techniques such as FESEM and TEM were used to visualize the HHP effects on cell structures. Therefore, osteoblasts were treated with HHP (100–150 MPa and 250–300 MPa for 10 min each) and subsequently prepared for FESEM and TEM analysis.

Untreated human osteoblasts could be detected via FESEM with an uneven cell surface with distinct elevations which appeared to be smooth (Figure 5a). In the sections, analyzed by TEM, cellular structures such as the nucleus, mitochondria, and the endoplasmic reticulum were observed (Figure 5b). Between the control group and the group treated with 100–150 MPa, only slight differences could be detected. Again, the cell surface appeared to be intact in FESEM, and TEM analyses revealed distinct structures as described before (Figure 5c,d). Only after treatment of human osteoblasts with 250–300 MPa, the cell surface appeared to be altered and became a sponge-like structure covered with small blebs (Figure 5e). While in the other two groups, the intracellular structures were clearly visible with TEM, these structures were destroyed following HHP treatment with 250–300 MPa (Figure 5f). Based on a comparison of the cell diameters of the differently treated cells, the untreated and 100–150 MPa treated groups had a diameter of approximately 20 µm while most of the cells treated with 250–300 MPa seemed to shrink approximately by one fifth.

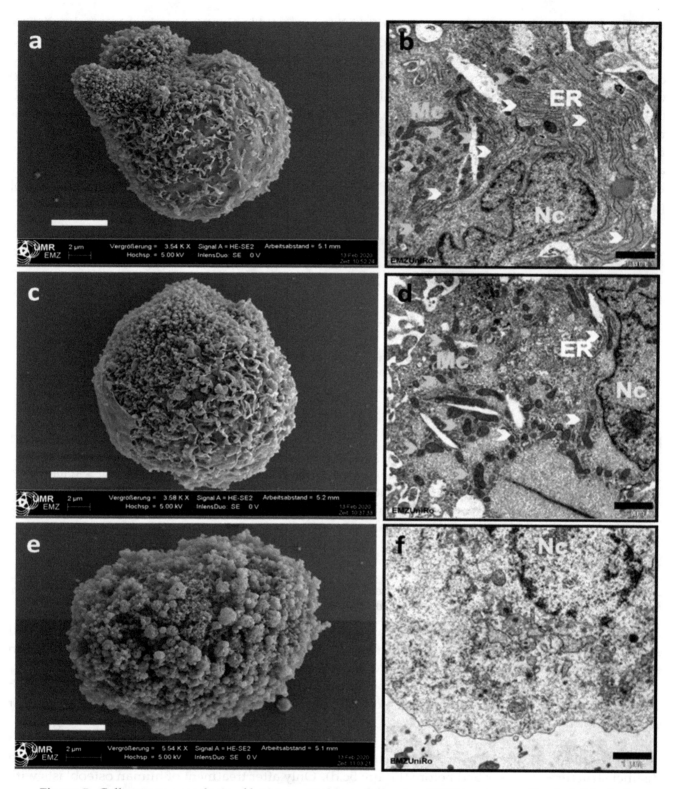

Figure 5. Cell structure analysis of human osteoblasts following HHP treatment. Treated cells (**c,d** 150–200 MPa; **e,f**: 250–300 MPa) and untreated cells (**a,b**) were fixed to carry out field emission electron (**a,c,e** scale bar: 5 μm) and transmission electron (**b,d,f**; scale bar: 1 μm) microscopy. Abbreviations: Nc = nucleus (light green), Mc = mitochondrion (orange), ER: endoplasmic reticulum (white).

3. Discussion

The need for tissue replacement materials, e.g., in the head and neck area, has increased in the last years, particularly with regard to the increasing amount of cancerous diseases [14]. The gold standard of using autologous material has the disadvantage of being limited, and patients have to undergo two operations. Allografts, which are also common, have to be treated with strong chemicals to reduce the immunogenic potential. At the same time, the matrix integrity is lost [15].

Synthetic materials, which are not limited in availability, often cannot withstand the biomechanical strain and are resorbed quickly, which makes them unsuitable for use in major bone defects [5]. The treatment of allogenic material with high hydrostatic pressure (HHP) could be a gentle and effective alternative to already existing methods for providing decellularized tissue replacement materials without negative effects on the tissue matrix [12]. The aim of this study was to evaluate an optimal HHP protocol for different cell types (osteoblasts and chondrocytes) regarding devitalization and induction of apoptosis or necrosis. Rivalain et al. already described the effect of different HHPs on mammalian cells and observed, that after treatment of around 200 MPa, cells underwent apoptosis whereas a treatment higher than 300 MPa resulted in necrosis [12].

We were able to characterize the influence of high hydrostatic pressure on cell survival and cell death in human osteoblasts and chondrocytes. We further aimed to distinguish differing HHP-effects when cells were exposed as cell pellets. The first observation was that the lowest HHP range of 100–150 MPa had no significant effect on the metabolic activity of all tested cell types. In contrast, treatment of cells with HHP ranging from either 250–300 MPa or 450–500 MPa had the same effect regarding the significant reduction of metabolic activity. This observation was a first hint that HHP treatment of 250 MPa and higher destroyed cells and resulted in devitalization. The previous work of Hiemer et al. [8] showed that an HHP of 480 MPa led to successful devitalization of hyaline cartilage cells. However, the limitation of the work was that this devitalizing effect with respect to apoptosis and necrosis was not investigated. As mentioned before, this point is of high relevance because cell death is decisive for the induction of immunological processes. The requirement for tissue replacement materials is low to negligible immunological potential for the acceptor. Therefore, the devitalization process should not have a necrotic effect because this would lead to inflammatory responses regarding the release of, e.g., damage-associated molecular patterns (DAMPS) which further stimulate the pro-inflammatory response [13]. To analyze the type of cell death induced by HHP, necrosis and apoptosis were first detected by ELISA. The collected data showed that the untreated control groups also underwent apoptosis and necrosis partly, which can be attributed to the fact that the controls were exposed to non-standard conditions (e.g., room temperature) during this time that the treated groups underwent the HHP treatment. Aside from this finding, HHD treatment of 100–150 MPa had no significant effect in comparison to the control group. A balanced apoptosis- and necrosis-inducing effect on osteoblasts was observed at 250–300 MPa, whereas an HHP treatment of 450–500 MPa induced primarily necrosis. In contrast, an HHP range of 100–150 MPa induced apoptosis in chondrocytes derived from hyaline cartilage whereby necrosis was already induced by a range from 250–300 MPa. The most effective HHP treatment should first and foremost lead to metabolically inactive cells that have, ideally, not yet undergone necrosis. Thus, taking the results of both, the WST-1 assay and ELISA-derived apoptosis and necrosis analysis into consideration, these findings suggest that an HHP of 200–300 MPa may be the most effective for gentle and noninflammatory cell devitalization. Analysis via flow cytometry, providing a more precise separation of vital, necrotic, apoptotic, and late apoptotic/necrotic cells, revealed that most osteoblasts were vital after HHP treatment of 100–150 MPa, while exposure to 250–300 MPa led to necrosis and apoptosis only in a small number of cells. Moreover, most of these cells were detected as being in a late apoptotic/necrotic condition. Identical observations were made for chondrocytes. We assumed that cells treated with the higher HHP range were apoptotic at an earlier time point and underwent necrosis. This result can be explained, among other things, by the fact that there was a time delay between the HHP treatment and the subsequent FACS analysis due to the availability of equipment at different locations. Therefore, it cannot be excluded that necrosis was induced in the cells after a short time. Additionally, the absence of any phagocytic cells in the experimental setting led to the absence of cell debris removal. Consequently, the membrane of the treated cells became permeable over time, and intracellular contents could be released, which might further stimulate the immunological response [13]. Additionally, permeable and destroyed cell membranes as well as cellular components were found in TEM analysis of osteoblasts treated with 250–300 MPa. In summary, the results support the idea that in the early phase after HHP, apoptosis was induced following a pressure of 250–300 MPa, but the initiation of intracellular material packaging was incomplete or aborted. As confirmed by FACS analyses, the cells were in a late apoptotic/necrotic state, which led to an uncontrolled destruction of cell components in the intracellular cavity, which was additionally supported by TEM. Interestingly, these results were not observed for the lower HHP range of 150–200 MPa. Therefore, we assumed that either a reduction of the HHP level or a shortening of the HHP exposure time could initiate apoptotic devitalization. This aspect has to be confirmed in future studies.

The aim of this study was to examine the cell-typical reactions to various HHP treatments to assess the tissue-specific reactions to HHP in the next step. Furthermore, our results on devitalization efficiency and cell death detection will help to better evaluate the results of a previous animal study on devitalized autografts [10]. Applied pressure of 480 MPa to osteochondral cylinders led to an extensive tissue remodeling process following tissue re-transplantation. However, excellent integration and revitalization were detected [10]. Possible reasons for these findings might be the autograft transplantation in the experimental setting and the animal-specific immune system, which plays an important role in tissue integration. For better understanding, it is even more important to perform targeted analysis of HHP-mediated cell death which is of high relevance for the subsequent translation of specific tissue types into the clinical use of allografts. It is also important to study additional tissue-specific cells to prove whether the respective high pressure successfully devitalizes all resident cells in the tissue through apoptosis. Taking late apoptosis/necrosis into account, rapid purification of the HHP-processed tissue is still mandatory so that cell residues can be removed gently and in sufficient time. These aspects will be analyzed in more detail in further studies.

4. Materials and Methods

4.1. Cell Culture

Human osteoblasts from femoral heads ($n = 5$) and human chondrocytes from hyaline knee cartilage ($n = 5$) were isolated from patients undergoing total joint replacements. Prior to isolation of human osteoblasts and chondrocytes, informed consent and ethical approval (A2010-10A (osteoblasts) and A2009-17 (chondrocytes) were obtained from the ethics committee of the University of Rostock, Germany) including IRB information. Harvesting and isolation of cells took place after patients signed agreement forms following the previously described protocols [16–18]. We have got the written informed consent from all patients. The isolated osteoblasts were cultivated in a 25 cm^2 culture flask with a volume of 8 mL osteogenic medium (Dulbecco's modified Eagle's medium (DMEM), PAN-Biotech, Aidenbach, Germany) supplemented with 10% fetal calf serum (FCS, PAN-Biotech), 1% amphotericin B, 1% penicillin/streptomycin, 1% HEPES buffer, 50 µg/mL ascorbic acid, 10 mM β-glycerophosphate, and 100 nM dexamethasone (all: Sigma–Aldrich, Munich, Germany). Human chondrocytes were cultivated in 25 cm^2 culture flasks containing 8 mL DMEM medium (Gibco® Invitrogen, Paisly, UK) containing 10% FCS, 1% amphotericin B, and 1% penicillin/streptomycin. In addition, chondrocytes were supplemented with 50 µg/mL ascorbic acid. Incubation took place in a humidified atmosphere of 5% CO_2 and at 37 °C. The respective medium was changed every second day so non-adherent cells could be aspirated. At a confluence of 100%, which was checked by light microscopy, the cells were transferred in a 75 cm^2 culture flask, and further culture took place under the same conditions as described above. Cells in the third passage were used for the experiments.

4.2. High Hydrostatic Pressure Treatment

To treat cells with high hydrostatic pressure (HHP), the medium was discarded and adherent cells were detached with 2 mL trypsin per 75 cm^2 flask. After an incubation time of 3 min at 5%

CO_2 and 37 °C, the reaction was stopped with 6 mL DMEM. The cell suspension was collected and centrifuged at $118 \times g$ for 8 min. The medium was discarded, and the obtained cell sediment was resuspended in 1 mL DMEM. After cell counting, 50,000 cells were transferred as duplicates into 2 mL CryoTubes (ThermoFisher Scientific, Waltham, MA, USA), filled up with the corresponding medium, and centrifuged at $118 \times g$ for 8 min to generate freshly pelleted cells. These cell pellets were treated with different HHPs ranging from 100–150 MPa, 250–300 MPa, and 450–500 MPa for 10 min using an HHP device (HDR-100, RECORD GmbH, Koenigsee, Germany). After HHP treatment, the tubes were centrifuged again at $118 \times g$ for 8 min; 200 μL of the supernatants were discarded, and the cells were incubated for 24 h under standard cell conditions. HHP-untreated cells served as controls; these cells were exposed to the same test conditions.

4.3. Metabolic Activity Analysis

After an incubation time of 24 h, HHP-treated and untreated cells were centrifuged at $118 \times g$ for 8 min. The supernatants were discarded, and the metabolic activity cells was analyzed by incubating the cells for 1.5 h under standard culture conditions with 250 μL of a 1:10 dilution of WST-1 reagent (Takara, Bio, Saint-Germain-en-Laye, France) and the respective cell culture medium. The amount of formed formazan dye, which correlates directly with the number of metabolic active cells, was determined by an extinction measurement using a Tecan-Reader Infinite® 200 Pro (Tecan, Maennedorf, Switzerland); (absorption: 450 nm, reference: 600 nm).

4.4. Analysis of Cell Death

For cell death detection, supernatants of exposed and unexposed cells were collected and stored at −20 °C for further necrosis detection by Cell Death Detection ELISA (Roche, Penzberg, Germany). For apoptosis detection, the residual cells were incubated with 200 μL lysis buffer provided by the kit for 30 min at room temperature. The cell lysates were centrifuged at $200 \times g$ for 10 min, and the supernatants without the cell debris were stored at −20 °C for further use.

To analyze the kind of cell death, all supernatants were thawed and transferred as duplicates into a streptavidin-coated microtiter plate with the related positive, negative, and background controls. In each well, 80 μL immunoreagent, containing incubation buffer, anti-histone-biotin, and anti-DNA-POD, all made available by the kit, were added to each well. The microtiter plate was covered with the cover foil and incubated for 2 h at room temperature on the shaker at 300 rpm. The supernatants were then discarded, and the plate was washed three times with 250 μL incubation buffer. Afterwards, the buffer was aspirated thoroughly, and 100 μL ABTS Solution was added. After an incubation period of 10 min at room temperature on the shaker at 250 rpm, a color reaction was visible and it was stopped by adding 100 μL ABTS Stop Solution. The quantification took place in a Tecan-Reader Infinite® 200 Pro at a wavelength of 405 nm and a reference wave length of 490 nm.

For precise analysis of cell death after HHP treatment, human osteoblasts and chondrocytes were analyzed by flow cytometry with the APC Annexin-V Apoptotis Detection Kit with propidium iodide (PI) (Biolegend, San Diego, CA, USA). Cells were cultured as described above (all cell types $n = 5$) and a cell number of 2×10^5 cells were treated with an HHP of 100–150 MPa and 250–300 MPa for 10 min, whereas cell pellets were treated and analyzed. Afterwards, cells were centrifuged at $118 \times g$ for 8 min. The supernatants were discarded and the cells were washed with 1 mL autoMACS® Running Buffer (Myltenyi Biotec, Bergisch Gladbach, Germany) at 1400 rpm for five minutes. A mastermix of 5 μL Annexin V FITC and 10 μL PI per sample was added, and the cells were incubated for 15 min at room temperature in darkness. Afterwards, 100 μL Annexin Binding buffer was added, and analysis took place with the FACS Calibur (BD, Franklin Lakes, NJ, USA). For the evaluation of the results, FloJo (BD, Franklin Lakes, NJ, USA) was used, and the percentage of positive cells was assessed.

4.5. Analysis of Cellular Damage by Field Emission Electron Microscopy (FESEM) and Transmission Electron Microscopy (TEM)

Human osteoblasts and human chondrocytes (100,000 cells each) were transferred into 2 mL CryoTubes (ThermoFisher), filled up with the corresponding medium, and centrifuged at 118× g for 8 min. These cell pellets were treated with different pressure ranges (w/o HHP, 100–150 MPa and 250–300 MPa). After treatment, tubes were again centrifuged at 118× g for eight minutes and the supernatant was discarded. The cells where then fixed with fixation buffer (1% paraformaldehyde, 2.5% glutaraldehyde, 0.1 M sodium phosphate buffer, pH 7.3) and stored at 4 °C. Afterwards, cells were washed with sodium phosphate buffer, adhered to glass coverslips previously coated with poly-L-lysine (Co. Sigma) for one hour and then dehydrated with an acetone series, followed by critical point drying using CO_2 (Emitech K850/Quorum Technologies Ltd., East Sussex, UK). The glass coverslips were mounted on aluminum sample holders and coated with gold under vacuum (SCD 004, Baltec, Balzers, Liechtenstein). With a field emission scanning electron microscope (MERLIN VP Compact, Carl Zeiss, Oberkochen Germany), images were taken from the selected regions (applied detector: HE-SE2; accelerating voltage: 5.0 kV; working distance: 5.2 mm).

For TEM preparation, fixed specimens were washed in 0.1 M sodium phosphate buffer (pH 7.3), and cell pellets were enclosed in 3% low melting agarose (Fluka, Munich, Germany) in water at 40 °C by centrifugation in Eppendorf tubes. After staining with 1% osmiumtetroxide (Roth GmbH, Karlsruhe, Germany) for 2 h, the specimens were dehydrated through an ascending series of acetone prior to embedding in Epon resin (Serva, Heidelberg, Germany). Resin infiltration started with a 1:1 mixture of acetone and resin overnight, followed by pure resin for 4 h. After transfer to rubber casting moulds, specimens were cured in an oven at 60 °C for at least 48 h. Resin blocks were trimmed using a Leica EM Trim 2 (Leica Microsystems, Wetzlar, Germany). Ultrathin sections (approx. 70–90 nm) were cut with a Leica UC7 ultramicrotome using a diamond knife (Diatome, Nidau, Switzerland). Ultrathin sections were mounted on formvar-coated copper grids and were contrasted with uranyl acetate and lead citrate. Ultrastructure was inspected with a Zeiss EM902 electron microscope operated at 80 kV (Carl Zeiss, Oberkochen, Germany). Digital images were acquired with a side-mounted 1x2k FT-CCD Camera (Proscan, Scheuring, Germany) using iTem Camera control imaging software (Olympus Soft Imaging Solutions, Muenster, Germany).

4.6. Statistics

All results are expressed as medians with interquartile ranges (25–75%) and whiskers. As this is the first study on this topic, no a priori power calculation could be performed. Testing for statistical significance was done by one- or two-way ANOVA. p-values ≤ 0.05 were seen as significant. All statistical analyses were performed with GraphPad Prism Version 7 (GraphPad Software, San Diego, CA, USA).

Author Contributions: Conceptualization, J.W., M.D. and A.J.-H.; Data curation, J.W., A.S. (Armin Springer) and M.F.; Funding acquisition, M.D., R.B. and A.J.-H.; Investigation, J.W., A.S. (Anika Seyfarth) and A.S. (Armin Springer); Methodology, J.W., A.S. (Anika Seyfarth) and M.F.; Project administration, R.B. and A.J.-H.; Resources, M.D., M.F. and R.B.; Supervision, M.D., R.B. and A.J.-H.; Writing—original draft, J.W., M.D. and A.J.-H.; Writing—review & editing, A.S. (Anika Seyfarth), A.S. (Armin Springer), M.F. and R.B., J.W. takes responsibility for the integrity of the work as a whole, from inception to finished article. All authors have read and agreed to the published version of the manuscript.

Acknowledgments: We thank Brigitte Mueller-Hilke, Michael Mueller, and Wendy Bergmann (Core Facility for Cell Sorting and Cell Analysis, Center for Medical Research, University Medical Center Rostock, Germany) for instruction and for providing the devices. We also thank Karoline Schulz and Ute Schulz (Medical Biology and Electron Microscopy Center, University Medical Center Rostock) for excellent technical support.

Abbreviations

HHP	High Hydrostatic Pressure
MPa	Mega pascal
FESEM	Field Emission Scanning Electron Microscopy
TEM	Transmission Electron Microscopy
ELISA	Enzyme Linked Immunosorbent Assay
SSC-H	Side Scatter Height
FSC-H	Forward Scatter Height
FL	Fluorescence
PI	Propidium Iodide
APC	Allophycocyanine

References

1. RKI. Muskuloskelletale Erkrankungen. 2017. Available online: https://www.rki.de/DE/Content/ Gesundheitsmonitoring/Themen/Chronische_Erkrankungen/Muskel_Skelett_System/Muskel_Skelett_ System_node.html (accessed on 25 March 2020).

2. Shrivats, A.R.; Alvarez, P.; Schutte, L.; Hollinger, J.O. *Chapter 24: Bone Regeneration*; Elsevier Inc.: Amsterdam, The Netherlands, 2014.

3. Ansari, M. Bone tissue regeneration: Biology, strategies and interface studies. *Prog. Biomater.* **2019**, *8*, 223–237. [CrossRef] [PubMed]

4. Checa, S. Multiscale agent-based computer models in skeletal tissue regeneration. *Numer. Methods Adv. Simul. Biomech. Biol. Process.* **2018**, 239–244.

5. Fernandez de Grado, G.; Keller, L.; Idoux-Gillet, Y.; Wagner, Q.; Musset, A.M.; Benkirane-Jessel, N.; Bornert, F.; Offner, D. Bone substitutes: A review of their characteristics, clinical use, and perspectives for large bone defects management. *J. Tissue Eng.* **2018**, *9*, 2041731418776819. [CrossRef] [PubMed]

6. Mistry, A.S.; Mikos, A.G. Tissue engineering strategies for bone regeneration. *Adv. Biochem. Eng. Biotechnol.* **2005**, *94*, 1–22. [PubMed]

7. Merkely, G.; Ackermann, J.; Lattermann, C. Articular Cartilage Defects: Incidence, Diagnosis and Natural History. *Oper. Tech. Sports Med.* **2018**, *26*, 156–161. [CrossRef]

8. Hiemer, B.; Genz, B.; Jonitz-Heincke, A.; Pasold, J.; Wree, A.; Dommerich, S.; Bader, R. Devitalisation of human cartilage by high hydrostatic pressure treatment: Subsequent cultivation of chondrocytes and mesenchymal stem cells on the devitalised tissue. *Sci. Rep.* **2016**, *6*, 1–12. [CrossRef] [PubMed]

9. Krueger, S.; Achilles, S.; Zimmermann, J.; Tischer, T.; Bader, R.; Jonitz-Heincke, A. Re-Differentiation Capacity of Human Chondrocytes in Vitro Following Electrical Stimulation with Capacitively Coupled Fields. *J. Clin. Med.* **2019**, *8*, 1771. [CrossRef] [PubMed]

10. Hiemer, B.; Genz, B.; Ostwald, J.; Jonitz-Heincke, A.; Wree, A.; Lindner, T.; Tischer, T.; Dommerich, S.; Bader, R. Repair of cartilage defects with devitalized osteochondral tissue: A pilot animal study. *J. Biomed. Mater. Res. Part B Appl. Biomater.* **2019**, *107*, 2354–2364. [CrossRef] [PubMed]

11. Shang, X.; Wang, H.; Li, J.; Li, O. Progress of sterilization and preservation methods for allografts in anterior cruciate ligament reconstruction. *Zhongguo Xiu Fu Chong Jian Wai Ke Za Zhi* **2019**, *33*, 1102–1107. [PubMed]

12. Rivalain, N.; Roquain, J.; Demazeau, G. Development of high hydrostatic pressure in biosciences: Pressure effect on biological structures and potential applications in Biotechnologies. *Biotechnol. Adv.* **2010**, *28*, 659–672. [CrossRef] [PubMed]

13. Kono, H.; Rock, K.L. How dying cells alert the immune system to danger. *Nat. Rev. Immunol.* **2008**, *8*, 279–289. [CrossRef] [PubMed]

14. Cancer Research UK. Head Neck Cancer Stat. Available online: https://www.cancerresearchuk.org/health-professional/cancer-statistics/statistics-by-cancer-type/head-and-neck-cancers#heading-Four (accessed on 30 March 2020).

15. Sohn, H.S.; Oh, J.K. Review of bone graft and bone substitutes with an emphasis on fracture surgeries. *Biomater. Res.* **2019**, *23*, 4–10. [CrossRef] [PubMed]

16. Lochner, K.; Fritsche, A.; Jonitz, A.; Hansmann, D.; Mueller, P.; Mueller-Hilke, B.; Bader, R. The potential role of human osteoblasts for periprosthetic osteolysis following exposure to wear particles. *Int. J. Mol. Med.* **2011**, *28*, 1055–1063. [CrossRef] [PubMed]

17. Jonitz, A.; Lochner, K.; Peters, K.; Salamon, A.; Pasold, J.; Mueller-Hilke, B.; Hansmann, D.; Bader, R. Differentiation capacity of human chondrocytes embedded in alginate matrix. *Connect. Tissue Res.* **2011**, *52*, 503–511. [CrossRef] [PubMed]

18. Jonitz-Heincke, A.; Tillmann, J.; Ostermann, M.; Springer, A.; Bader, R.; Høl, P.J.; Cimpan, M.R. Label-Free Monitoring of Uptake and Toxicity of Endoprosthetic Wear Particles in Human Cell Cultures. *Int. J. Mol. Sci.* **2018**, *19*, 3468. [CrossRef] [PubMed]

8

Multilineage Differentiation Potential of Human Dental Pulp Stem Cells—Impact of 3D and Hypoxic Environment on Osteogenesis In Vitro

Anna Labedz-Maslowska [1,†], Natalia Bryniarska [1,2,†], Andrzej Kubiak [1,3,†], Tomasz Kaczmarzyk [4], Malgorzata Sekula-Stryjewska [5], Sylwia Noga [1,5], Dariusz Boruczkowski [6], Zbigniew Madeja [1] and Ewa Zuba-Surma [1,*]

[1] Department of Cell Biology, Faculty of Biochemistry, Biophysics and Biotechnology, Jagiellonian University, 30-387 Krakow, Poland; anna.labedz-maslowska@uj.edu.pl (A.L.-M.); natalia.bryniarska@outlook.com (N.B.); andrzej.kubiak@ifj.edu.pl (A.K.); sylwia.noga@uj.edu.pl (S.N.); z.madeja@uj.edu.pl (Z.M.)

[2] Department of Experimental Neuroendocrinology, Maj Institute of Pharmacology, Polish Academy of Sciences, 31-343 Krakow, Poland

[3] Institute of Nuclear Physics, Polish Academy of Sciences, 31-342 Krakow, Poland

[4] Department of Oral Surgery, Faculty of Medicine, Jagiellonian University Medical College, 31-155 Krakow, Poland; tomasz.kaczmarzyk@uj.edu.pl

[5] Laboratory of Stem Cell Biotechnology, Malopolska Centre of Biotechnology, Jagiellonian University, 30-387 Krakow, Poland; malgorzata.sekula@uj.edu.pl

[6] Polish Stem Cell Bank, 00-867 Warsaw, Poland; dariusz.boruczkowski@pbkm.pl

* Correspondence: ewa.zuba-surma@uj.edu.pl

† These authors contributed equally to this work.

Abstract: Human dental pulp harbours unique stem cell population exhibiting mesenchymal stem/stromal cell (MSC) characteristics. This study aimed to analyse the differentiation potential and other essential functional and morphological features of dental pulp stem cells (DPSCs) in comparison with Wharton's jelly-derived MSCs from the umbilical cord (UC-MSCs), and to evaluate the osteogenic differentiation of DPSCs in 3D culture with a hypoxic microenvironment resembling the stem cell niche. Human DPSCs as well as UC-MSCs were isolated from primary human tissues and were subjected to a series of experiments. We established a multiantigenic profile of DPSCs with $CD45^-/CD14^-/CD34^-/CD29^+/CD44^+/CD73^+/CD90^+/CD105^+/Stro-1^+/HLA-DR^-$ (using flow cytometry) and confirmed their tri-lineage osteogenic, chondrogenic, and adipogenic differentiation potential (using qRT-PCR and histochemical staining) in comparison with the UC-MSCs. The results also demonstrated the potency of DPSCs to differentiate into osteoblasts in vitro. Moreover, we showed that the DPSCs exhibit limited cardiomyogenic and endothelial differentiation potential. Decreased proliferation and metabolic activity as well as increased osteogenic differentiation of DPSCs in vitro, attributed to 3D cell encapsulation and low oxygen concentration, were also observed. DPSCs exhibiting elevated osteogenic potential may serve as potential candidates for a cell-based product for advanced therapy, particularly for bone repair. Novel tissue engineering approaches combining DPSCs, 3D biomaterial scaffolds, and other stimulating chemical factors may represent innovative strategies for pro-regenerative therapies.

Keywords: stem cells; dental pulp stem cells; osteogenesis; biomaterials; tissue engineering; regenerative medicine

1. Introduction

The mesenchymal stem/stromal cells (MSCs) represent an adult stem cell (SCs) population that exhibits tri-lineage differentiation potential [1]. The MSCs can be isolated from various tissues, including bone marrow (BM), umbilical cord blood and Wharton's jelly of the umbilical cord (UC), adipose tissue, peripheral blood, synovium or dental pulp [2]. For distinguishing better between MSCs of various origins, the International Society for Cellular Therapy (ISCT) proposed minimal criteria for defining MSCs in 2006 that included the following: (i) an ability to adhere to plastic surfaces under standard culture conditions in vitro; (ii) a specific antigenic profile including at least 95% positivity for CD105, CD73 and CD90 antigen expression with parallel negativity (less than 2%) for CD45, CD34, CD14 or CD11b, CD79α or CD19 and HLA-DR), and (iii) a capacity to differentiate into the cells of mesodermal lineages, such as osteoblasts, chondroblasts and adipocytes in vitro [3]. Moreover, MSCs derived from various tissues may cross the boundaries of the germ layers and differentiate into cardiomyocyte- or endothelial-like cells (with parallel ability to form capillary-like structures on Matrigel) upon exposure to lineage-specific growth factor cocktails [4–7]. Recent evidence indicates that murine BM-MSCs may differentiate into hepatocytes, lung epithelial cells, myofibroblasts, and renal tubular cells in vivo [8]. Thus, based on their wide differentiation potential and their pro-regenerative potential in injured tissues, MSCs are promising "candidates" as medicinal products employed in in vivo applications for advanced therapy in humans. MSCs have been widely evaluated in preclinical studies as well as with a multitude of clinical trials focused on several tissue injuries, including bone defects [9] and other defects such as cartilage injury [10]. Of note, recent evidence demonstrates the biological variability (including differentiation capacity) within the MSCs of various origins [11]. Thus, the selection of the appropriate source of MSCs might be critical for the effective regeneration of injured tissues following their application in vivo.

BM-derived MSCs represent the gold standard for MSCs used in research and clinical applications (predominantly in autologous transplantations) in recent years [12]. However, the application of autologous MSCs may have some limitations. BM-derived MSCs isolated from elderly patients exhibit decreased biological activity, including osteogenic potential [13], which may result in limited outcomes of the treatment with these cells. Moreover, MSCs isolated from the tissues of patients suffering from chronic conditions like diabetes or systemic lupus erythematosus may also exhibit altered biological properties impairing their pro-reparative functions in autologous applications [14,15]. Nowadays, growing evidence indicates that umbilical cord tissue represents an attractive alternative source of MSCs for allogenic applications, which eliminates the limitations of autologous MSCs as well as the need for the invasive procedure of tissue harvesting that is required for BM [16,17]. Moreover, UC tissue possesses higher accessibility and lower ethical concerns in comparison with BM. Importantly, UC-MSCs have been demonstrated to exhibit similar transcriptomic profiles with respect to the expression of stemness- and bone development-related genes to BM-MSCs (harvested from healthy donors), demonstrating the great promise for umbilical cord tissue as an alternative source of MSCs [17].

In 2000, Gronthos et al. described a unique population of SCs residing in human dental pulp referred to as dental pulp stem cells (DPSCs). They exhibit mesenchymal characteristics, including adhesion to plastic surfaces, fibroblast-like morphology, lack of expression of CD14, CD34, and CD45, as well as the potential to differentiate into osteoblasts [18]. Moreover, DPSCs show antigenic profile similar to BM-MSCs, along with greater proliferation capacity when compared to that of BM-MSCs [18]. Although DPSCs predominantly originate from the neural crest, which is an ectodermal structure, they are considered as MSCs or MSC-like cells [19–21]. Thus, the unique developmental origin of DPSCs may indicate their ability to produce neural cells and their impact on neural tissue regeneration, conferring certain advantages in comparison with other sources of MSCs [22–26]. DPSCs exhibit capability not only to differentiate into neural-like cells [25] but also to secret factors enhancing endogenous repair mechanisms [27]. The unique features of DPSCs resulted in the first clinical application for the treatment of neurodegenerative diseases [28]. Moreover, DPSCs, when cultured under appropriate conditions, are known to exhibit osteogenic potential [18,29]. While long bones develop from mesoderm during the process of endochondral ossification, flat bones of the skull, similar to the mandible, are formed during the intramembranous ossification from ectomesenchymal cells derived from neural crest SCs [30,31]. Once DPSCs were identified as MSCs derived from the neural crest, they were recognized as a potential promising candidate for the repair of defects of the jaws [20]. Interestingly, DPSCs may also be used in the innovative strategies for dentin regeneration [32].

Following numerous clinical trials employing MSCs of various origins, some concerns about their efficacy and safety have been raised [33]. In many cases, results of clinical applications were less favourable than expected, with the observation of certain side effects upon the administration of MSCs into a suboptimal niche [33]. One of the reasons for the limited pro-regenerative capacity of MSCs may be the lack of well-established protocols for MSC isolation, expansion, and preparation of a cell therapy product for in vivo applications [33] as well as the use of an inappropriate cell source or differentiation strategy [32]. Importantly, the standard culture conditions with plastic or glass surfaces (representing two-dimensional culture, 2D) do not fully mimic the three-dimensional (3D) microarchitecture of SC niches [34]. In this context, the use of novel 3D scaffolds (created by f.e. hydrogels) for cells ex vivo propagation, differentiation, or as a carrier for the administration of cells, provides a new exciting approach to enhance the biological potential of MSCs for the repair of the injured tissue. MSCs encapsulated in alginate beads exhibit greater osteogenic differentiation potential than MSCs in traditional cell culture, which was confirmed by the higher expression of mRNAs corresponding to osteoblast-specific genes (e.g., *Runx2, Col1A1*, sclerostin or dental matrix protein-1) [35]. Similar observations have been described for DPSCs differentiated in scaffolds, such as Matrigel or collagen-based sponges, in vitro [36].

Oxygen (O_2) concentration in the culture environment represents another important factor that should be taken into consideration during the differentiation of MSCs both in vitro and in vivo. It has been established that the normoxic environment does not reflect the oxygen concentration in SC niches [37]. Importantly, in numerous reports, MSCs were shown to proliferate faster when cultured in hypoxic conditions with 1% [38], 2% [39], or 3% O_2 [40], compared to that in the normoxic atmosphere. In contrast, a decrease in the proliferation of rat BM-MSCs in the presence of 1% O_2 was reported owing to the upregulation of p27 protein correlating with the downregulation of cyclin D expression [41]. Hsu et al. demonstrated that the ex vivo culture of human MSCs in the hypoxic (1% O_2) environment decreases their osteogenic differentiation. The hypoxic environment at the site of cell transplantation in the injured tissues may be a critical factor impairing the efficacy of MSCs to differentiate following their transplantation [42], which, along with the spatial niche organization, provides a unique environment for functional behaviour of the cells in vivo.

Thus, this study aimed to examine the differentiation potential of ectoderm-derived human DPSCs in vitro, in comparison with other human MSC fractions, such as the Wharton's jelly-derived MSCs from the umbilical cord (UC-MSCs). We also evaluated osteogenic differentiation of human DPSCs in 3D cultures and hypoxic microenvironment in vitro as well as examined their selected functional properties, including proliferation and metabolic activity.

2. Results

2.1. Human Dental Pulp Harbours a Population of Adherent Cells with MSC Characteristics

Accumulating evidence indicates that the dental pulp contains a unique population of adherent cells with mesenchymal characteristics, including adhesion to plastic surfaces, fibroblast-like morphology, lack of expression of CD14, CD34 or CD45 and potential to differentiate into osteoblasts in vitro [18,43]. Thus, in the current study, we developed and optimized the protocol for the isolation of such cells with the corresponding characteristics of MSCs (Figure 1a). The dental pulp extracted from the permanent teeth of healthy human donors was subjected to enzymatic digestion to isolate a mixture of different populations of cells, which were further seeded on cell culture plates. After 10 days, we obtained a population of adherent cells that were proliferated further. Moreover, we also isolated

UC-MSCs from human UC Wharton's Jelly tissue (Figure 1b) by employing a similar approach, which were used as a "classic" control MSCs for comparison with the dental pulp- derived cells in vitro.

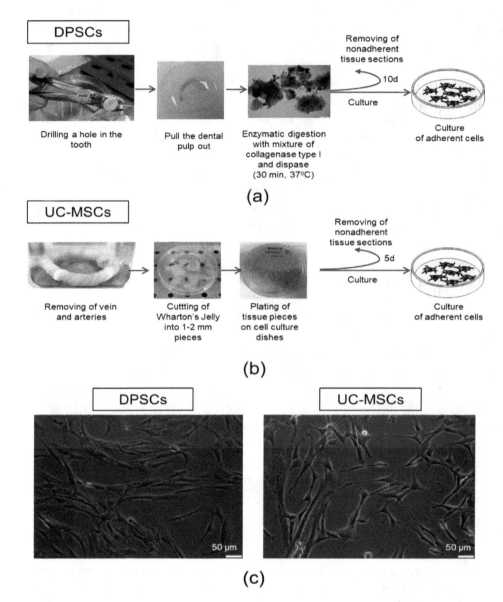

Figure 1. Isolation procedure and morphology of dental pulp stem cells (DPSCs) and umbilical cord Wharton's jelly-derived mesenchymal stem/stromal cells (UC-MSCs). (**a**) Isolation of DPSCs from pulp tissue. The upper part of the tooth was drilled and the dental pulp was extracted. The dental pulp was enzymatically digested by a mixture of collagenase I and dispase. The isolated cells and tissue sections were seeded onto cell culture plates in a complete cell culture medium. On day 10 post-seeding, non-adherent cells and tissue pieces were removed. (**b**) Isolation of UC-MSCs from Wharton's jelly. The umbilical cord was washed with PBS to remove residual cord blood, and arteries and vein were further dissected. Wharton's Jelly tissue was cut into 12 mm pieces and placed on the tissue culture dishes in a complete cell culture medium. On day five post-seeding, non-adherent cells and tissue pieces were removed. (**c**) Representative images of the morphology of DPSCs (left) and UC-MSCs (right). Scale bars: 50 μm.

We observed that both DPSCs, as well as UC-MSCs, demonstrated adhesion to plastic surfaces when maintained under standard culture conditions in vitro. Dental pulp-derived cells exhibit a morphology of spindled-shaped, fibroblast-like cells, similar to UC-MSCs (Figure 1c). By using a flow cytometry platform, we analysed the antigenic profile of isolated cells following the minimal criteria for defining multipotent mesenchymal stem/stromal cells published by the ISCT [3].

We demonstrated that the population of dental pulp-derived cells isolated in this study exhibited a high expression of MSC-specific markers, such as CD29, CD44, CD73, CD90, CD105 and Stro-1, and does not express markers specific to hematopoietic cells, such as CD45, CD14, CD34 or HLA-DR antigen (Figure 2a). UC-MSCs also express antigens typical for MSCs, such as CD29, CD44, CD73, CD90 and Stro-1, and, in parallel, do not possess CD45, CD14, CD34 and HLA-DR antigens on their surface that are considered markers of hematopoietic cells (Figure 2b). Thus, the multiantigenic phenotype of the dental pulp-derived cells is similar to the antigenic profile of UC-MSCs as shown in Figure 2c. Thus, based on their ability to adhere to plastic surfaces and antigenic profiles, we confirmed the identity of the isolated dental pulp-derived cells as previously described, representing the subpopulation of MSCs.

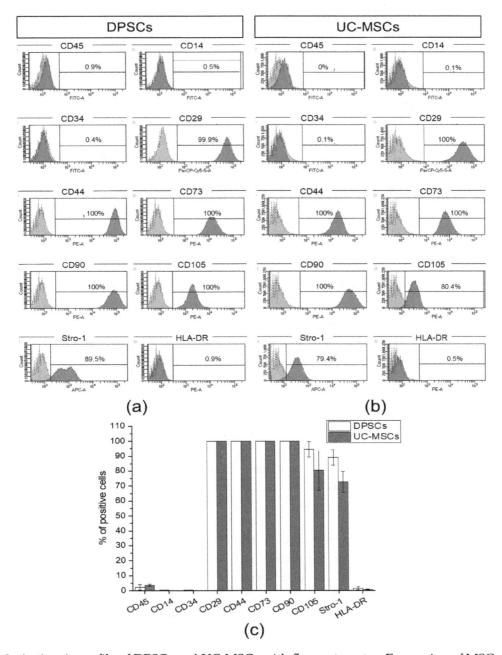

Figure 2. Antigenic profile of DPSCs and UC-MSCs with flow cytometry. Expression of MSC-negative markers (D45, CD14, CD34), MSC- positive markers (CD29, CD44, CD73, CD90, CD105, Stro-1), and HLA-DR antigen on DPSCs and UC-MSCs. (**a**) Representative histograms of the expression of analysed

antigens on DPSCs. (**b**) Representative histograms of the expression of analysed antigens on UC-MSCs. The peaks of unstained cells (grey) were overlaid with the peak for analysed antigen (violet). (**c**) Quantitative data representing the percentage content of DPSCs or UC-MSCs positive for analysed antigens. Results are presented as mean ± SD, $n = 3$.

2.2. DPSCs Exhibit Wide Differentiation Potential In Vitro

In the next step, to answer the question about the biological potential of DPSCs with respect to their pro-regenerative ability in injured tissues, we first analysed the tri-lineage differentiation potential of such cells compared to UC-MSCs in vitro. For that purpose, the DPSCs and UC-MSCs were differentiated into osteoblasts, chondroblasts, and adipocytes after 7, 14 and 21 days in tissue-specific differentiation media. We observed that both DPSCs and UC-MSCs exhibit tri-lineage differentiation potential (as shown in Figures 3 and 4, respectively), which also confirmed their MSC phenotype as defined by minimal criteria recommended by ISCT [3].

In the case of osteogenic differentiation, we analysed the expression of osteogenesis-related genes during the differentiation process of both MSC populations, such as Runx2, osteocalcin and osteopontin, in comparison with the control (undifferentiated) cells, which were cultured under standard culture conditions. We observed that the expression levels of transcription factor Runx2 and osteocalcin (a marker of bone formation) were comparable between DPSCs and UC-MSCs, whereas the fold change in expression of osteopontin (a protein expressed in maturated bone tissue) was elevated in UC-MSCs, notably on the 14th-day post-stimulation (Figure 3a, Table S1). Real-time RT-PCR results obtained for both MSC populations were compared with those of the control (undifferentiated) cells cultured in a standard cell culture medium (mRNA levels in such cells were calculated as 1.0).

The histochemical staining of cells differentiated into osteoblasts demonstrated larger deposits of calcium phosphate (indicated by red-coloured deposits of calcium phosphate) that were observed following DPSC differentiation when compared to the differentiation of UC-MSCs. Moreover, the deposits were observed earlier (at 14 days) in the case of DPSC osteogenic differentiation compared to those with differentiation of UC-MSCs (Figure 4). The comparable expression of the genes between DPSCs and UC-MSCs along with the higher formation of calcium phosphate deposits following DPSC differentiation may demonstrate a higher osteogenic differentiation potential of the DPSCs compared to that of the UC-MSCs.

The DPSCs, as well as UC-MSCs, were successfully differentiated into chondroblasts in vitro (Figures 3b and 4, respectively). In the case of DPSCs, we observed increased expression of *Sox9* transcription factor mRNA on days 7 and 14 of differentiation, compared to that in the undifferentiated cells, which confirmed their chondrogenic differentiation potential. However, the expression of *Sox9* gene was higher in UC-MSCs in comparison with DPSCs. We did not observe any significant change in the expression of *Col2A1* between both types of cells, while the fold change in the expression of *Col10A1* was higher in the UC-MSCs compared to that in the DPSCs (Figure 3b, Table S2). Recent evidence indicates that *Col10A1* is a marker of hypertrophic chondrocytes, which may be implicated as the principal factor driving bone growth. It has also been observed in skeletal dysplasia and osteoarthritis disorders [44]. The histochemical staining of DPSCs and UC-MSCs that were differentiated into chondroblasts indicated extracellular secretion of sulphated proteoglycans (indicated by blue coloured staining) by both types of MSCs, and the kinetics of differentiation seemed to be similar between both populations of SCs (Figure 4).

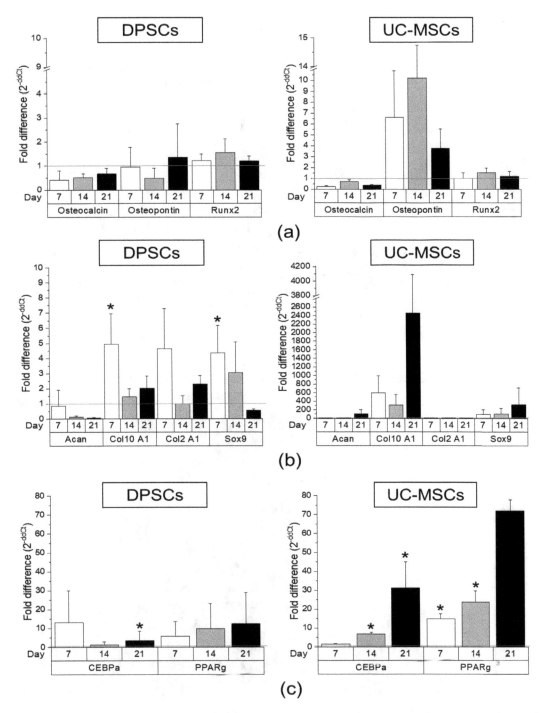

Figure 3. Comparison of tri-lineage differentiation potential of DPSCs and UC-MSCs by real-time RT-PCR. (**a**) Quantitative analysis of mRNA expression for osteogenesis related genes (osteocalcin, osteopontin, *Runx2*) in DPSCs (left) and UC-MSCs (right). (**b**) Quantitative analysis of mRNA expression for chondrogenesis related genes (*Acan, Col10A1, Col2A1, Sox9*) in DPSCs (left) and UC-MSCs (right). (**c**) Quantitative analysis of mRNA expression for adipogenesis related genes (*CEBPα, PPARγ*) in DPSCs (left) and UC-MSCs (right). Cells were cultured in a StemPro osteogenesis differentiation kit, StemPro chondrogenesis differentiation kit, and StemPro adipogenesis differentiation kit for 7, 14, and 21 days, respectively. Fold differences in expression (2^{-ddCt}) of analysed genes in control cells cultured in standard cell culture medium (undifferentiated) were calculated as 1.0 and marked by a solid line. Graphs present different scales. Results are presented as mean ± standard error of the mean (SEM), $n = 3$ (every sample prepared for each DPSCs line derived from each donor were run in duplicates); t-test, (*) $p < 0.05$ vs. undifferentiated cells.

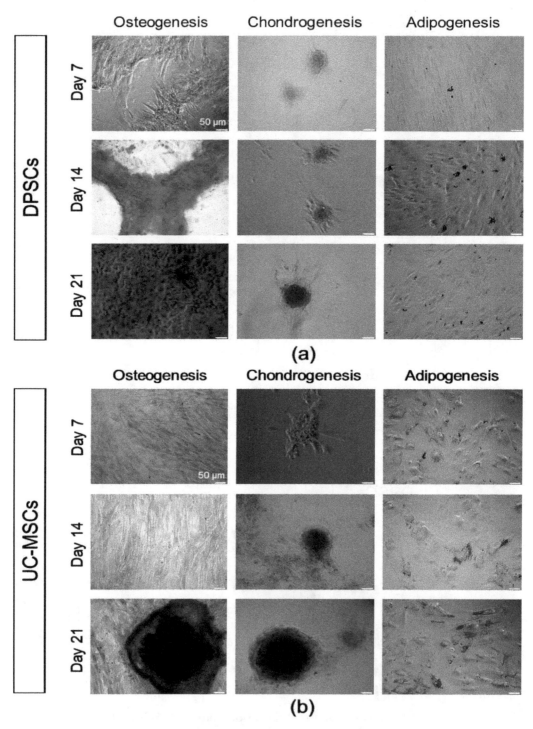

Figure 4. Tri-lineage differentiation potential of DPSCs and UC-MSCs in an in vitro culture demonstrated by histochemical staining. (**a**) Representative images of DPSCs differentiated into osteoblasts, chondroblasts and adipocytes. (**b**) Representative images of UC-MSCs differentiated into osteoblasts, chondroblasts, and adipocytes. DPSCs and UC-MSCs were cultured in a StemPro osteogenesis differentiation kit, StemPro chondrogenesis differentiation kit, or StemPro adipogenesis differentiation kit. On days 7, 14, and 21 of differentiation, DPSCs and UC-MSCs were fixed with paraformaldehyde and stained with Alizarin Red S (red staining of calcium phosphate deposits that are a characteristic of osteogenic differentiation), Alcian Blue (blue staining of sulphated proteoglycans that are a characteristic of chondrogenic differentiation) or Oil Red O (brownish red oil droplets that are a characteristic of adipogenic differentiation). Scale bars: 50 μm.

When focusing on adipogenic differentiation, the mRNA expression corresponding to *PPARγ* adipogenesis-related transcription factor on day 21 as well as *CEBPα* protein expression on day 14 was significantly higher in the UC-MSCs in comparison with that in the DPSCs (Figure 3c, Table S3). Moreover, the presence of oil droplets indicating an ongoing process of adipogenesis was typically observed in both cell fractions on the 14th day of differentiation. The results considering the level of demonstrate higher adipogenic differentiation of UC-MSCs compared to DPSCs.

Taken together, our first analyses confirmed that DPSCs may be successfully differentiated into osteoblasts, chondroblasts, and adipocytes similar to other "classic" MSC populations such as UC-MSCs. The histochemical analyses of the final cell phenotypes confirmed a higher ability of DPSCs to differentiate into osteoblasts compared to UC-MSCs, which are primarily restricted to chondrogenic and adipogenic differentiation.

To establish whether DPSCs exhibit any cardiomyogenic potential in vitro, they were cultured in cardiomyogenesis stimulating medium as previously described [6]. We analysed mRNA expression of cardiac markers such as *Gata-4*, *Nkx2.5* and *Myl2c* after 7, 14 and 21 days following the induction of differentiation. We observed that the expression of these genes was markedly elevated after 7 and 21 days of cardiomyogenic differentiation induction in UC-MSCs than that in DPSCs (Figure 5a and Table S4). Moreover, both types of MSCs express intranuclear cardiac transcription factor *Gata-4* as well as cytoplasmic structural protein troponin T-C after seven days of differentiation (Figure 6). This may suggest that DPSCs possess the lower capacity for cardiac cell phenotypes, when culture under lineage-specific conditions, in comparison with other MSC fractions, such as UC-MSCs.

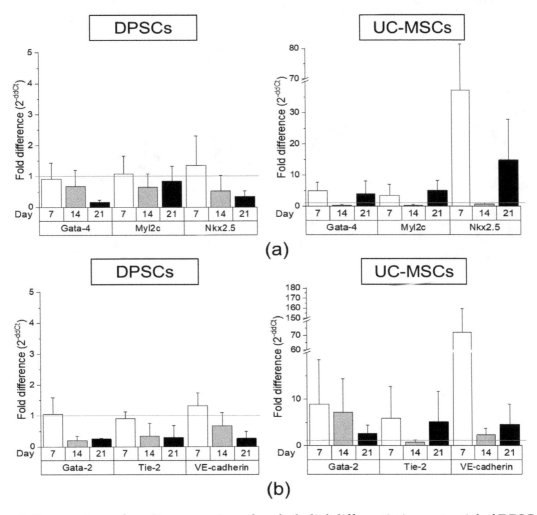

Figure 5. Comparison of cardiomyogenic and endothelial differentiation potential of DPSCs and UC-MSCs by Real-Time PCR. (**a**) Quantitative analysis of mRNA expression of cardiomyogenesis

related genes (*Gata-4, Nkx2.5, Myl2c*) in DPSCs (left) and UC-MSCs (right). Cells were cultured in DMEM/F12 supplemented with 2% FBS and 10 ng/mL basic fibroblast growth factor (bFGF), 10 ng/mL vascular endothelial growth factor (VEGF) and 10 ng/mL transforming growth factor β1 (TGF-β1) for 7, 14 and 21 days. (**b**) Quantitative analysis of mRNA expression for endothelial related genes (*Gata-2, Tie-2,* VE-cadherin) in DPSCs (left) and UC-MSCs (right). Cells were cultured in EGM-2MV endothelial cell growth medium for 7, 14 and 21 days. Fold differences in the expression (2^{-ddCt}) of analysed genes in control cells cultured in standard cell culture medium (undifferentiated) were calculated as 1.0 and marked by a solid line. Graphs present different scales. Results are presented as mean ± SEM, $n = 3$ (every sample prepared for each DPSCs line from each donor was run in duplicate); *t*-test, $p < 0.05$ vs. undifferentiated cells.

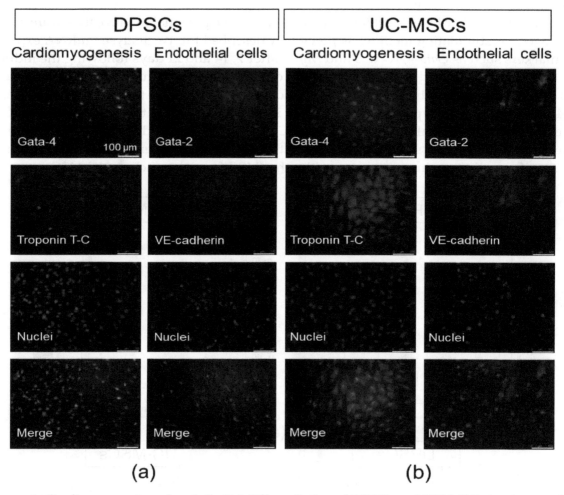

(a)   (b)

Figure 6. Cardiomyogenic and endothelial differentiation of DPSCs and UC-MSCs in vitro on day 7. (**a**) Representative images of cardiomyogenic and endothelial marker expression in DPSCs. (**b**) Representative images of cardiomyogenic and endothelial marker expression in UC-MSCs. In the case of cardiomyogenic differentiation, cells were cultured in DMEM/F12 supplemented with 2% FBS and 10 ng/mL bFGF, 10 ng/mL VEGF and 10 ng/mL TGF-β1. On day 7, cells were fixed, permeabilized, and stained against intranuclear transcription factor *Gata-4* (Alexa Fluor 488, green) and troponin T-C (Alexa Fluor 546, red), whereas nuclei were co-stained with DAPI (blue). In the case of endothelial differentiation, cells were cultured in EGM-2MV. On day 7, cells were fixed, permeabilized, and stained against intranuclear transcription factor *Gata-2* (Alexa Fluor 488, green) and VE-cadherin (Alexa Fluor 546, red), whereas nuclei were co-stained with DAPI (blue). Cells were analysed with Leica DMI6000B ver. AF7000 fluorescent microscope. Scale bars: 100 μm.

To assess the potential angiogenic differentiation capacity of DPSCs, in comparison with UC-MSCs, we launched a long-term culture of these cells in proangiogenic medium containing VEGF (EGM-2MV).

The angiogenic potential was examined both at the mRNA and protein levels. The highest expression of angiogenesis-related genes (*Gata-2, Tie2*, VE-cadherin) was observed in UC-MSCs on day 7 of differentiation, whereas the DPSCs were unresponsive to proangiogenic stimulation (expression of the proangiogenic genes was at the same level as in unstimulated DPSCs; Figure 5b, Table S5). Interestingly, the enhanced expression of these genes was also observed in UC-MSCs on the 14th and 21st days of proangiogenic stimulation. However, in the immunocytochemical staining, we did not observe any prominent expression of proangiogenic transcription factor Gata-2 or cell membrane protein VE-cadherin supporting angiogenesis (Figure 6).

Collectively, we observed the following features of DPSCs as cells that (i) exhibit higher osteogenic differentiation capacity, (ii) demonstrate comparable chondrogenic and adipogenic differentiation potential, and (iii) possess limited ability for cardiac or endothelial phenotype, in comparison with other "classic" MSCs (UC-MSCs).

2.3. 3D Encapsulated DPSCs Exhibit Higher Differentiation Capacity into Osteoblasts in Vitro

As shown previously, the DPSCs exhibit a higher osteogenic potential compared to the UC-MSCs. Based on the fact that osteogenic differentiation leading to bone formation is a process that takes place in the regular in vivo 3D niche of developing organism [45], we encapsulated the DPSCs in a hydrogel matrix to mimic such 3D niche in vitro and further analysed their morphology, proliferation, metabolic activity, and osteogenic potential in both normoxic or hypoxic culture environment.

Young's moduli of hydrogel matrix measured using AFM were normally distributed (assessed by the Shapiro–Wilk test) with a mean value of E = 3.69 ± 1.49 kPa (Figure 7a). Elasticity maps demonstrate a heterogeneous distribution of elastic modulus (Figure 7b,c), thus providing more realistic conditions for cell growth. The resulting Young's modulus was in the range of physiological tissue elasticity (~1–100 kPa, [34]) demonstrating rather highly deformable substrate properties. For proliferating cells encapsulated inside hydrogels, such gel does not constitute a barrier. Cells may be able to generate protrusive forces during cellular divisions and can be released into the surrounding environment [46].

We further observed that the DPSCs encapsulated in the 3D hydrogel exhibited a round shape 24 h post mixing, whereas certain flattened cells exhibiting a spindle-shaped morphology were observed after 48 h, indicating a phenotypical change.

In contrast, DPSCs seeded on cell culture plates (2D) exclusively exhibited spindle-shaped morphology at both time points as expected in standard 2D conditions (Figure 7d,e). On the first day of 2D and 3D cultures, the relative proliferation of DPSCs in an environment containing 2% or approximately 18% of O_2 was the same (Figure 7f). Between days 2 and 7 after culture initiation, we observed a greater relative proliferation of DPSCs in 2D culture compared to that in the 3D culture. However, we did not observe any influence of O_2 concentration on the proliferation ratio of DPSCs in both 2D and 3D culture conditions at fixed-time intervals (Figure 7f).

The increased proliferation of DPSCs in 2D culture was correlated with the higher metabolic activity of these cells (Figure 7g). DPSCs in 3D culture exhibited lower metabolic activity along with their lower proliferation. We also did not observe any significant impact of O_2 concentration on the levels of ATP production in DPSCs in 2D and 3D cell culture (Figure 7g).

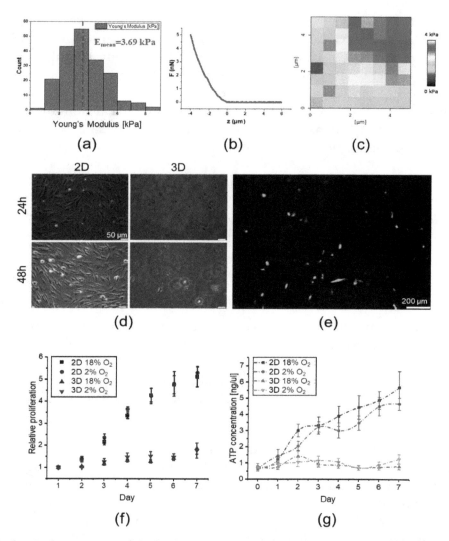

Figure 7. Mechanical properties of the hydrogel matrix and their impact on morphology, proliferation, and metabolic activity of DPSCs. (**a**) Young's modulus distributions of hydrogel matrix by AFM. (**b**) Exemplary force curve recorded on the hydrogel. (**c**) Exemplary elasticity map of peptide hydrogel. (**d**) Morphology of DPSCs encapsulated in hydrogel (3D) or seeded on the surfaces coated with gelatin (2D) at 24 and 48 h post-seeding. DPSCs were cultured in DMEM/F12 supplemented with 10% FBS and cultured under standard culture conditions (5% CO_2, normoxia). (**e**) Morphology of DPSCs encapsulated in hydrogel (3D culture) on day 4 post-seeding. Cells were stained with fluorescein diacetate and analysed with Leica DMI6000B ver. AF7000 fluorescent microscope to visualize their morphology. (**f**) The proliferation of DPSCs in 2D and 3D cultures in the environment containing about 18% or 2% O_2 analysed every 24 h until day 7. The analyses were conducted using the Cell Counting Kit-8. (**g**) Metabolic activity of DPSCs in 3D and 2D culture in the environment containing about 18% or 2% O_2 measured every 24 h until day 7. The analyses were conducted using the ATP Lite Luminescence assay kit. The results from proliferation and metabolic activity are presented as mean ± SEM, $n = 3$ (every sample prepared for each DPSCs line from each donor was analysed in triplicate).

In the next step, we conducted in vitro osteogenic differentiation of DPSCs in 2D or 3D culture in the presence of 2% or approximately 18% of O_2. After seven days of stimulation, the concentration of mRNA for Runx2 was elevated by three times in DPSCs cultured in both 2D and 3D conditions in the presence of hypoxia (2% O_2) in comparison with undifferentiated cells. The elevated expression of Runx2 in 3D culture in an environment containing 2% O_2 was sustained up to 14 days after stimulation. Importantly, under such culture conditions, we observed the highest fold change in the expression of mRNA corresponding to Runx2 as well as Col1A, at every analysed experimental time point (Figure 8a).

Hypoxic microenvironment (2% O_2) also stimulated the expression of a gene encoding osteopontin on day 7 in DPSCs in 2D culture (Figure S2). Recent evidence indicates that osteopontin regulates matrix remodelling and tissue calcification and may also be implicated in the pathophysiological process such as osteoporosis [47]. After seven days of differentiation, despite the increased expression of Runx2, we did not observe prominent calcium phosphate deposits in cells cultured in an environment containing 2% O_2. The most explicit deposits of calcium phosphate were observed after 21 days of differentiation: in case of 3D culture, we observed large rounded deposits of calcium phosphate especially in the presence of 2% O_2, whereas in case of 2D culture, these deposits were more prominent in the presence of approximately 18% O_2 (Figure 8b).

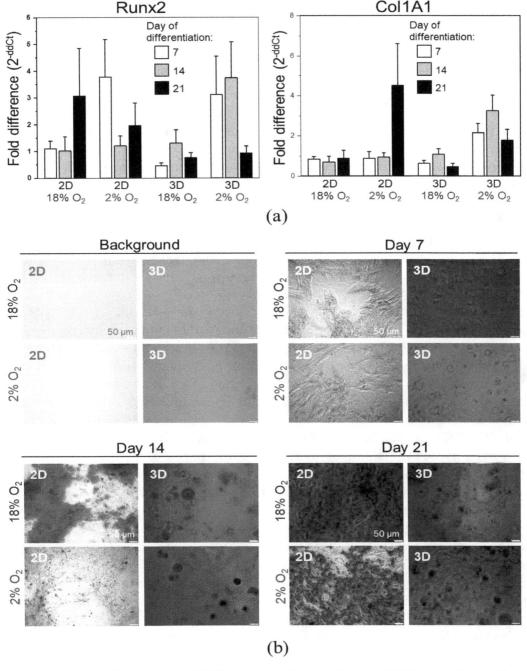

Figure 8. Osteogenic differentiation of DPSCs encapsulated in hydrogel (3D) or seeded on the surface coated with gelatin (2D) cultured in hypoxic (2% O_2) or normoxic (about 18% O_2) environment. (**a**) Quantitative analysis of mRNA expression for osteogenesis associated genes (*Col1A, Runx2*) in DPSCs on days 7, 14, and 21 of differentiation. Fold change in the expression of analysed genes in control cells

before differentiation was calculated as 1.0 and marked by a solid line. Results are presented as mean ± SEM, $n = 3$ (every sample was analysed in duplicate). $p < 0.05$ (t-test). (b) Representative images of DPSCs differentiated into osteoblasts on days 7, 14, and 21 days of differentiation. Panel "Background" contains the representative images of 2D and 3D surfaces (standards plastic dish and hydrogel without cells, respectively) stained with Alizarin Red S solution to visualize background staining (on day 7 post surface preparation). Panels "Day 7", "Day 14", "Day 21" demonstrate representative images of DPSCs cultured in StemPro osteogenesis differentiation kit. On days 7, 14 and 21, DPSCs were fixed with paraformaldehyde and stained with Alizarin Red S (red staining of deposits of calcium phosphate is a characteristic for osteogenic differentiation).

Taken together, the results indicate that 3D cell encapsulation as well as the low concentration of O_2 resembling conditions in the stem cell niches may favour osteogenic differentiation of DPSCs in an in vitro environment.

3. Discussion

Recent evidence indicates that dental pulp harbours a population of stem cells, such as DPSCs [43]. Nowadays, specific markers that may allow to uniquely define the immunophenotype of DPSCs are still unknown [48]. However, these cells express antigens that are characteristic of "classic" MSCs derived from mesodermal tissues. Such cells do not express hematopoietic markers [49]. Importantly, DPSCs were also shown to exhibit the expression of stemness-associated markers, such as *Oct-4, Nanog* and *Sox-2* [50] along with vimentin, as well as neural-specific markers such as nestin, N-tubulin and neurogenin-2 [51]. Due to the previously described unique features of such cells, one of the aims of this study was to analyse differentiation potential and other essential functional and morphological features of DPSCs in comparison with UC-MSCs. UC-MSCs may be used for allogenic or autologous applications. Moreover, accumulating evidence indicates that the transcriptomic profile of UC-MSCs indicating the expression of stemness- and bone development-related genes is similar to BM-MSCs. There is a possibility of elimination of restrictions that are characteristic for autologous BM-MSC. This might result in the replacement of these cells that are considered as gold standard for MSC research and clinical applications [16,17,52]. In the second phase of the current study, we focused on the osteogenic differentiation of DPSCs in 3D cultures in the hypoxic microenvironment resembling the stem cell niche.

We isolated DPSCs from dental pulp extracted from permanent healthy teeth of donors, whereas UC-MSCs were isolated from Wharton's Jelly of UC. Isolated DPSCs represent an adherent fraction of dental pulp tissue and are spindle-shaped, fibroblast-like cells similar in morphology to UC-MSCs. By employing flow cytometry platform, we established the antigenic profiles of DPSCs containing the expression of $CD45^-/CD14^-/CD34^-/CD29^+/CD44^+/CD73^+/CD90^+/CD105^+/Stro-1^+/HLA-DR^-$. The immunophenotype of DPSCs is similar to the antigenic profile of UC-MSCs. Subsequently, by employing quantitative RT-PCR and histochemical staining, we confirmed that DPSCs, as well as UC-MSCs exhibit the capacity for tri-lineage mesenchymal differentiation. Collectively, DPSCs, as well as UC-MSCs, fulfil the classification criteria of MSCs published by ISCT [3]. It is important to emphasize that some investigators have pointed out that the exact status of DPSCs as MSCs is still not fully defined and requires further investigation [48]. Based on the fact that the typical characteristics and immunophenotype of DPSCs fulfil MSC classification criteria, DPSCs may be considered as a population of MSC with certain unique properties resulting from their developmental origin.

Due to the ectomesenchymal origin [53], the DPSCs may be effectively differentiated into functionally active neurons [51,54] demonstrating an attractive prospect for the treatment of injuries of the nervous system. However, the comparisons between the differentiation potential of DPSCs and other "classic" MSCs toward mature cell types other than neural cells have never been performed. Thus, in the current study, we compared the differentiation capacity of DPSCs with UC-MSCs isolated from human neonatal tissue, focusing on osteogenic, chondrogenic, adipogenic, cardiomyogenic, and angiogenic differentiation in vitro. In the case of osteogenic differentiation, the expression levels of mRNA for *Runx2* and osteocalcin were comparable between the DPSCs and UC-MSCs, whereas the expression of osteopontin was elevated in the UC-MSCs notably on the latter days following stimulation (14d). Other studies have shown that osteopontin may be expressed in many organs, with or without matrix, suggesting that this molecule may act as a structural molecule or humoral factor or cytokine and may not be a specific marker of osteogenesis [45], which may explain the elevated expression observed in UC-MSCs. Despite the comparable expression level of osteogenesis associated genes between DPSCs and UC-MSCs, histochemical staining demonstrated larger deposits of calcium phosphate following DPSC osteogenic stimulation in comparison with those in the stimulated UC-MSCs, which may indicate higher osteogenic capacity of DPSCs. Deposition of calcium phosphate, a bone mineral, is a characteristic for osteoblasts [55]. Since some investigators indicate a poor correlation between mRNA and protein expression levels, it may be necessary to extend the studies examining mRNA and protein expression in DPSCs in comparison with UC-MSCs to explain the observed higher deposition of calcium phosphate in DPSCs following osteogenic stimulation despite the low expression of mRNA [56]. Osteogenic differentiation of DPSCs was also confirmed by other investigators [57]. A higher alkaline phosphatase activity has been detected in human DPSCs cultured in osteogenic differentiation medium than that in BM-MSCs [11].

Interestingly, we also confirmed the chondrogenic potential of DPSCs and UC-MSCs. We observed higher expression of *Sox9, Acan,* and *Col10A1* mRNA in UC-MSCs, whereas the expression of *Col2A1* was comparable between both groups of SCs. Histochemical staining of DPSCs and UC-MSCs indicated the extracellular secretion of sulphated proteoglycans by both types of cells. Kinetics of differentiation was also similar between the analysed populations of SCs. Recent evidence indicates that *Col10A1* is a marker of hypertrophic chondrocytes, which may be implicated in bone growth. In contrast, it has also been observed in skeletal dysplasia and osteoarthritis disorders [44]. Thus, further investigation may be necessary to confirm the proper chondrogenic differentiation of UC-MSCs as well as DPSCs. Some investigators, however, have confirmed the higher chondrogenic potential of BM-MSCs in comparison to DPSCs [11]. Moreover, we observed a higher level of adipogenesis in UC-MSCs than that in DPSCs following stimulation, as measured by adipogenic gene expression at mRNA level. Both MSC populations secrete oil droplets, a characteristic of the ongoing adipogenic differentiation [58]. A lower adipogenic potential of DPSCs in comparison with BM-MSCs has also been demonstrated by Guo et al. [11].

In the next step, we analysed differentiation potential of DPSCs toward cells with cardiac and endothelial phenotype. After cardiomyogenic stimulation, we observed higher expression of *Gata-4, Nkx2.5,* and *Myl2c* mRNA in UC-MSCs than in DPSCs, which was confirmed by immunocytochemical staining. In the case of endothelial cell formation, elevated expression of angiogenesis-related genes (*Gata-2, Tie2,* VE-cadherin) was observed in UC-MSCs after seven days of stimulation with VEGF, whereas DPSCs were typically unresponsive toward such proangiogenic stimulation. The results suggested q limited switch towards cardiac and endothelial phenotype in UC-MSCs, while the signs of differentiation were hardly detected in DPSCs. However, other investigators have shown that in the co-culture of DPSCs with neonatal rat cardiomyocytes, the expression of cardiac-specific markers (including troponin I and β-myosin heavy chain) was increased in DPSCs shortly after co-incubation, suggesting a possible transient phenotypical switch into cardiac lineage under certain circumstances [59].

Since DPSCs may be effectively differentiated into osteoblasts, we also encapsulated these cells in the hydrogel matrix to create a 3D culture niche mimicking tissue and organ-specific microarchitecture to conduct osteogenic differentiation in vitro. Nevertheless, we observed a greater relative proliferation of DPSCs in 2D culture in comparison with 3D culture. Moreover, contrary to our initial expectations, we did not observe any influence of oxygen concentration on the proliferation ratio of DPSCs in both 2D and 3D culture at predetermined intervals. However, the increased proliferation of DPSCs in 2D culture was correlated with higher metabolic activity, whereas DPSCs in 3D culture exhibited lower metabolic activity correlating with lower proliferation and higher differentiation capacity in such conditions. It has been demonstrated that 3D culture allows the cells to maintain a better cell-to-cell contact and intercellular signalling network [60]. In the case of a hypoxic environment, we set up oxygen concentration equal to 2%, which was in the range of O_2 concentration measured by other investigators in murine BM (1.54–2.0%) [61]. For a normoxic culture environment, Wegner et al. showed that the O_2 concentration should be approximately 18% [62]. Hence, we assumed that the O_2 concentration in the normoxic environment was approximately 18%. The typical mammalian cell culture conditions inside an incubator also include the temperature of 37 °C, 95% relative humidity and CO_2 concentration of approximately 5%.

Kwon et al. has shown increased proliferation rate and percentage of DPSCs in S-phase under 5% hypoxic culture environment from day 5 of the culture accompanied with enhanced osteogenic differentiation capacity of these cells in comparison with the normoxic environment [63], whereas Iida et al. observed the inhibition of the osteo/odontogenic differentiation capacity of such cells, despite the enhanced proliferation rate of DPSCs under hypoxic environment [64]. Results related to the proliferation of DPSCs cultured under hypoxia have been controversial, which may be attributed to the lack of standardization of cell culture protocols [65]. In this study, we cultured DPSCs in the presence of 2% O_2, whereas other investigators cultured cells in the presence of 5% or 3% O_2 [63,64].

A decrease in cell proliferation might often be accompanied by the induction of cell differentiation [66] and 3D culture may facilitate the developmental processes allowing the cells to differentiate into more complex structures [67]. Thus, we analysed the osteogenic differentiation of DPSCs in 3D cultures under hypoxic or normoxic conditions.

We also analysed the osteogenic differentiation of DPSCs in 3D culture under hypoxic and normoxic conditions. The expression of mRNA corresponding to Runx2–osteogenesis regulating transcription factor-was elevated by approximately thrice the initial amount in DPSCs in 2D and 3D culture in the presence of 2% O_2, compared to that in the undifferentiated cells. The elevated expression of *Runx2* in a 3D culture in an environment containing 2% O_2 persisted up to day 14, suggesting long-term osteogenic induction. We observed the highest increase in the expression of early and late osteogenic markers such as *Runx2* and *Col1A*, respectively, at every interval in 3D culture in microenvironment under hypoxia than that in the 3D culture in a normoxic microenvironment. In the case of 3D culture, we also observed large rounded deposits of calcium phosphate, primarily in the presence of hypoxia.

Taken together, our results indicate that 3D cell encapsulation as well as low concentrations of oxygen may favour the osteogenic differentiation of DPSCs in vitro. The culture of human MSCs under hypoxic conditions enhances the differentiation into osteocytes. These cells produce more osteomatrix and the expression of osteogenesis-related genes (e.g., *Runx2, ALP*) is up-regulated [68,69]. Burakova et al. suggested that the effect of hypoxia on MSCs may be driven by specific factors explaining the opposite influence of low oxygen concentration observed by many investigators [70]. Recent evidence indicates that the 3D scaffold may have a major impact on the proliferation and activation of MSCs, whereas the 3D architecture supports the osteogenic differentiation of MSCs and enhances the production of mineralized bone matrix [71]. Increased expression of *Runx2* and *Col1A1* characterizing osteogenic differentiation of DPSCs in 3D culture under hypoxic environment observed in our study is consistent with recent evidence demonstrated by other investigators, confirming that the differentiation of MSCs encapsulated in alginate beads significantly increases the osteogenic differentiation of these cells in vitro [35]. Moreover, in the novel bone development or bone regeneration, vasculature delivers nutrients, growth factors, minerals, and oxygen. The scaffolds for bone regeneration should allow endothelial cell migration and the growth of new vessels, which may induce VEGF expression promoting the migration and proliferation of endothelial cells and positively influence the bone regeneration process [72]. After the co-culture of DPSCs and endothelial cell line EA.hy926 on chitlac-coated thermosets, an increased proliferation with enhanced DPSC osteogenic differentiation and EA.hy926 vessel formation was observed [73], confirming positive feedback of both cell fractions.

In our experiments, we used primary human DPSCs isolated from dental pulp derived from the teeth of three donors, whereas UC-MSCs were isolated from two umbilical cords. Moreover, for qRT-PCR analysis, every sample (prepared for each DPSCs line from each donor) was run in duplicate, whereas for the proliferation and metabolic activity assessment, every sample was run in triplicate. In some cases, we observed a high error of the mean (SEM), which may be the result of biological variability related to primary human tissues obtained from various donors. Data published by other investigators indicate that the DPSCs isolated from the patients at different ages exhibit different doubling time, expression of CD29 antigen, and percentage of apoptotic cells during propagation along with differentiation potential [74], which may have an additional impact on the variability of the cell response. Moreover, the DPSCs isolated from different teeth may possess varied differentiation capacity or immunomodulatory potential [75]. The high variation of the response and data presented in our study is not unusual during the analyses of SCs, including DPSCs. Other investigators have also demonstrated highly variable data from analysis of different populations of SCs, including DPSCs (results with high SEM values) [76]. Therefore, further investigations are required to find a correlation between the type of dental pulp derived from different teeth, the age of the donor, and the biological potential of DPSCs. Proper protocols might be developed for the preparation of optimal DPSC fraction as a product for advanced therapy with high pro-regenerative potential.

To summarize, we observed that human DPSCs possess a higher potential for osteogenic differentiation and similar chondrogenic and adipogenic differentiation potential compared to the "classic" MSCs such as UC-MSCs but a limited ability for the cardiac or endothelial phenotype in vitro. Moreover, the hypoxic environment especially in 3D culture conditions may enhance the osteogenic capacity of DPSCs in vitro. DPSCs encapsulated in injectable hydrogel might be a candidate for innovative advanced therapy for the treatment of bone (including alveolar bone) loss.

4. Materials and Methods

4.1. Isolation of Human DPSCs

The DPSCs were isolated from the dental pulp of human permanent healthy teeth, which were extracted based on the orthodontic indications in the Department of Oral Surgery, Faculty of Medicine, Jagiellonian University Medical College in Krakow and then donated for scientific research following approval by the Bioethics Committee at the Jagiellonian University in Krakow (approval number: 1072.6120.41.2017).

In the first step of isolation of DPSCs, the pulp chamber was exposed using the pulp drill (size: D Ø: 0.25 mm and 0.30 mm) to extract the dental pulp. Subsequently, the pulp chamber was gently rinsed with phosphate-buffered saline (PBS; GE Healthcare Life Sciences HyClone Laboratories, Malborough, MA, USA) containing 100 IU/mL penicillin and 100 µg/mL streptomycin solutions (Gibco, ThermoFisher Scientific, Waltham, MA, USA) to wash out the remaining pulp tissue. Following the mechanical disruption, the isolated pulp tissue was subjected to further enzymatic digestion using a mixture of collagenase I (3 mg/mL, Sigma-Aldrich, St. Louis, MO, USA) and dispase (4 mg/mL, Sigma-Aldrich) for 30 min at 37 °C. The enzymes were inactivated by adding a complete cell culture medium (DMEM/F12 supplemented with 10% FBS, Sigma-Aldrich; and 100 IU/mL penicillin, 100 µg/mL streptomycin, Gibco, ThermoFisher Scientific). Released cells as well as the larger pieces of dental pulp were washed and seeded on 24-well culture plates favouring primary cell adhesion (Corning Primaria Culture Plate; Falcon Coring, Tewksbury, MA, USA) and were further cultured under standard conditions (37 °C, 5% CO_2, 95% humidity). Fresh culture medium was added after 24 h post-seeding. The tissue pieces were removed after 10 days and the adherent cells were washed with PBS and cultured in the complete cell culture medium. The scheme of isolation of DPSCs is presented in Figure 1a. DPSCs were passaged with 0.25% trypsin/EDTA (Gibco, ThermoFisher Scientific) when the confluence of cells reached close to 80–90%. The cells were isolated from three separate teeth derived from three individual donors and were further cultivated individually as separate isolates of DPSCs. DPSCs collected from a single human donor were used in every experiment. Thus, three independent DPSC lines derived from three teeth that were harvested from three individual human donors were used to replicate the experiments.

4.2. Isolation of Human UC-MSCs

The UC-MSCs were isolated from human umbilical cords obtained from The Polish Stem Cell Bank, with permissions from the Polish Ministry of Health (MZ-PZ-TSZ-025-15906-36/AB/14). The umbilical cord was washed with PBS to remove the remaining blood. Subsequently, the vein and arteries were dissected, the Wharton's jelly tissue was cut into 1–2 mm pieces and placed on tissue culture dishes (Falcon) containing complete cell culture medium (DMEM/F12 supplemented with 10% FBS, Sigma-Aldrich,; and 100 IU/mL penicillin, 100 μg/mL streptomycin, Gibco, ThermoFisher Scientific). The tissue pieces were removed after 5 days, the adherent cells were washed with PBS and cultured in the complete cell culture medium. The scheme of isolation of UC-MSCs is presented in Figure 1b. The cells were cultured under standard cell culture conditions. UC-MSCs were passaged with 0.25% trypsin/EDTA (Gibco, ThermoFisher Scientific) when the confluence of cells reached close to 80–90%. The cells isolated from the Wharton's Jelly tissue from the UC of every donor were cultivated separately. In the three replicated experiments, two independent UC-MSC lines derived from the two Wharton's jelly tissues obtained from UCs collected from two separate human donors were used.

4.3. Cell Counting and Viability Assessment

To assess the number of cells and examine their viability, the cell suspension was mixed with 0.2% Trypan Blue Stain (ThermoFisher Scientific), placed in cell counting slides, and analysed by Countess Automated Cell Counter (ThermoFisher Scientific).

4.4. Antigenic Phenotyping by Flow Cytometry

To compare the phenotype of DPSCs and UC-MSCs, the cells of both fractions were resuspended in standard staining medium (DMEM/F12 supplemented with 2% FBS; both from Sigma-Aldrich, and were further immunolabelled with the following monoclonal antibodies against human antigens: anti-CD45 (FITC, clone: HI30, Biolegend, San Diego, CA, USA), anti-CD14 (FITC, clone: MφP9, BD Biosciences, Franklin Lakes, NJ, USA), anti-CD34 (FITC, clone: 581, BD Biosciences, Franklin Lakes NJ, USA), anti-CD29 (PE/Cy5, clone: TS2/16, Biolegend), anti-CD44 (PE, clone: BJ18, Biolegend), anti-CD73 (PE, clone: AD2, Biolegend), anti-CD90 (PE, clone: 5E10, Biolegend), anti-CD105 (PE, clone: 43A3, Biolegend), anti-Stro-1 (Alx647, clone: STRO-1, Biolegend,) and anti-HLA-DR (PE, clone: L243, Biolegend). For each analysed antigen, appropriate isotype control was used as following: mouse IgG1 (FITC, clone: MOPC-21, BD Biosciences, mouse IgG2 (Alx488, clone: 133303, R&D Systems, Minneapolis, MN, USA), mouse IgG1 (PE/Cy5, clone: MOPC-21, BD Biosciences,), mouse IgG1 (PE, clone: MOPC-21, BD Biosciences,), mouse IgG2 (PE, clone: IS6-11E5.11, Miltenyi Biotec, Bergisch Gladbach, Germany) and mouse IgG1 (APC, clone: IS5-21F5, Miltenyi BiotecStaining was performed for 30 min at 4 °C according to the manufacturer's protocols. Cells were further washed and analysed using LSR Fortessa flow cytometer and FACS Diva software (Becton Dickinson, Franklin Lakes, NJ, USA).

4.5. Differentiation of DPSCs and UC-MSCs

4.5.1. Osteogenic, Chondrogenic, and Adipogenic Differentiation

Plates with 12-wells were coated with 0.1% gelatin (Sigma-Aldrich). In the case of osteogenic and adipogenic differentiation, 2.0×10^4 cells were seeded per well in the complete cell culture medium (DMEM/F12 supplemented with 10% FBS, Sigma-Aldrich; and 100 IU/mL penicillin, 100 μg/mL streptomycin, Gibco, ThermoFisher Scientificto reach 60% confluence. Subsequently, the medium was replaced with complete StemPro Osteogenesis Differentiation Kit or StemPro Adipogenesis Differentiation Kit (Gibco, ThermoFisher Scientific), stimulating osteogenic or adipogenic differentiation, respectively. The cultures were refed every 3–4 days. In the case of chondrogenic differentiation, micro mass cultures were generated by seeding 5 μL droplets of cell solution (1.6×10^7 viable cells/mL) and were incubated for 2 h under high humidity conditions. Next, the micro masses were flooded with the complete cell culture medium. After 24 h, the medium was changed to StemPro chondrogenesis differentiation medium (Gibco, ThermoFisher Scientific). The cultures were re-fed every 2–3 days.

Cells were examined for osteogenic, chondrogenic, and adipogenic differentiation on days 7, 14 and 21 of culture, following histochemical staining performed for identifying the cell phenotype.

4.5.2. Cardiomyogenic Differentiation

12-well plates were coated with 50 μg/mL collagen type I (Sigma-Aldrich), and 2.0×10^4 cells were seeded per well in complete cell culture medium (DMEM/F12 supplemented with 10% FBS, Sigma-Aldrich; and 100 IU/mL penicillin, 100 μg/mL streptomycin, Gibco, ThermoFisher Scientific) for 24 h. The following day, the culture medium was changed with a cardiomyogenesis-stimulating medium containing DMEM/F12 supplemented with 2% FBS (both from Sigma-Aldrich) and 10 ng/mL bFGF, 10 ng/mL VEGF and 10 ng/mL TGF-β1 (all growth factors were from Peprotech, London, UK). The growth factors were supplemented daily and the whole medium was replaced every two days. Cells were examined for cardiac differentiation on days 7, 14 and 21 of culture, following the staining for cardiac-specific markers.

4.5.3. Endothelial Differentiation

Twelve-well plates were coated with a solution containing 50 μg/mL fibronectin (BD Bioscience) and 0.1% gelatin (Sigma-Aldrich). Cells were seeded with the density of 2.0×10^4 cells per well in the complete cell culture medium (DMEM/F12 supplemented with 10% FBS, Sigma-Aldrich; and 100 IU/mL penicillin, 100 μg/mL streptomycin, Gibco, ThermoFisher Scientific) for 24 h. The following day, the culture medium was changed with EGM-2MV Endothelial Cell Growth Medium (Lonza, Basel, Switzerland). EGM-2MV was replaced every two days. Cells were examined for endothelial differentiation on days 7, 14 and 21 days of culture, following staining for endothelial markers.

The expression of selected osteogenic, chondrogenic, adipogenic, cardiomyogenic and angiogenic-specific markers was evaluated by measuring the levels of corresponding mRNAs as well as by histochemical or immunocytochemical staining.

4.6. Gene Expression Analysis by Real-Time RT-PCR

Total RNA was isolated using the GeneMATRIX Universal RNA Purification Kit (Eurx, Gdansk, Poland) according to the manufacturer's protocol for RNA isolation from cell cultures. Briefly, to lyse the cells, lysis buffer (Eurx) with 1% Bond-Breaker reagent (ThermoFisher Scientificwas used. RNAs were treated with Turbo DNase (Ambion, ThermoFisher Scientificto remove DNA contamination. RNA concentration was determined by Nano Photometer® device (Implen, Munich, Germany). Purified RNAs were stored at −80 °C.

cDNA synthesis was performed using NG dART RT kit (EURx, Gdansk, Poland) according to the manufacturer's protocol with the following conditions: one cycle at 25 °C for 10 min, one cycle at 50 °C for 40 min, and one cycle at 85 °C for 5 min by employing C1000 Touch™ Thermal Cycler (Bio-Rad, Hercules, CA, USA). The samples of cDNAs were stored at −20 °C for further analysis. Expression of selected human genes associated with osteogenic (Osteocalcin, Osteopontin, *Runx2*), chondrogenic (*Acan, Col10A1, Col1A1, Sox9*), adipogenic (*CEBPα, PPARγ*), cardiomyogenic (*Gata-4, Myl2c, Nkx2.5*) or endothelial (*Gata-2, Tie-2*, VE-cadherin) differentiation were examined by real-time PCR using an ABI PRISM 7000 sequence detection system (Applied Biosystems, ThermoFisher Scientific, Waltham MA, USA). β2-microglobulin was used as a control housekeeping gene.

Real-time PCR was performed using SYBR Green qPCR Master Mix (EURx), cDNA template (10 ng), forward primer (1 μM) and reverse primer (1 μM; both from Genomed, Warsaw, Poland). The sequences of primers used are included in Table 1. Reactions were performed under the following conditions: one cycle at 50 °C for 2 min, one cycle at 95 °C for 10 min, followed by 40 cycles at 94 °C for 15 s, 60 °C for 30 s and 72 °C for 30 s. Relative quantification of genes expression was calculated using the comparative ddC$_t$ method.

Table 1. List of primers employed in real-time RT-PCR.

Gene	Sequences
β2-microglobulin	(F) 5′ AATGCGGCATCTTCAAACCT 3′ (R) 5′ TGACTTTGTCACAGCCCAAGATA 3′
ACAN	(F) 5′ AGGCAGCGTGATCCTTACC 3′ (R) 5′ GGCCTCTCCAGTCTCATTCTC 3′
Sox-9	(F) 5′ TGGGCAAGCTCTGGAGACTTC 3′ (R) 5′ ATCCGGGTGGTCCTTCTTGTG 3′
Col10A1-F	(F) 5′ GCAACTAAGGGCCTCAATGG 3′ (R) 5′ CTCAGGCATGACTGCTTGAC 3′
Col2A1	(F) 5′ CGTCCAGATGACCTTCCTACG 3′ (R) 5′ TGAGCAGGGCCTTCTTGAG 3′
Osteocalcin	(F) 5′ CGCTGGTCTCTTCACTAC 3′ (R) 5′ CTCACACTCCTCGCCCTATT 3′
Osteopontin	(F) 5′ ACTCGAACGACTCTGATGATGT 3′ (R) 5′ GTCAGGTCTGCGAAACTTCTTA 3′
Runx2	(F) 5′ GGAGTGGACGAGGCAAGAGTTT 3′ (R) 5′ AGCTTCTGTCTGTGCCTTCTGG 3′
PPARγ	(F) 5′ AGGCGAGGGCGATCTTGACAG 3′ (R) 5′ GATGCGGATGGCCACCTCTTT 3′
CEBPα	(F) 5′ AGGTTTCCTGCCTCCTTCC 3′ (R) 5′ CCCAAGTCCCTATGTTTCCA 3′
GATA-4	(F) 5′ AACGACGGCAACAACGATAAT 3′ (R) 5′ GTTTTTTCCCCTTTGATTTTTGATC 3′
Nkx2.5	(F) 5′ TGCTGCTCACAGGGCCCGATACTTC 3′ (R) 5′ TCCTTTCGAGCTCAGTGCACCACAAAC 3′
hMyl2A-F	(F) 5′ GGGCCCCATCAACTTCACCGTCTTCC 3′ (R) 5′ TGTAGTCGATGTTCCCCGCCAGGTCC 3′
Tie-2	(F) 5′ TCCCGAGGTCAAGAGGTGTA 3′ (R) 5′ AGGGTGTGCCTCCTAAGCTA 3′
GATA-2	(F) 5′ GCTCGTTCCTGTTCAGAAGG 3′ (R) 5′ GCCATAAGGTGGTGGTTGTC 3′
VE-cadherin	(F) 5′ TTTTCCAGCAGCCTTTCTACCA 3′ (R) 5′ GCGGATGGAGTATCCAATGCTA 3′

4.7. Histochemical Staining

On days 7, 14 and 21 of osteogenic, chondrogenic and adipogenic differentiation, the cells were washed with PBS and fixed with 4% paraformaldehyde (POCH, Avantor Performance Materials Poland S.A., Gliwice, Poland) for 30 min at RT.

To evaluate calcium phosphate deposition in the cells differentiated into osteoblasts, the cells were rinsed twice with distilled water following fixation and stained with 2% Alizarin Red S solution with pH 4.2 (Sigma-Aldrich) for 3 min.

To visualize sulphated proteoglycans that are characteristic for chondrogenic differentiation, fixed cells were rinsed with PBS and stained with 1% Alcian Blue solution (Sigma-Aldrichprepared in 0.1 N HCl (POCH, Avantor Performance Materials Poland S.A) for 30 min. The wells were rinsed thrice with 0.1 N HCl (POCH) and then distilled water was added to neutralize the acidity.

To visualize the presence of oil droplets during adipogenic differentiation, the cells were rinsed with distilled water and incubated with 60% isopropanol (POCH, Avantor Performance Materials Poland S.A) for 5 min. After the removal of isopropanol, the cells were stained with 1% Oil Red O solution (Sigma-Aldrich) for 5 min.

After histochemical staining, the cells were rinsed with distilled water and visualized using an Olympus IX81 microscope equipped with MicroPublisher 3.3 RTV camera (Olympus, Tokyo, Japan).

4.8. Immunocytochemistry

Immunocytochemistry staining was performed to evaluate (i) Gata-4 and troponin T-C expression in DPSCs and UC-MSCs differentiated into cardiomyocytes as well as (ii) Gata-2 and VE-cadherin expression in angiogenic differentiation. For this purpose, on days 7, 14 and 21 of the differentiation culture, the medium was removed, cells were washed with PBS (GE Healthcare Life Sciences HyClone Laboratories, Malborough, MA, USA), and fixed with 4% paraformaldehyde (POCH, Avantor Performance Materials Poland S.A.) for 20 min at RT. Cells were subsequently washed thrice with PBS, permeabilized with 0.1% Triton X-100 solution (Sigma-Aldrich) for 8 min at RT, and washed again thrice with PBS. Subsequently, cells were stained against: (i) cardiac-specific proteins–with primary anti-Gata-4 antibody (mouse monoclonal IgG$_{2a}$, 1:50; Santa Cruz Biotechnology, Dallas, TX, USA) and anti-troponin T-C antibody (goat polyclonal IgG, 1:20; Santa Cruz Biotechnology) (ii) endothelial-specific proteins–with primary anti-Gata-2 (rabbit polyclonal IgG, 1:50; Santa Cruz Biotechnology) and anti-VE-cadherin (mouse monoclonal IgG$_1$, 1:20; Santa Cruz Biotechnology) for 16 h at 4 °C. For (i) cardiac and (ii) endothelial markers detection, the following secondary antibodies were subsequently added, respectively: (i) donkey anti-mouse IgG antibody conjugated with Alexa Fluor 488 (1:250; Jackson ImmunoResearch, Cambridgeshire, UK) and donkey anti-goat antibody conjugated with Alexa Fluor 546 (1:250; ThermoFisher Scientific) (ii) goat anti-mouse IgG antibody conjugated with Alexa Fluor 546 (1:250; ThermoFisher Scientific) and goat anti-rabbit IgG antibody conjugated with Alexa Fluor 488 (1:250; ThermoFisher Scientific). The concentrations of the used primary and secondary antibodies were selected based on the optimization experiments conducted previously. Moreover, appropriate IgG controls were used to confirm the specificity of antibodies prior to the current study. Staining with secondary antibodies was performed for 2 h in 37 °C protected from light. Cells were further washed with PBS and nuclei were stained with DAPI (2 µM, ThermoFisher Scientific) for 15 min in 37 °C protected from light. VECTASHIELD Mounting Medium (Vector Laboratories, Burlingame, CA, USA) was used to mount coverslips. The preparations were analysed with Leica DMI6000B ver. AF7000 fluorescent microscope (Leica Microsystems GmbH, Welzlar, Germany). The control (undifferentiated) DPSCs and UC-MSCs were also stained with primary and secondary antibodies according to the protocol described above, and the results are presented in Figure S1.

4.9. Mechanical Characterization of the Hydrogel Matrix

Mechanical properties of BD PuraMatrix Peptide Hydrogel (Corning, Tewskbury, MA, USA) were investigated using atomic force microscopy (AFM, CellHesion head, JPK Instruments, Berlin, Germany) in force mapping mode. To probe hydrogel samples, commercially-available silicon nitride cantilevers (MLCT-C, Bruker, Billerica, MA, USA) with a nominal spring constant of 0.01 N/m were applied. Force curves, i.e., dependencies between cantilever deflection and relative sample position, were acquired over a grid of 8 × 8 pixels within a scan area of 50 × 50 µm. The maximum load force (F) was 5 nN and load speed of 8 µm/s was maintained. Young's modulus was determined using Hertz contact mechanics as described previously [77]. Briefly, the following relation between the load force (F) and resulting indentation (Δz) for the paraboloidal assumption of the probing tip was applied:

$$F = \frac{4 \cdot \sqrt{R} \cdot E}{3 \cdot (1 - v^2)} \cdot \Delta z^{\frac{3}{2}} \tag{1}$$

Theorem 1. *Maximum load force (F). In this theorem, R is the radius of tip curvature, E is Young's modulus and v is the Poisson ratio of the material (here assumed to be 0.5 treating hydrogels as incompressible material).*

JPK Data Processing software was used to apply this equation to the experimental data to obtain the value of Young's modulus for each force curves. The final Young's modulus was obtained by averaging all force curves and was expressed as a mean and standard deviation.

4.10. 3D Encapsulation of DPSCs within the Hydrogel Matrix

BD PuraMatrix Peptide Hydrogel (Corning) with a concentration of 1% (*w/v*) after decreasing the viscosity by vortexing was diluted with a cell culture-grade water (PAA,) to the concentration of 0.3%. The cells re-suspended in a 20% sucrose solution (Sigma-Aldrich) were centrifuged at 320× *g* for 7 min at RT to wash out residual salts from the culture medium (to prevent early hydrogel gelation). After centrifugation, the cells were re-suspended in a 20% sucrose solution, and the cell number and viability were assessed by Countess Automated Cell Counter as described in Section 4.3. To obtain a final hydrogel concentration of 0.15%, 0.3% hydrogel was diluted with the same volume of cell suspension in 20% sucrose solution (at 2× the final desired cell concentration). Thus, the final concentrations of hydrogel and sucrose were equal to 0.15% and 10%, respectively. The whole volume was carefully mixed and added to the centre of each well without introducing bubbles. The following volumes of the mixtures were used: 50 μL/well in a 96-well plate, 250 μL/well in a 24-well plate, and 400 μL/well in a 12-well plate (Corning) and are presented in Table 2. The gelation of the BD PuraMatrix hydrogel was initiated by the addition of the complete cell culture medium (DMEM/F12 supplemented with 10% FBS, Sigma-Aldrich; and 100 IU/mL penicillin, 100 μg/mL streptomycin, Gibco, ThermoFisher Scientific) and by gently running culture media down the side of the well on top of the hydrogel. Within 1 h post 3D cell encapsulation in hydrogel and following their gelation, 70% of the medium was changed twice to stabilize pH. The DPSCs were cultured for two days under standard culture conditions.

Table 2. Preparing of 3D encapsulation of DPSCs within the hydrogel matrix.

	3D Cell Encapsulation		
Type of Cell Culture Plate	Volume of Mixture: 0.15% Hydrogel + 10% Sucrose (Per Well)	Number of Encapsulated Cells	Volume of Added Cell Culture Media (for Gelation)
96-well	50 μL	2.0×10^3	100 μL
24-well	250 μL	5.0×10^4	500 μL
12-well	400 μL	1.0×10^5	800 μL

4.11. 3D and 2D Culture of DPSCs in Vitro

To establish a 2D culture of the DPSCs, cell culture plates (Corning) were coated with 0.1% gelatin (Sigma-Aldrich). The cell suspension was prepared in 20% sucrose solution and seeded at a concentration of 2×10^4/well in a 12-well plate in the complete cell culture medium (DMEM/F12 supplemented with 10% FBS, Sigma-Aldrich; and 100 IU/mL penicillin, 100 μg/mL streptomycin, Gibco, ThermoFisher Scientific, Waltham, MA, USA). DPSCs were cultured for two days under standard culture conditions. DPSCs encapsulated in the 3D hydrogel or seeded on 2D gelatin-coated cell culture plates were further cultured at 37 °C in a humidified atmosphere containing 5% CO_2 and the following oxygen concentrations were used: (i) 2% of O_2 (hypoxia) or (ii) about 18% of O_2 (normoxia) for 7, 14 and 21 days.

4.12. Assessment of DPSC Proliferation and Metabolic Activity In Vitro

The proliferation and metabolic activity of DPSCs in 3D or 2D culture in the environment containing 2% (hypoxia) or about 18% of O_2 (normoxia) were measured by MTS assay or by analysing ATP content, respectively. The tests were performed every 24 h until seven days post cell encapsulation/seeding.

For MTS assay, DPSCs were encapsulated in the hydrogel or seeded onto gelatin-coated surfaces of 96-well transparent plates (Corning, Tewskbury MA, USA) at a density of 10^3 cells/well for 2D culture conditions and 2×10^3 cells/well for 3D culture conditions. The analysis was performed using the Cell Counting Kit-8 (Sigma-Aldrich, St. Louis MO, USA). For this purpose, 50 μL of WST-8 reagent was added into each well and incubated for 4 h. The absorbance was measured at 450 nm wavelength using the Multiskan FC Microplate Photometer (ThermoFisher Scientific, Waltham MA, USA). For the measurement of ATP concentrations, DPSCs were encapsulated in the hydrogel or seeded onto gelatin-coated surfaces of 96-well white plates (Perkin Elmer, Waltham, MA, USA). Subsequently, the assay was conducted using the ATP Lite Luminescence assay kit according to the manufacturer's instructions (Perkin ElmerLuminescence was measured using the Infinite® M200PRO plate reader (Tecan, Mannedorf, Zurich, Switzerland).

4.13. Osteogenic Differentiation of DPSCs in 3D or 2D in Vitro Culture

Osteogenic differentiation of DPSCs in 3D or 2D cultures was initiated 1–2 days after the encapsulation of DPSCs in hydrogel (3D culture) or seeding of DPSCs into cell culture plates coated with 0.1% gelatin (2D culture). For this purpose, a StemPro Osteogenesis Differentiation Kit (ThermoFisher Scientific) was used. The differentiation medium was changed every 3–4 days. On days 7, 14 and 21 of osteogenic differentiation, DPSCs from 3D and 2D culture were recovered for the analysis of expression of osteogenesis-associated genes by Real-Time RT-PCR. Moreover, the presence of calcium phosphate deposits was confirmed by the staining of cells in the hydrogel or those seeded on the 2D surface by Alizarin Red S solution as described in Section 4.7. For 3D culture, a larger number of washes (approx. 4–6) were performed to reduce the red background staining of the hydrogel.

4.14. Recovery of DPSCs from the Hydrogel Matrix

The DPSCs were isolated from the hydrogel according to the Puramatrix Peptide Hydrogel manufacturer's protocol titled "Cell recovery for sub-culturing or biochemical analyses" (Corning). Briefly, the hydrogel with the culture medium was mechanically disrupted by pipetting. The suspension was then transferred to a 15 mL centrifuge tube. The wells were washed with PBS (GE Healthcare Life Sciences HyClone Laboratories) to collect the remaining hydrogel fragments, transferred into a centrifuge tube, and centrifuged at 320× g for 7 min at RT. The hydrogel pellet was re-suspended in PBS (GE Healthcare Life Sciences HyClone Laboratories) and centrifuged at 320× g for 7 min at RT. Subsequently, the hydrogel pellet was re-suspended in 0.25% Trypsin/EDTA (Gibco; ThermoFisher Scientific) following digestion for approx. 10 min at 37 °C. The trypsin was inactivated with complete cell culture medium and the suspension was centrifuged at 320× g for 7 min at RT.

4.15. Statistical Analysis

Data are represented as mean ± SD or SEM as indicated. Statistical analyses were performed with Student's t-test. $p < 0.05$ was considered statistically significant.

5. Conclusions

In conclusion, DPSCs exhibit: (i) higher osteogenic differentiation capacity, (ii) comparable chondrogenic and adipogenic differentiation potential and (iii) limited ability for the cardiac or endothelial phenotype in comparison with the other "classic" MSCs (UC-MSCs). The current results may help determine the future direction of the application of these cells in regenerative therapies.

Importantly, 3D cell encapsulation as well as the low concentration of O2 resembling conditions in the stem cell niches may favour osteogenic differentiation of DPSCs in an in vitro environment. The positive impact of hypoxia on the osteogenic potential of DPSCs was visible notably in 3D culture conditions.

Thus, tissue engineering approaches combining DPSCs, 3D biomaterial scaffolds, and other stimulating chemical factors may represent new innovative paths in the development of tissue repair.

Supplementary Materials
Figure S1. Immunocytochemical staining of DPSCs and UC-MSCs at 7 day of standard cell culture. Figure S2. Expression of osteopontin during osteogenic differentiation of DPSCs encapsulated in hydrogel (3D) or seeded on the surface coated with gelatin (2D) cultured in hypoxic (2% O2) or normoxic (18% O2) environment. Table S1. Fold change in mRNA expression for osteogenesis related genes (osteocalcin, osteopontin, Runx2) in DPSCs and UC-MSCs by Real-Time RT-PCR. Table S2. Fold change in mRNA expression for chondrogenesis related genes

(Acan, Col10A1, Col2A1, Sox9) in DPSCs and UC-MSCs by Real-Time RT-PCR. Table S3. Fold change in mRNA expression for adipogenesis related genes (*CEBPα, PPARγ*) in DPSCs and UC-MSCs by Real-Time RT-PCR. Table S4. Fold change in mRNA expression for cardiomyogenesis related genes (*Gata-4, Nkx2.5, Myl2c*) in DPSCs and UC-MSCs by Real-Time RT-PCR. Table S5. Fold change in mRNA expression for endothelial related genes (*Gata-2, Tie-2*, VE-cadherin) in DPSCs and UC-MSCs by Real-Time RT-PCR.

Author Contributions: Conceptualization: A.L.-M. and E.Z.-S.; methodology: A.L.-M., T.K., M.S.-S., S.N. and E.Z.-S.; validation: A.L.-M., Z.M. and E.Z.-S.; formal analysis: A.L.-M., N.B. and A.K.; investigation: A.L.-M., N.B. and A.K.; resources: T.K., D.B., M.S.-S., S.N., E.Z.-S. and Z.M.; writing—original draft preparation: A.L.-M., N.B. and A.K.; writing—review and editing: E.Z.-S. and T.K.; visualization: A.L.-M., N.B. and A.K.; supervision, financial and logistical support: E.Z.-S. and Z.M.; project administration: E.Z.-S.; funding acquisition: E.Z.-S. All authors have read and agreed to the published version of the manuscript.

Acknowledgments: The authors would like to thank Sylwia Bobis-Wozowicz from Department of Cell Biology, Faculty of Biochemistry, Biophysics and Biotechnology, Jagiellonian University, Krakow, Poland for designing primer sequences used in this study. We would also to thank Małgorzata Lekka from the Institute of Nuclear Physics, Polish Academy of Sciences, Krakow, Poland for access to AFM and providing conceptual support during analyses of AFM data.

Abbreviations

BM	Bone marrow
BM-MSCS	Bone marrow-derived mesenchymal stem/stromal cells
DPSCs	Dental pulp stem cells
ISCT	International Society for Cellular Therapy
MSCs	Mesenchymal stem/stromal cells
O_2	Oxygen
SCs	Stem cells
UC-MSCs	Umbilical cord Wharton's jelly-derived mesenchymal stem/stromal cells
2D	Two-dimensional
3D	Three-dimensional

References

1. Uccelli, A.; Moretta, L.; Pistoia, V. Mesenchymal stem cells in health and disease. *Nat. Rev. Immunol.* **2008**, *8*, 726–736. [CrossRef]

2. Berebichez-Fridman, R.; Montero-Olvera, P.R. Sources and clinical applications of mesenchymal stem cells state-of-the-art review. *Sultan Qaboos Univ. Med. J.* **2018**, *18*, e264–e277. [CrossRef]

3. Dominici, M.; Le Blanc, K.; Mueller, I.; Slaper-Cortenbach, I.; Marini, F.C.; Krause, D.S.; Deans, R.J.; Keating, A.; Prockop, D.J.; Horwitz, E.M. Minimal criteria for defining multipotent mesenchymal stromal cells. The International Society for Cellular Therapy position statement. *Cytotherapy* **2006**, *8*, 315–317. [CrossRef] [PubMed]

4. Labovsky, V.; Hofer, E.L.; Feldman, L.; Fernández Vallone, V.; García Rivello, H.; Bayes-Genis, A.; Hernando Insúa, A.; Levin, M.J.; Chasseing, N.A. Cardiomyogenic differentiation of human bone marrow mesenchymal cells: Role of cardiac extract from neonatal rat cardiomyocytes. *Differentiation* **2010**, *79*, 93–101. [CrossRef] [PubMed]

5. Guo, X.; Bai, Y.; Zhang, L.; Zhang, B.; Zagidullin, N.; Carvalho, K.; Du, Z.; Cai, B. Cardiomyocyte differentiation of mesenchymal stem cells from bone marrow: New regulators and its implications. *Stem Cell Res. Ther.* **2018**, *9*, 1–12. [CrossRef] [PubMed]

6. Labedz-Maslowska, A.; Lipert, B.; Berdecka, D.; Kedracka-Krok, S.; Jankowska, U.; Kamycka, E.; Sekula, M.; Madeja, Z.; Dawn, B.; Jura, J.; et al. Monocyte chemoattractant protein-induced protein 1 (MCPIP1) enhances angiogenic and cardiomyogenic potential of murine bone marrow-derived mesenchymal stem cells. *PLoS ONE* **2015**, *10*, e0133746. [CrossRef] [PubMed]

7. Janeczek Portalska, K.; Leferink, A.; Groen, N.; Fernandes, H.; Moroni, L.; van Blitterswijk, C.; de Boer, J. Endothelial Differentiation of Mesenchymal Stromal Cells. *PLoS ONE* **2012**, *7*, e46842. [CrossRef] [PubMed]
8. Anjos-Afonso, F.; Siapati, E.K.; Bonnet, D. In vivo contribution of murine mesenchymal stem cells into multiple cell-types under minimal damage conditions. *J. Cell Sci.* **2004**, *117*, 5655–5664. [CrossRef]
9. Jin, Y.Z.; Lee, J.H. Mesenchymal stem cell therapy for bone regeneration. *CiOS Clin. Orthop. Surg.* **2018**, *10*, 271–278. [CrossRef]
10. Murphy, J.M.; Fink, D.J.; Hunziker, E.B.; Barry, F.P. Stem Cell Therapy in a Caprine Model of Osteoarthritis. *Arthritis Rheum.* **2003**, *48*, 3464–3474. [CrossRef]
11. Alge, D.L.; Zhou, D.; Adams, L.L.; Wyss, B.K.; Shadday, M.D.; Woods, E.J.; Chu, T.M.G.; Goebel, W.S. Donor-matched comparison of dental pulp stem cells and bone marrow-derived mesenchymal stem cells in a rat model. *J. Tissue Eng. Regen. Med.* **2010**, *4*, 73–81. [CrossRef] [PubMed]
12. Chu, D.-T.; Phuong, T.N.T.; Tien, N.L.B.; Tran, D.K.; Van Thanh, V.; Quang, T.L.; Truong, D.T.; Pham, V.H.; Ngoc, V.T.N.; Chu-Dinh, T.; et al. An Update on the Progress of Isolation, Culture, Storage, and Clinical Application of Human Bone Marrow Mesenchymal Stem/Stromal Cells. *Int. J. Mol. Sci.* **2020**, *21*, 708. [CrossRef] [PubMed]
13. Mueller, S.M.; Glowacki, J. Age-related decline in the osteogenic potential of human bone marrow cells cultured in three-dimensional collagen sponges. *J. Cell. Biochem.* **2001**, *82*, 583–590. [CrossRef] [PubMed]
14. Cianfarani, F.; Toietta, G.; Di Rocco, G.; Cesareo, E.; Zambruno, G.; Odorisio, T. Diabetes impairs adipose tissue-derived stem cell function and efficiency in promoting wound healing. *Wound Repair Regen.* **2013**, *21*, 545–553. [CrossRef] [PubMed]
15. Nie, Y.; Lau, C.S.; Lie, A.K.W.; Chan, G.C.F.; Mok, M.Y. Defective phenotype of mesenchymal stem cells in patients with systemic lupus erythematosus. *Lupus* **2010**, *19*, 850–859. [CrossRef]
16. Marmotti, A.; Mattia, S.; Castoldi, F.; Barbero, A.; Mangiavini, L.; Bonasia, D.E.; Bruzzone, M.; Dettoni, F.; Scurati, R.; Peretti, G.M. Allogeneic Umbilical Cord-Derived Mesenchymal Stem Cells as a Potential Source for Cartilage and Bone Regeneration: An in Vitro Study. *Stem Cells Int.* **2017**, *2017*, 1732094. [CrossRef]
17. El Omar, R.; Beroud, J.; Stoltz, J.F.; Menu, P.; Velot, E.; Decot, V. Umbilical cord mesenchymal stem cells: The new gold standard for mesenchymal stem cell-based therapies? *Tissue Eng. Part B Rev.* **2014**, *20*, 523–544. [CrossRef]
18. Gronthos, S.; Mankani, M.; Brahim, J.; Robey, P.G.; Shi, S. Postnatal human dental pulp stem cells (DPSCs) in vitro and in vivo. *Proc. Natl. Acad. Sci. USA* **2000**, *97*, 13625–13630. [CrossRef]
19. Dupin, E.; Sommer, L. Neural crest progenitors and stem cells: From early development to adulthood. *Dev. Biol.* **2012**, *366*, 83–95. [CrossRef]
20. Kaukua, N.; Shahidi, M.K.; Konstantinidou, C.; Dyachuk, V.; Kaucka, M.; Furlan, A.; An, Z.; Wang, L.; Hultman, I.; Ährlund-Richter, L.; et al. Glial origin of mesenchymal stem cells in a tooth model system. *Nature* **2014**, *513*, 551–554. [CrossRef]
21. D'aquino, R.; de Rosa, A.; Laino, G.; Caruso, F.; Guida, L.; Rullo, R.; Checchi, V.; Laino, L.; Tirino, V.; Papaccio, G. Human dental pulp stem cells: From biology to clinical applications. *J. Exp. Zool. Part B Mol. Dev. Evol.* **2009**, *312*, 408–415. [CrossRef]
22. Arthur, A.; Rychkov, G.; Shi, S.; Koblar, S.A.; Gronthos, S. Adult Human Dental Pulp Stem Cells Differentiate Toward Functionally Active Neurons Under Appropriate Environmental Cues. *Stem Cells* **2008**, *26*, 1787–1795. [CrossRef] [PubMed]
23. Young, F.; Sloan, A.; Song, B. Dental pulp stem cells and their potential roles in central nervous system regeneration and repair. *J. Neurosci. Res.* **2013**, *91*, 1383–1393. [CrossRef] [PubMed]
24. Pisciotta, A.; Carnevale, G.; Meloni, S.; Riccio, M.; De Biasi, S.; Gibellini, L.; Ferrari, A.; Bruzzesi, G.; De Pol, A. Human Dental pulp stem cells (hDPSCs): Isolation, enrichment and comparative differentiation of two sub-populations Integrative control of development. *BMC Dev. Biol.* **2015**, *15*, 14. [CrossRef] [PubMed]
25. Gervois, P.; Struys, T.; Hilkens, P.; Bronckaers, A.; Ratajczak, J.; Politis, C.; Brône, B.; Lambrichts, I.; Martens, W. Neurogenic Maturation of Human Dental Pulp Stem Cells Following Neurosphere Generation Induces Morphological and Electrophysiological Characteristics of Functional Neurons. *Stem Cells Dev.* **2015**, *24*, 296–311. [CrossRef] [PubMed]
26. Bianco, P.; Cao, X.; Frenette, P.S.; Mao, J.J.; Robey, P.G.; Simmons, P.J.; Wang, C.Y. The meaning, the sense and the significance: Translating the science of mesenchymal stem cells into medicine. *Nat. Med.* **2013**, *19*, 35–42. [CrossRef] [PubMed]

27. Leong, W.K.; Henshall, T.L.; Arthur, A.; Kremer, K.L.; Lewis, M.D.; Helps, S.C.; Field, J.; Hamilton-Bruce, M.A.; Warming, S.; Manavis, J.; et al. Human Adult Dental Pulp Stem Cells Enhance Poststroke Functional Recovery Through Non-Neural Replacement Mechanisms. *Stem Cells Transl. Med.* **2012**, *1*, 177–187. [CrossRef]

28. Nagpal, A.; Kremer, K.L.; HamiltonBruce, M.A.; Kaidonis, X.; Milton, A.G.; Levi, C.; Shi, S.; Carey, L.; Hillier, S.; Rose, M.; et al. TOOTH (The Open study Of dental pulp stem cell Therapy in Humans): Study protocol for evaluating safety and feasibility of autologous human adult dental pulp stem cell therapy in patients with chronic disability after stroke. *Int. J. Stroke* **2016**, *11*, 575585. [CrossRef]

29. Bryniarska, N.; Kubiak, A.; Labędz-Maslowska, A.; Zuba-Surma, E. Impact of developmental origin, niche mechanics and oxygen availability on osteogenic differentiation capacity of mesenchymal stem/stromal cells. *Acta Biochim. Pol.* **2019**, *66*, 491–498. [CrossRef]

30. Kini, U.; Nandeesh, B.N. Physiology of bone formation, remodeling, and metabolism. In *Radionuclide and Hybrid Bone Imaging*; Springer: Berlin\Heidelberg, Germany, 2012; Volume 9783642024, pp. 29–57. ISBN 9783642024009.

31. Theveneau, E.; Mayor, R. Neural crest migration: Interplay between chemorepellents, chemoattractants, contact inhibition, epithelial-mesenchymal transition, and collective cell migration. *Wiley Interdiscip. Rev. Dev. Biol.* **2012**, *1*, 435–445. [CrossRef]

32. Baranova, J.; Büchner, D.; Götz, W.; Schulze, M.; Tobiasch, E. Tooth Formation: Are the Hardest Tissues of Human Body Hard to Regenerate? *Int. J. Mol. Sci.* **2020**, *21*, 4031. [CrossRef] [PubMed]

33. Lukomska, B.; Stanaszek, L.; Zuba-Surma, E.; Legosz, P.; Sarzynska, S.; Drela, K. Challenges and Controversies in Human Mesenchymal Stem Cell Therapy. *Stem Cells Int.* **2019**, *37*, 855–864. [CrossRef] [PubMed]

34. Barnes, J.M.; Przybyla, L.; Weaver, V.M. Tissue mechanics regulate brain development, homeostasis and disease. *J. Cell Sci.* **2017**, *130*, 71–82. [CrossRef] [PubMed]

35. Westhrin, M.; Xie, M.; Olderøy, M.; Sikorski, P.; Strand, B.L.; Standal, T. Osteogenic differentiation of human mesenchymal stem cells in mineralized alginate matrices. *PLoS ONE* **2015**, *10*, e0120374. [CrossRef] [PubMed]

36. Riccio, M.; Resca, E.; Maraldi, T.; Pisciotta, A.; Ferrari, A.; Bruzzesi, G.; de Pol, A. Human dental pulp stem cells produce mineralized matrix in 2D and 3D cultures. *Eur. J. Histochem.* **2010**, *54*, 205–213. [CrossRef]

37. Mohyeldin, A.; Garzón-Muvdi, T.; Quiñones-Hinojosa, A. Oxygen in Stem Cell Biology: A Critical Component of the Stem Cell Niche. *Cell Stem Cell* **2010**, *7*, 150–161. [CrossRef]

38. Kakudo, N.; Morimoto, N.; Ogawa, T.; Taketani, S.; Kusumoto, K. Hypoxia enhances proliferation of human adipose-derived stem cells via HIF-1α activation. *PLoS ONE* **2015**, *10*, e0139890. [CrossRef]

39. Grayson, W.L.; Zhao, F.; Bunnell, B.; Ma, T. Hypoxia enhances proliferation and tissue formation of human mesenchymal stem cells. *Biochem. Biophys. Res. Commun.* **2007**, *358*, 948–953. [CrossRef]

40. Sakdee, J.B.; White, R.R.; Pagonis, T.C.; Hauschka, P. V Hypoxia-amplified Proliferation of Human Dental Pulp Cells. *J. Endod.* **2009**, *35*, 818–823. [CrossRef]

41. Kumar, S.; Vaidya, M. Hypoxia inhibits mesenchymal stem cell proliferation through HIF1α-dependent regulation of P27. *Mol. Cell. Biochem.* **2016**, *415*, 29–38. [CrossRef]

42. Hsu, S.-H.; Chen, C.-T.; Wei, Y.-H. Inhibitory effects of hypoxia on metabolic switch and osteogenic differentiation of human mesenchymal stem cells. *Stem Cells* **2013**, *31*, 2779–2788. [CrossRef] [PubMed]

43. Noda, S.; Kawashima, N.; Yamamoto, M.; Hashimoto, K.; Nara, K.; Sekiya, I.; Okiji, T. Effect of cell culture density on dental pulp-derived mesenchymal stem cells with reference to osteogenic differentiation. *Sci. Rep.* **2019**, *9*, 5430. [CrossRef] [PubMed]

44. Gu, J.; Lu, Y.; Li, F.; Qiao, L.; Wang, Q.; Li, N.; Borgia, J.A.; Deng, Y.; Lei, G.; Zheng, Q. Identification and characterization of the novel Col10a1 regulatory mechanism during chondrocyte hypertrophic differentiation. *Cell Death Dis.* **2014**, *5*, e1469. [CrossRef] [PubMed]

45. Bilezikan, J.P.; Raisz, L.G.; Martin, T.J. Principles of Bone Biology, 3rd ed. *Am. J. Neuroradiol.* **2009**, *30*, E139. [CrossRef]

46. Nam, S.; Chaudhuri, O. Mitotic cells generate protrusive extracellular forces to divide in three-dimensional microenvironments. *Nat. Phys.* **2018**, *14*, 621–628. [CrossRef]

47. De Fusco, C.; Messina, A.; Monda, V.; Viggiano, E.; Moscatelli, F.; Valenzano, A.; Esposito, T.; Chieffi, S.; Cibelli, G.; Monda, M.; et al. Osteopontin: Relation between Adipose Tissue and Bone Homeostasis. *Stem Cells Int.* **2017**, *2017*. [CrossRef] [PubMed]

48. Lan, X.; Sun, Z.; Chu, C.; Boltze, J.; Li, S. Dental pulp stem cells: An attractive alternative for cell therapy in ischemic stroke. *Front. Neurol.* **2019**, *10*, 824. [CrossRef]

49. Sakai, K.; Yamamoto, A.; Matsubara, K.; Nakamura, S.; Naruse, M.; Yamagata, M.; Sakamoto, K.; Tauchi, R.; Wakao, N.; Imagama, S.; et al. Human dental pulp-derived stem cells promote locomotor recovery after complete transection of the rat spinal cord by multiple neuro-regenerative mechanisms. *J. Clin. Investig.* **2012**, *122*, 80–90. [CrossRef]

50. Cheng, P.H.; Snyder, B.; Fillos, D.; Ibegbu, C.C.; Huang, A.H.C.; Chan, A.W.S. Postnatal stem/progenitor cells derived from the dental pulp of adult chimpanzee. *BMC Cell Biol.* **2008**, *9*, 20. [CrossRef]

51. Király, M.; Porcsalmy, B.; Pataki, Á.; Kádár, K.; Jelitai, M.; Molnár, B.; Hermann, P.; Gera, I.; Grimm, W.D.; Ganss, B.; et al. Simultaneous PKC and cAMP activation induces differentiation of human dental pulp stem cells into functionally active neurons. *Neurochem. Int.* **2009**, *55*, 323–332. [CrossRef]

52. Davis, J. *Tissue Regeneration—From Basic Biology to Clinical Application*; IntechOpen: London, UK, 2012; ISBN 978-953-51-0387-5. [CrossRef]

53. Ibarretxe, G.; Crende, O.; Aurrekoetxea, M.; García-Murga, V.; Etxaniz, J.; Unda, F. Neural crest stem cells from dental tissues: A new hope for dental and neural regeneration. *Stem Cells Int.* **2012**, *2012*, 103503. [CrossRef] [PubMed]

54. Ellis, K.M.; O'Carroll, D.C.; Lewis, M.D.; Rychkov, G.Y.; Koblar, S.A. Neurogenic potential of dental pulp stem cells isolated from murine incisors. *Stem Cell Res. Ther.* **2014**, *5*, 30. [CrossRef] [PubMed]

55. Camci-Unal, G.; Laromaine, A.; Hong, E.; Derda, R.; Whitesides, G.M. Biomineralization Guided by Paper Templates. *Sci. Rep.* **2016**, *6*, 1–12. [CrossRef]

56. Koussounadis, A.; Langdon, S.P.; Um, I.H.; Harrison, D.J.; Smith, V.A. Relationship between differentially expressed mRNA and mRNA-protein correlations in a xenograft model system. *Sci. Rep.* **2015**, *5*, 10775. [CrossRef]

57. Fujii, Y.; Kawase-Koga, Y.; Hojo, H.; Yano, F.; Sato, M.; Chung, U.-I.; Ohba, S.; Chikazu, D. Bone regeneration by human dental pulp stem cells using a helioxanthin derivative and cell-sheet technology. *Stem Cell Res. Ther.* **2018**, *9*, 215. [CrossRef]

58. Rizzatti, V.; Boschi, F.; Pedrotti, M.; Zoico, E.; Sbarbati, A.; Zamboni, M. Lipid droplets characterization in adipocyte differentiated 3T3-L1 cells: Size and optical density distribution. *Eur. J. Histochem.* **2013**, *57*. [CrossRef]

59. Armiñán, A.; Gandía, C.; Bartual, M.; García-Verdugo, J.M.; Lledó, E.; Mirabet, V.; Llop, M.; Barea, J.; Montero, J.A.; Sepúlveda, P. Cardiac differentiation is driven by nkx2.5 and gata4 nuclear translocation in tissue-specific mesenchymal stem cells. *Stem Cells Dev.* **2009**, *18*, 907–917. [CrossRef]

60. Abbott, A. Biology's new dimension. *Nature* **2003**, *424*, 870–872. [CrossRef]

61. Spencer, J.A.; Ferraro, F.; Roussakis, E.; Klein, A.; Wu, J.; Runnels, J.M.; Zaher, W.; Mortensen, L.J.; Alt, C.; Turcotte, R.; et al. Direct measurement of local oxygen concentration in the bone marrow of live animals. *Nature* **2014**, *508*, 269–273. [CrossRef]

62. Wenger, R.; Kurtcuoglu, V.; Scholz, C.; Marti, H.; Hoogewijs, D. Frequently asked questions in hypoxia research. *Hypoxia* **2015**, *3*, 35–43. [CrossRef]

63. Kwon, S.Y.; Chun, S.Y.; Ha, Y.S.; Kim, D.H.; Kim, J.; Song, P.H.; Kim, H.T.; Yoo, E.S.; Kim, B.S.; Kwon, T.G. Hypoxia Enhances Cell Properties of Human Mesenchymal Stem Cells. *Tissue Eng. Regen. Med.* **2017**, *14*, 595–604. [CrossRef] [PubMed]

64. Iida, K.; Takeda-Kawaguchi, T.; Tezuka, Y.; Kunisada, T.; Shibata, T.; Tezuka, K.I. Hypoxia enhances colony formation and proliferation but inhibits differentiation of human dental pulp cells. *Arch. Oral Biol.* **2010**, *55*, 648–654. [CrossRef] [PubMed]

65. Werle, S.B.; Chagastelles, P.; Pranke, P.; Casagrande, L. The effects of hypoxia on in vitro culture of dental-derived stem cells. *Arch. Oral Biol.* **2016**, *68*, 13–20. [CrossRef] [PubMed]

66. Ikezoe, T.; Daar, E.S.; Hisatake, J.I.; Taguchi, H.; Koeffler, H.P. HIV-1 protease inhibitors decrease proliferation and induce differentiation of human myelocytic leukemia cells. *Blood* **2000**, *96*, 3553–3559. [CrossRef]

67. Cukierman, E.; Pankov, R.; Yamada, K.M. Cell interactions with three-dimensional matrices. *Curr. Opin. Cell Biol.* **2002**, *14*, 633–640. [CrossRef]

68. Valorani, M.G.; Montelatici, E.; Germani, A.; Biddle, A.; D'Alessandro, D.; Strollo, R.; Patrizi, M.P.; Lazzari, L.; Nye, E.; Otto, W.R.; et al. Pre-culturing human adipose tissue mesenchymal stem cells under hypoxia increases their adipogenic and osteogenic differentiation potentials. *Cell Prolif.* **2012**, *45*, 225–238. [CrossRef]

69. Basciano, L.; Nemos, C.; Foliguet, B.; de Isla, N.; de Carvalho, M.; Tran, N.; Dalloul, A. Long term culture of mesenchymal stem cells in hypoxia promotes a genetic program maintaining their undifferentiated and multipotent status. *BMC Cell Biol.* **2011**, *12*, 12. [CrossRef]

70. Buravkova, L.B.; Andreeva, E.R.; Gogvadze, V.; Zhivotovsky, B. Mesenchymal stem cells and hypoxia: Where are we ? Mitochondrion Mesenchymal stem cells and hypoxia: Where are we? *MITOCH* **2014**, *19*, 105–112. [CrossRef]

71. Persson, M.; Lehenkari, P.P.; Berglin, L.; Turunen, S.; Finnilä, M.A.J.; Risteli, J.; Skrifvars, M.; Tuukkanen, J. Osteogenic Differentiation of Human Mesenchymal Stem cells in a 3D Woven Scaffold. *Sci. Rep.* **2018**, *8*, 1–12. [CrossRef]

72. Diomede, F.; Marconi, G.D.; Fonticoli, L.; Pizzicanella, J.; Merciaro, I.; Bramanti, P.; Mazzon, E.; Trubiani, O. Functional relationship between osteogenesis and angiogenesis in tissue regeneration. *Int. J. Mol. Sci.* **2020**, *21*, 3242. [CrossRef]

73. Rapino, M.; Di Valerio, V.; Zara, S.; Gallorini, M.; Marconi, G.D.; Sancilio, S.; Marsich, E.; Ghinassi, B.; Di Giacomo, V.; Cataldi, A. Chitlac-coated thermosets enhance osteogenesis and angiogenesis in a co-culture of dental pulp stem cells and endothelial cells. *Nanomaterials* **2019**, *9*, 928. [CrossRef] [PubMed]

74. Wu, W.; Zhou, J.; Xu, C.T.; Zhang, J.; Jin, Y.J.; Sun, G.L. Derivation and growth characteristics of dental pulp stem cells from patients of different ages. *Mol. Med. Rep.* **2015**, *12*, 5127–5134. [CrossRef] [PubMed]

75. Yildirim, S.; Zibandeh, N.; Genc, D.; Ozcan, E.M.; Goker, K.; Akkoc, T. The comparison of the immunologic properties of stem cells isolated from human exfoliated deciduous teeth, dental pulp, and dental follicles. *Stem Cells Int.* **2016**, *2016*. [CrossRef]

76. Monterubbianesi, R.; Bencun, M.; Pagella, P.; Woloszyk, A.; Orsini, G.; Mitsiadis, T.A. A comparative in vitro study of the osteogenic and adipogenic potential of human dental pulp stem cells, gingival fibroblasts and foreskin fibroblasts. *Sci. Rep.* **2019**, *9*, 1–13. [CrossRef] [PubMed]

77. Lekka, M. Discrimination Between Normal and Cancerous Cells Using AFM. *Bionanoscience* **2016**, *6*, 65–80. [CrossRef] [PubMed]

9

Modeling Rheumatoid Arthritis In Vitro: From Experimental Feasibility to Physiological Proximity

Alexandra Damerau [1,2] **and Timo Gaber** [1,2,*]

[1] Charité—Universitätsmedizin Berlin, corporate member of Freie Universität Berlin, Humboldt-Universität zu Berlin, and Berlin Institute of Health, Department of Rheumatology and Clinical Immunology, 10117 Berlin, Germany; alexandra.damerau@charite.de

[2] German Rheumatism Research Centre (DRFZ) Berlin, a Leibniz Institute, 10117 Berlin, Germany

* Correspondence: timo.gaber@charite.de

Abstract: Rheumatoid arthritis (RA) is a chronic, inflammatory, and systemic autoimmune disease that affects the connective tissue and primarily the joints. If not treated, RA ultimately leads to progressive cartilage and bone degeneration. The etiology of the pathogenesis of RA is unknown, demonstrating heterogeneity in its clinical presentation, and is associated with autoantibodies directed against modified self-epitopes. Although many models already exist for RA for preclinical research, many current model systems of arthritis have limited predictive value because they are either based on animals of phylogenetically distant origin or suffer from overly simplified in vitro culture conditions. These limitations pose considerable challenges for preclinical research and therefore clinical translation. Thus, a sophisticated experimental human-based in vitro approach mimicking RA is essential to (i) investigate key mechanisms in the pathogenesis of human RA, (ii) identify targets for new therapeutic approaches, (iii) test these approaches, (iv) facilitate the clinical transferability of results, and (v) reduce the use of laboratory animals. Here, we summarize the most commonly used in vitro models of RA and discuss their experimental feasibility and physiological proximity to the pathophysiology of human RA to highlight new human-based avenues in RA research to increase our knowledge on human pathophysiology and develop effective targeted therapies.

Keywords: in vitro models; rheumatoid arthritis; cytokines; mesenchymal stromal cells; co-culture; tissue engineering; 3D cell culture; explants; joint-on-a-chip

1. Introduction

Rheumatoid arthritis (RA) is a progressive systemic, chronic, and inflammatory autoimmune disease with an average prevalence of 0.5–1.0% in the population worldwide, demonstrating ethnic and geographic differences [1]. Its pathogenesis is characterized by immune cell infiltration into the synovial membrane and the joint cavity and the formation of hyperplastic and invasive synovium, resulting in progressive cartilage destruction and subchondral bone erosion in late stages of disease if not treated (Figure 1). Along with the joints, RA can affect many of the body's organs, including the heart, eyes, skin, intestine, kidney, lung, and brain, as well as the skeleton [2,3]. A disease most likely RA was first recognized more than 20 centuries ago as a disease that painfully affects the body's joints [4]. It is the most common inflammatory joint disease affecting both individuals and society. The affected patients suffer a considerable loss of quality of life and a decline in productivity, and the effort and costs of health care increase, ultimately resulting in a major economic and social burden [5]. Symptoms of RA most commonly include pain, swelling, and morning stiffness in the affected joints. It is a multifactorial disorder and recent studies have identified multiple genetic and environmental factors associated with an increased risk of RA, e.g., female sex, smoking, and major histocompatibility complex (MHC) regions encoding human leukocyte antigen (HLA) proteins (amino acids at positions 70 and 71) [2,6]. Years before first clinical symptoms of RA occur, autoimmunity against modified self-proteins is initiated, which results in the onset of the disease [1].

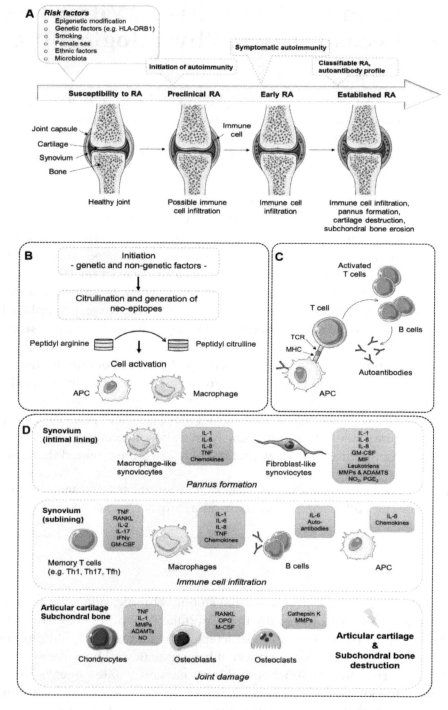

Figure 1. Establishment of rheumatoid arthritis (RA): Mechanisms of disease initiation, development, and progression. (**A**) Multiple risk factors, including both genetic and non-genetic influences, are required to induce the development of RA in susceptible individuals. Years before first clinical symptoms of RA occur, autoimmunity against modified self-proteins is initiated, which results in the onset of a subclinical inflamed synovium (symptomatic autoimmunity) propagated by immune cell infiltration and pannus formation. Once established, RA can be classified according to the clinical symptoms. (**B**) Onset of autoimmunity is supposed to occur in the mucosa (e.g., mouth, lung, and gut) by the creation of neo-epitopes as a result of post-translational modifications, e.g., by citrullination. These neo-epitopes can

be recognized by antigen-presenting cells (APCs) of the adaptive immune system and (**C**) are presented to adaptive immune cells in lymphoid tissues, activate an immune response, and induce autoantibody formation (e.g., ACPA and RF). (**D**) Activated immune cells and immune complexes can activate synovial cells, such as fibroblast-like synoviocytes (FLS) and macrophage-like synoviocytes of the intimal lining and APCs in the sublining area, to produce a range of inflammatory factors and expand and form the cartilage- and bone-invasive pannus. Autoimmune activation and immune cell infiltration (T cells, B cells, macrophages) of the sublining area further contribute to the excessive production of inflammatory factors, autoantibodies, and synovial vascular leakage, ultimately leading to articular cartilage and subchondral bone destruction as a result of matrix-degrading enzymes and a de-balanced bone homeostasis characterized by an imbalanced RANKL/RANK/OPG system and activated osteoclasts. ADAMTS, a disintegrin and metalloproteinase with thrombospondin motifs; APCAs, anti-citrullinated protein antibodies; RF, rheumatoid factor; GM-CSF, granulocyte–macrophage colony-stimulating factor; M-CSF, macrophage colony-stimulating factor; MHC, major histocompatibility complex; MMP, matrix metalloproteinase; NO, nitric oxide; OPG, osteoprotegerin; RANKL, receptor activator of nuclear factor-κB ligand; RANK, receptor activator of nuclear factor-κB; TCR, T cell receptor; TNF, tumor necrosis factor. Figure contains graphics from Servier Medical Art, licensed under a Creative Common Attribution 3.0 Generic License. http://smart.servier.com/.

As the course of RA within the individual patients may differ with regard to pathogenesis, clinical symptoms, and diseases subtypes, personalized precision medicine must be the ultimate goal to achieve disease remission. To date, we are far from curing RA in part due to the need for (i) objective patient-related biomarkers to identify disease subtypes and treatment response and (ii) the management of patients who are refractory or resistant to available treatments. Having both will enable us to understand the disease and their pathogenic processes to optimize and introduce personalized precision health care.

2. The Course of RA Pathogenesis

The course of RA pathogenesis involves several stages. Before clinical symptoms are established, a certain level of RA susceptibility (e.g., genetic factors) coupled with the accumulation of risk factors proceed through the pre-clinical stage of the disease, leading to synovial inflammation, which, if not resolved, ultimately leads to the development of RA. During the early development of RA, post-translational modifications of a wide range of cellular (e.g., collagen) and nuclear proteins (e.g., histones) occur, including the conversion of the amino acid arginine to citrulline, a process called citrullination. Citrullination may be a result of smoking on mucosa, induced by microbiota (e.g., *Porphyromonas gingivalis*) or by an overarching neutrophil reaction. Altered modified self-proteins engage professional antigen presenting cells (APCs), such as macrophages, as foreign and induce a normal immune response via the help of T cells, thereby stimulating B cells to produce a wide range of (auto)antibodies recognizing self-proteins, such as rheumatoid factor (RF) and anti-citrullinated protein antibodies (ACPAs). The presence of autoantibodies often occurs before the onset of clinical synovitis, leading to the assumption that a second not-fully-understood mechanism seems to be necessary for the transition of autoimmunity to local synovial inflammation [1,2,6].

However, during the progression of RA, increase in vascular permeability, a disrupted extracellular matrix, and synovial immune cell infiltration transform the paucicellular synovium into chronically inflamed tissue. This process includes the expansion of the intimal lining and activation of macrophage- and stromal-fibroblast-like synoviocytes (FLSs), which then produce a variety of pro-inflammatory humoral mediators, such as cytokines and chemokines, including interleukin (IL)-1β, IL-6, IL-8, tumor necrosis factor (TNF), granulocyte–macrophage colony-stimulating factor (GM-CSF), macrophage migration inhibitory factor (MIF), and matrix-degrading enzymes, e.g., matrix metallopeptidases (MMPs) and a disintegrin and metalloproteinase with thrombospondin motifs (ADAMTs), prostaglandins, leukotrienes, and reactive nitric oxide. The aggressive and invasive phenotype of expanding FLSs, forming the hyperplastic pannus tissue, contributes to cartilage damage but may also be responsible for propagation and systemic spreading of inflammation by migrating from joint to joint and other organs [7–9]. The inflammation-induced expansion of FLSs and the infiltration of inflammatory cells into the usually paucicellular synovium lead to an enhanced metabolic need and, therefore, to an undersupply of both nutrients and oxygen to the synovial tissue. Due to the resulting local hypoxia, new vessels are formed that further facilitate the inflammatory process by increasing the amount of adaptive immune cells, and especially CD4+ memory T (Th) cells infiltrating the synovial sublining. Lymphocyte infiltrates accumulate and form aggregates ranging from small and loosely arranged lymphocyte clusters to large and organized ectopic lymphoid structures, which, in some cases, develop germinal centers that facilitate local T cell–B cell interactions. In these ectopic germinal structures, specific pathologic follicular helper T cells (Tfh) promote B-cell responses and (auto)antibody production within pathologically-inflamed non-lymphoid tissues. Apart from pathogenic Tfh cells, Th1 and Th17 cells have been identified in the pathogenesis of RA. Although the evidence of the pathogenic function of Th1 cells in RA is controversial due to the lack of therapeutic efficiency targeting of interferon (IFN)-γ [10,11], it should be noted that biologic targeting of TNF-α, which is a Th1 cytokine, are successful treatments in RA [11]. An effect that can be explained by the suppressive nature of Th1 on Th17 is that the responses contribute to tissue damage through production of TNF and GM-CSF [12].

IL-17-producing CD4+ T cells have been identified in synovial tissues from patients with RA, including their inducing cytokines IL-6 IL-1β, IL-21, transforming growth factor (TGF)-β, and IL-23 [13–17], and have been demonstrated to be increased/maintained in the peripheral blood of RA patients [18–21], whereas IL-17 was shown to induce bone resorption and contribute to neutrophil recruitment, and particularly into the synovial fluid, a hallmark of RA [22,23]. Besides effector T helper cells, antigen-presenting follicular dendritic cells, macrophages, and mast cells are present in the synovial sublining and contribute to the chronic inflammation by a large number of inflammatory mediators, such as cytokines, chemokines, and reactive oxygen and nitrogen species, as well as matrix-degrading enzymes. In contrast, neutrophils are lacking in the inflamed synovial lining and sublining but are abundantly present in the synovial fluid. Recent studies proposed that distinct subtypes of synovial histology displaying inflammatory versus non-inflammatory patterns are associated with different clinical phenotypes and a concurring response to novel targeted therapeutic interventions [24,25]. Technical progression and the development and combination of state-of-the-art methods from single cell genomics to mass cytometry have provided new insights into the complex interplay of cells and soluble immune mediators, particularly cytokines and chemokines [2]. Thus, specific pathogenic infiltrating immune cell subsets—such as IL-1β positive pro-inflammatory monocytes, autoimmune-associated B cells, and peripheral helper T (Tph) cells sharing similarities with Tfh cells, distinct subsets of CD8+ T cells, as well as mast cells—contribute to the inflammatory pattern of the RA synovial lining/sublining [26–31].

Invading immune cells and FLSs of the synovial lining produce large amounts of pro-inflammatory cytokines and express high levels of MMPs, while the expression of endogenous MMP inhibitors remains insufficiently low. Finally, the invasive and destructive FLS-front of synovial tissue, called the pannus, attaches to the articular surface and contributes to local matrix destruction and cartilage degradation. The chondrocytes of the damaged articular cartilage contribute to the vicious cycle of cartilage degeneration by inducing inflammatory cytokines, such as IL-1β and TNF-α, as well as MMPs and nitric oxide (NO). Additionally, FLSs negatively affect the subchondral bone by activation and maturation of bone-resorbing osteoclasts. Osteoclasts are highly responsive to autoantibodies; pro-inflammatory cytokines, in particular TNF-α, IL-1β, and IL-6; and more importantly, receptor activator of nuclear factor kappa B ligand (RANKL), which is the key regulator of osteoclastogenesis. RANKL binds to its receptor, the receptor activator of nuclear factor-κB (RANK), and activates osteoclasts, leading to an enhancement of bone resorption. Conversely, osteoblasts that play a key role in the regulation of anabolic bone metabolism produce bone matrix constituents, induce bone matrix mineralization, and modulate osteoclasts through the production of osteoprotegerin (OPG) [32]. Although osteoblasts producing OPG, which is a decoy receptor for RANKL, results in protection from bone destruction by osteoclasts, they also generate RANKL and M-CSF, both of which contribute to osteoclastogenesis. Imbalanced bone remodeling both in the subchondral and periarticular bone of joints leads to bone erosions and periarticular osteopenia; generalized bone loss is a general feature of established RA.

3. Lessons from Animal Models of Arthritis: None are Truly RA

Animal models represent an integral part of the preclinical drug discovery process and are used to study pathophysiological mechanisms of RA. Despite their extreme usefulness for testing new approaches of intervention in many cases, concerns about low clinical development success rates for investigational drugs have been raised [33], "Dozens of preclinical arthritis models have been developed . . . none of these, however, is truly RA, and none consistently predicts the effect of a therapeutic agent in patients" [33].

Importantly, animals do not naturally develop autoimmune disorders, such as RA, which is an inherent limitation of these arthritis models (Table 1). Instead, animal models can be used to study certain specific pathophysiological aspects of human disease, such as destructive pathways involved in the erosion of articular cartilage and bone. To this end, arthritis can be chemically induced in these animals by soluble agents (e.g., type II collagen-induced arthritis model) or develop spontaneously after genetic manipulation (e.g., human TNF transgene model) (Table 1) [34–36]. Although most of these models display features of human RA, such as inflammatory cell infiltrate, synovial hyperplasia, pannus formation, cartilage destruction, and bone erosions, they also demonstrate specific limitations, such as the development of self-limiting arthritis, development of arthritis only in susceptible strains of rodents, and a pathophysiology that does not recapitulate the endogenous breach of tolerance and excludes systemic components of disease [34–36]. The mutations used in genetically engineered arthritis models have not been identified in human RA [36]. When comparing transcriptional programs of mice and humans overlapping but notably different gene expression patterns have been observed [37]. Therefore, therapeutic approaches, such as the application of biologics highly specific for human target proteins, cannot be proven using non-humanized rodent models [38]. Finally, mice and humans differ in their locomotion, life span, evolutionary pressures, ecological niches, circadian rhythms, weight bearing, and blood leukocyte population ratios. Thus, none of the animal models is capable of fully replicating human pathogenesis of RA, which provides an explanation for the observed challenges in clinical translation [33].

Modern management guidelines recommend early and rigorous treatment to achieve low disease activity or remission targets as rapidly as possible. Thus, RA is currently treated with a wide variety of therapeutic drugs ranging from steroidal/nonsteroidal anti-inflammatory drugs (NSAID), glucocorticoids (GCs), and disease-modifying anti-rheumatic drugs (DMARDs) of synthetic origin, such as conventional synthetic DMARDs (e.g., methotrexate), biological, and biosimilar DMARDs (e.g., TNF inhibitors or IL-6 inhibitors), as well as targeted synthetic DMARDs (the Janus kinase (JAK) inhibitors) targeting specific immune cells, cytokines, or pro-inflammatory pathways [2,26,39]. Today's therapeutic approaches using state-of-the-art biologicals or JAK inhibitors have been proven to be highly successful and effective in most patients with RA, including those with severe disease progression. Despite major progress in the treatment of RA, a strong unmet medical need remains, as not all patients reach sustained clinical remission (less than half of patients with RA) and about 25% still suffer from moderate or even high disease activity [2,40]. Defining patients with RA (i) refractory to available treatments among patients with RA who are undertreated or non-adherent to treatment, (ii) identifying objective biomarkers for disease states (e.g., early versus established RA) and/or (iii) 'refractory' states and finally (iv) for states treatment response is still the greatest unmet need in RA [40]. The lack of therapeutic efficacy in the true refractory patients may be due to the nature of the "one-fits-it-all" approach of standardized therapeutic regimes. Thus, clinical management of patients often neglects their heterogeneity with regard to the endogenous circadian rhythms, disease states, subtypes and duration, as well as autoantibody, cytokine, and infiltrating immune cell pattern. Identifying objective biomarkers to delineate disease subtypes and treatment response will be necessary to provide a 'precise' customized treatment strategy for each individual patient enhancing our repertoire in the battle against this potentially devastating disease.

Table 1. Selected rodent models for rheumatoid arthritis (as reviewed in Reference [34-36]).

Animal Models for Rheumatoid Arthritis	Species	Induction/Genetic Alteration	Limitations	References
Induced Arthritis Models				
Collagen-induced arthritis (CIA)	Mouse, rat	Inoculation with type II heterologous or homologous collagen in complete Freund's adjuvant in strains expressing major histocompatibility complex (MHC) Class II I-Aq haplotypes	■ General variable incidence, severity, and inter-group inconsistency ■ Only inducible in susceptible strains of rodents ■ Low incidence, as well as variability, of arthritis severity in c57bl/6 mice ■ Acute and self-limiting polyarthritis in contrast to human RA ■ Greater incidence in males in contrast to human RA	[36,41–43]
Collagen-antibody-induced arthritis (CAIA)	Mouse	Anti-collagen antibodies have been demonstrated to induce arthritis	■ Pathogenesis is not mediated via T and B cell response in contrast to human RA ■ Pathogenesis is inducible irrespective of the presence of MHC class II haplotype in contrast to human RA	[44,45]
Adjuvant-induced arthritis (AA)	Mouse, rat	Mixture of mineral oils, heat-killed mycobacteria, and emulsifying agent, which was termed complete Freund's adjuvant (CFA); when omitting mycobacteria, also known as incomplete Freund's adjuvant (IFA); see also pristane-induced arthritis (PIA)	■ Acute and self-limiting polyarthritis in contrast to human RA ■ Not antigenic but displays an autoimmune pathophysiology	[34,35,46,47]
Zymosan-induced arthritis	Mouse, rat	Intra-articular injection of zymosan, a polysaccharide from the cell wall of *Saccharomyces cerevisiae*, into the knee joints of mice causes proliferative arthritis, including immune cell infiltration, synovial hypertrophy, and pannus formation	■ Technical skill required for an intra-articular injection in mice ■ Monoarthritis in contrast to human RA	[48,49]
Streptococcal cell-wall-induced arthritis (SCWIA)	Mouse, rat	*Streptococcus pyogenes* synthesize a peptidoglycan-polysaccharide (PG-PS) polymer	■ Pathogenesis is inducible in selected susceptible strains of rodents ■ Germ-free conditions are necessary to reach susceptibility in rats ■ Multiple injections are needed; otherwise, acute and self-limiting arthritis develops, in contrast to human RA ■ Tumor necrosis factor (TNF)-α is less important in SCW-induced arthritis but not in human RA ■ Rheumatoid factor is missing in polyarticular arthritis in rats	[35,50,51]

Table 1. *Cont.*

Animal Models for Rheumatoid Arthritis	Species	Induction/Genetic Alteration	Limitations	References
Cartilage oligomeric matrix protein (COMP)-induced arthritis	Mouse, rat	Immunization with IFA combined with native and denatured COMP, which is a large protein that is synthesized by chondrocytes (see also adjuvant-induced arthritis)	■ Acute and self-limiting polyarthritis in contrast to human RA ■ Not antigenic but displays an autoimmune pathophysiology	[52,53]
Pristane-induced arthritis (PIA)	Mouse, rat	Injection of the hydrocarbon pristane intraperitoneally into mice	■ No evidence of autoimmune reactions ■ Inflammation is restricted to the joints but systemic abnormalities are absent in rats	[47,54,55]
Antigen-induced arthritis (AIA)	Mouse	Inoculation with antigen by intra-articular injection	■ Intra-articular injection in mice requires advanced technical skills ■ Does not recapitulate the endogenous breach of tolerance in contrast to human ra ■ Excludes systemic component of disease	[36,56,57]
Proteoglycan-induced arthritis	Mouse	Intraperitoneal injection of proteoglycan that is emulsified with an adjuvant	■ Only inducible in susceptible strains of mice ■ Incidence of ankylosing spondylitis without any exacerbations and remissions in contrast to human RA	[34–36]
Glucose-6-phosphate isomerase (G6PI)-induced arthritis	Mouse	Immunization using the ubiquinone containing glycolytic enzyme G6PI with CFA for induction of RA	■ Only inducible in susceptible strains of mice ■ Low prevalence of antibodies against G6PI in patients with RA	[34–36,58]
Genetically manipulated spontaneous arthritis models				
K/BxN model	Mouse	K/BxN mice were generated by crossing mice expressing the MHC class II molecule Ag7 with the T cell receptor (TCR) transgenic KRN line expressing a TCR specific for a G6PI-peptide	■ Mutations have only been identified in mice ■ Low prevalence of antibodies to g6pi in patients with ra ■ Without systemic manifestations or production of rheumatoid factor in contrast to human RA	[58–60]
SKG model	Mouse	Induction of arthritis due to point mutation in ZAP-70	■ Mutations have only been identified in mice ■ Disease manifestations in germ-free mice only upon induction	[34–36,61]
Human TNF transgene model	Mouse	Transgene for human TNF-α	■ Mutations have only been identified in mice ■ No production of rheumatoid factor in contrast to human RA	[34–36,62,63]

Therefore, preclinical models are essential to help improve our understanding of pathological mechanisms and to develop and verify new therapeutic approaches with the aim of meeting this unmet medical need. This includes the investigation of human-specific alternatives to identify objective biomarkers to delineate disease subtypes and treatment response, and novel targets to manipulate the function of immune cells involved in the pathogenesis of RA. The purpose of this review was to summarize the most commonly used and often cytokine-based in vitro models of RA, and discuss how they reflect human pathophysiology to further understand the underlying mechanisms of RA.

4. Lessons from In Vitro Models of Arthritis: An Alternative without Alternatives

During the last decade, promising in vitro techniques have been improved by advances in tissue engineering. Thus, the pathogenesis of RA has been simulated and studied using a variety of in vitro and in vivo models. Cell-based in vitro assays range from tissue explants and relatively simplified (co)-culture systems to complex engineered three-dimensional (multi)component tissue systems using a variety of cell types from cell lines, primary cells, or patient-derived cells, such as mesenchymal stromal cells (MSCs) or pluripotent stem cells (iPSCs), to study, e.g., cell migration, activation, antigen presentation, and cell–cell interaction, as well as cell- and matrix-related changes. Additionally, organoids incubated on microfluidic chips, as well as using in silico models, show promise as an approach to further studying the mechanisms underlying RA pathophysiology and to identify potential new targets. Thus, next-generation preclinical in vitro screening systems will be based on microphysiological in vitro human-joint-on-a-chip systems using primary cells from patients with RA and from different organs, mimicking the systemic nature of the disease and fostering the translational process to humans, while reducing the number of animal experiments. Ultimately, the main goal for all in vitro approaches is to achieve the greatest possible physiological proximity to the disease, while ensuring experimental feasibility, breaking down the barrier to translational medicine and thus to conducting high-quality, reproducible research (Figure 2).

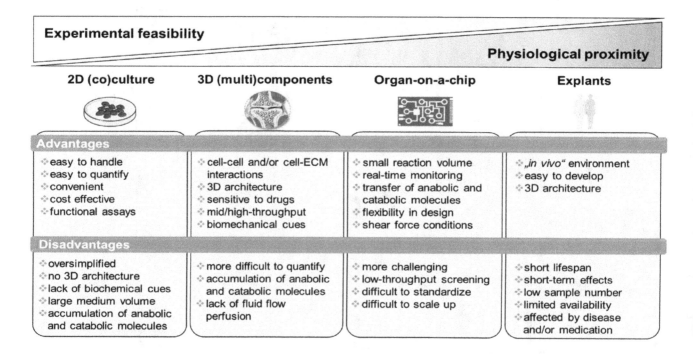

Figure 2. Overview of state-of-the-art in vitro models classified according to experimental feasibility and physiological proximity. Figure contains graphics from Servier Medical Art, licensed under a Creative Common Attribution 3.0 Generic License. http://smart.servier.com/.

4.1. Tissue Explants: Close Physiological Proximity but Low Experimental Feasibility

Ex vivo culture models or tissue explants represent the closest physiological similarity to pathological tissue due to the nature of their origin. If ethically and clinically available, these models can be easily obtained, are easy to develop, and allow the semi-controlled study of the behavior of cells cultured. Although tissue explants reflect the human physiology in terms of 3D structure and environment, they are often affected by individual health status and medication, as well as sample preparation. However, tissue explant approaches are still a powerful tool in, e.g., osteochondral bone research due to the ability to retain native bone cell communication and to study cellular responses and extracellular matrix remodeling processes, including disease-specific matrix degradation in a (patho)physiological bone environment [64]. In addition to their limited availability (especially in terms of healthy human material), the main limitations of tissue explant models are shortened lifespan due to simultaneous disruption of the supplying vessels and, consequently, induced cell death and necrosis-induced cell death at the explant/wound edges [65]. With synovial tissue, explants can be obtained from patients with RA or osteoarthritis (OA) during joint replacement surgery, as well as by needle and arthroscopic biopsy. These types of samples have been comprehensively examined using molecular and immunohistochemical techniques leading to a better understanding of the pathogenic events that occur in the course of the disease [66]. For instance, when studying the association between synovial imaging activity by magnet resonance imaging or color Doppler ultrasound with the expression of synovial inflammatory mediators using tissue explants, Andersen et al. observed a correlation of distinct synovial cytokines with corresponding imaging pathology and disease activity [67].

Samples of synovial [68] and bone explants [69] have been used to study the efficacy and efficiency of therapeutic treatments on the (i) production of pro-inflammatory mediators, (ii) expression of matrix-degrading enzymes, and (iii) adhesion molecules. Of note, IL-1β, TNF-α, and IL-17 have been demonstrated to produce many additive and/or synergistic effects in vitro. Using synovial explants from patients with RA, therapeutic intervention with a combination of biologicals, e.g., anti-TNF-α antibodies and IL-1Ra, resulted in significantly decreased IL-6 and MMP-3 production, indicating the superior efficacy of combinatorial therapy over a single biological treatment [70]. Kirenol, which is a Chinese herbal active component, was demonstrated to inhibit FLS proliferation, migration, invasion, and secretion of pro-inflammatory IL-6 in explants from RA synovium [71].

To examine disease-related expression profiles, explants, like articular cartilage discs, have been obtained from patients with RA after knee arthroplasty. Using this approach, Gotoh et al. demonstrated that the interaction of CD40 with CD154 increased the expression of inflammatory cytokines and MMPs, resulting in an increased cartilage degradation in patients with RA [72]. Based on the aforementioned types of explants, Schultz et al. developed a 3D in vitro model to investigate destructive processes in RA. Although the explant co-culture system did not address all aspects of RA, such as the presence of immune cells, the authors confirmed the capability of their model to study FLS activity on destructive processes of established joint diseases in vitro [73]. More than 10 years later, Pretzel et al. established an in vitro that which closely reflects early processes in cartilage destruction caused by synovial fibroblasts via, e.g., the suppression of anabolic matrix synthesis highlighting the value and close proximity of tissue explant models [9].

4.2. Simplified 2D Culture and Co-Culture Approaches for High-Throughput Drug Screening

Closely mimicking physiological and pathophysiological biological complexity in terms of physiological or pathophysiological characteristics requires the use of tissue explants, using 3D architecture or the development of sophisticated complex 3D tissue models. However, achieving experimental feasibility and ensuring adequate nutrient and oxygen supply are more challenging tasks with 3D designs than with 2D cell cultures. Therefore, 2D monolayer cell cultures are a simple and cost-effective alternative, especially for high-throughput screening approaches, which are common in pharmaceutical, industrial, and toxicological research. They are still used to investigate the efficiency and efficacy of therapeutics, to determine their optimal concentration, to analyze disease-related gene expression profiles, and to study cell–cell, cell–microenvironment, or cell–humoral interactions using auto- and paracrine signals, such as in aggregate–cell interactions, in a simplified co-culture system [74–77]. Two-dimensional monolayer cell cultures are used for rapid in vitro cell expansion, despite the risk of cellular alterations in terms of morphology, genetic alteration, cell diversity, cell cycle progression, and cell differentiation capacity [78]. Accordingly, when 2D modeling cartilage, for instance, the phenotype of chondrocytes becomes unstable, which is indicated by a downregulation of type II collagen (COL2) with a simultaneous increase in the expression of type I collagen (COL1). To avoid these artificial changes, an optimized cultivation procedure is required using specific plate coatings, such as poly(L-lactic acid) [79]. When investigating the effects of RA-associated cytokines on cartilage, monolayer chondrocyte cultures are considered an optimal tool due to their easy handling in combination with the rapid response of chondrocytes to pro-inflammatory cytokines. In addition, chondrocytes, when stimulated with, e.g., IL-1β, TNF-α, or IFN-γ, show a classical RA-like phenotype as evidenced by decreased expressions of COL2 and aggrecan (ACAN) when MMP13 expression increases [80] and induced apoptosis in chondrocytes [81], reflecting the human in vivo situation [82,83]. Using the 2D approach, Teltow et al. demonstrated that the majority of IL-1β-treated chondrocytes are produced in collagenase 1 instead of collagenase 3, although the latter has been assumed to foster the destructive processes of RA joints by degrading collagen type II [84]. IL-1β was demonstrated to decrease the expression of COL2 in 2D monolayer cultures [85].

Expanding the 2D monolayer cultures using co-culture systems, the interaction between cells growing in the same environment can be either indirectly (physical barrier) cultivated by simple medium transfer and using a trans-well chamber or directly cultivated in a mixed culture system providing cell-to-cell contact. Using direct and indirect co-cultivation, Donlin et al. demonstrated that human RA synovial fibroblasts suppress the TNF-α-induced IFN-γ signature in macrophages under both conditions, indicating that no cell contact is required, but rather soluble fibroblast products inhibit the IFN-γ signature of macrophages [86]. To extend the co-culture systems, Pagani et al. developed an advanced tri-culture model to study the interaction between osteoblasts, osteoclasts, and endothelial cells and the cytokine-induced effects on bone homeostasis with respect to RA [87].

4.3. 3D tissue Engineering Approaches: Mimicking Structural Features of the Joint

In the field of musculoskeletal disorders, simplified 2D cell culture systems have been stepwise replaced by promising in vitro 3D tissue engineering approaches, including (i) scaffold-free 3D approaches, such as cell-sheet formation [88], self-assembly, or self-organization [89], (ii) natural scaffold-based 3D approaches, such as hyaluronic-acid-based scaffolds [90], and (iii) synthetic scaffold-based 3D approaches, such as poly-(lactide)-based scaffolds [91].

These 3D approaches offer considerable advantages compared to the above-mentioned 2D approaches because they facilitate cell–cell and cell–matrix interactions; cell proliferation, differentiation, and migration and they maintain the cell fate as a result of the physiological 3D structure. To mimic the structural features of the joint, which is a prerequisite for simulating the pathogenesis of RA, the various cell-based components, such as synovial membrane and the chondrogenic and osteogenic parts, must be developed for an in vitro 3D approach.

4.3.1. Synovial Membrane 3D In Vitro Models: From Monolayer to Micromass Culture

The synovial membrane, or synovia, lines the joint cavity and can be divided into the synovial intimal lining (intima) and subintimal lining (subintima). In the healthy state, the intima lining consists of one to four cell layers of type A (macrophages) and type B (FLSs) synoviocytes. The subintimal lining is based on fibrous, areolar, and fatty tissues [92]. As described above, activated FLSs are supposed to be key mediators of joint destruction and drivers of the inflammatory processes during the course of RA. Therefore, FLSs are receiving attention for creating 3D models of the synovial membrane. For this purpose, FLS are resuspended in gels to map a 3D micromass [93]. Karonitsch et al. used such an in vitro 3D micromass model of the synovial membrane to determine the individual effects of pro-inflammatory cytokines, such as IFN-γ and TNF-α, on mesenchymal tissue remodeling [94]. Whereas IFN-γ promotes the invasive potential of FLSs via JAK activation, TNF induces pronounced aggregation of FLSs, indicating that both cytokines affect synovial tissue remodeling in a different manner [94]. Using a similar 3D in vitro approach, Bonelli et al. recently observed that TNF regulates the expression of the transcription factor interferon regulatory factor 1 (IRF1), a key regulator of the IFN-mediated inflammatory cascade, which was confirmed by a TNF transgenic arthritis mouse model [95]. Although both studies relied on 3D models solely consisting of FLSs, they indicated that 3D in vitro approaches are sufficient to elucidate mechanistically cellular processes in the FLS-driven inflammation during RA.

Broeren et al. established a sophisticated, promising, and more complex in vitro 3D synovial membrane model by combining either primary RA-FLSs with peripheral CD14+ monocytes or using a complete human RA synovial cell suspension [96]. This model reflects the native 3D architecture of the synovium forming a lining layer at the outer surface consisting of fibroblast-like and macrophage-like synoviocytes. Long-term exposure to TNF-α led to hyperplasia of the lining layer, an altered macrophage phenotype, and an increase in pro-inflammatory cytokines, such as TNFA, IL6, IL8, and IL1B, reassembling key features of established RA, thereby confirming previous observations by Kiener et al. [93,96]. The findings of the latter study highlighted the unrestricted possibilities of 3D in vitro approaches to be an excellent alternative for drug testing and mechanistic research.

Although these models closely reflect the inflamed synovial membrane, they all rely on diseased FLSs, which are often limited in availability and are affected by different stages of disease, as well as current medication [97]. To mimic a healthy situation, which is essential to understanding pathogenic alterations of the synovium, an easy to handle and available cell source from different sources that shares properties of FLSs would be ideal for simulating the synovial tissue in vitro. Adult MSCs share most properties with FLSs, including surface markers, differentiation capacity, and the capability to produce hyaluronic acid, and are indistinguishable from each other. Thus, MSCs could be a promising cell source for the development of in vitro 3D models of the synovial membrane or even the other components of the joint [98].

4.3.2. Modeling Articular Cartilage: Scaffold Revisited

To mimic articular cartilage for a 3D in vitro model of arthritis, healthy hyaline cartilage is a relatively acellular and avascular tissue with limited regenerative capacity, nourished by the synovial fluid through diffusion [99]. Articular cartilage is characterized by an organized structure consisting of different layers (superficially tangential, transitional, and radial) that absorbs mechanical loads and forces within the joint and thus protects the underlying subchondral bone. Chondrocytes/-blasts are the only cell population that produce and maintain the highly organized extracellular matrix (ECM), consisting of collagens, mainly type II, type IX, and XI; non-collagen proteins; and proteoglycans, such as aggrecan [99]. During RA, pro-inflammatory stimuli, such as TNF-α or IFN-γ, result in the molecular activation of catabolic and inflammatory processes in human chondrocytes, which decreases their viability and proliferation and increases matrix degradation [81,100].

Due to the sensitivity of chondrocytes to the molecular and mechanical cues of the environment, the consensus is that 3D tissue models, using a matrix that corresponds to the natural tissue properties, are closer to the in vivo situation [101]. Therefore, most 3D approaches involve a scaffold to provide the cells with a predetermined 3D structure. These scaffolds include porous scaffolds made of collagen type II [102], natural gels, such as gelatine microspheres [103], alginate beads [104], hyaluronic acid, and chitosan [105]. Using gelatine microspheres, Peck et al. created a 3D cartilage model very closely mimicking human cartilage, as confirmed by the high expression of type II collagen and proteoglycans [103]. Using a tri-culture approach combining the gelatine microspheres-based 3D cartilage model with a synovial cell line and lipopolysaccharide (LPS)-activated monocytic THP-1 cells, the authors confirmed and validated the pathological alteration in the phenotype of chondrocytes characterized by increased apoptosis, decreased gene expression for matrix components, such as collagen type II and aggrecan, increased gene expression for tissue degrading enzymes (*MMP1*, *MMP3*, *MMP13*, and *ADAMTS4*, *ADAMTS5*) and upregulation of the expression of inflammatory mediator genes (*TNFA*, *IL1B*, and *IL6*), as observed in a disease state of RA [103]. Along this line, stimulation of alginate-based 3D cartilage tissue models with supernatant from RA synovial fibroblasts led to the activation of catabolic and inflammatory processes that could be reversed by anti-rheumatic drugs when used [106]. Ibold et al. developed a 3D articular cartilage model for RA based on the interactive co-culture of high-density scaffold-free porcine cartilage with a RA-derived synovial fibroblast cell line to provide a tool for high-throughput drug screening. For high-throughput purposes, automation of cell seeding was introduced, which improved the quality of the generated pannus cultures as assessed by the enhanced formation of cartilage-specific ECM [107]. However although the stiffness and absorption rate of these natural matrices cannot be adjusted to the specific requirements of each cartilage zone, Karimi et al., modeled the superficial, middle, and calcified zone using varying cell amounts, mechanical loading, and biochemical influences [108].

To establish scaffold-free 3D cartilage constructs, intrinsic processes, such as spontaneous self-assembly, or extrinsic processes, such as mechanical load-induced self-organization, have been described [109–112] and used as 3D in vitro models for, e.g., preclinical high-throughput screenings [113,114].

Since MSCs, which are progenitors of chondrocytes, can be forced in vitro to enter chondrogenic differentiation, they represent an ideal cell source for the development of in vitro cartilage models: MSCs are available from different tissue sources (even autologous), they are immune privileged, easy-to-handle, and highly expandable. Thus, MSCs have been the focus in numerous studies with and without the incorporation of scaffolds [115,116]. Using this approach, a chondrocyte-like morphology and cartilage-like matrix corresponding to that of native cartilage were reported, particularly with the aim of develop cartilage grafts for therapeutic purposes [115].

4.3.3. The Complexity of Mimicking 3D Subchondral Bone: Mission Impossible?

A key feature of RA is focal bone loss or bone erosion [117]. To address this feature, mimicking bone tissue is mandatory. However, bone tissue is complex in terms of cell composition, matrix organization, vascularization, and mechanical loading. Bone is a dynamic, highly vascularized, and connective tissue that undergoes lifelong remodeling processes in an adaptive response to mechanical stress. It provides a supporting function within the musculoskeletal system and consists of different cell types, such as osteoblasts, osteocytes, and osteoclasts embedded in the ECM, which consists of organic and inorganic phases. Osteoclasts and osteoblasts are key players during bone turnover, whereas osteocytes play a crucial role in bone homeostasis, responsible for mechanosensing and mechanotransduction [118]. Traditionally, bone tissue engineering has been used to produce implants for bone regeneration [119]. In recent years, however, bone tissue engineering has been increasingly applied to create artificial in vitro bone models to improve our understanding of bone-related (patho)physiological mechanisms, such as osteoporosis. Commonly, approaches used to mimic bone in vitro are scaffold-based. Thus, numerous innovative scaffolds (synthetic, natural, biodegradable, and non-biodegradable) have been developed that are capable of mimicking the mechanical stiffness and structural properties of bone; the latter includes mimicking porosity and pore sizes to provide cavities for cell penetration and nutrient supply [120]. These scaffolds are further optimized to have both osteoconductive and osteoinductive properties [121].

Apart from the scaffold-based approaches, scaffold-free organoids or spheroids and 3D printing, hydrogels, or beads have been used [90,122–125]. However, all of the aforementioned approaches commonly use MSCs capable of differentiating into the osteogenic lineage, osteoblasts, and a combination of either osteoblasts and osteocytes or osteoblasts and osteoclasts. To further support osteogenic properties, bioactive compounds, such as bone morphogenetic protein 2 (BMP-2) or vascular endothelial growth factor (VEGF), have been included [126,127]. To achieve the mechanical impact important for native bone, suitable bioreactors combined with bioceramics further support the in vitro osteogenesis in a defined, standardized, controlled, and reproducible manner [128]. Novel promising approaches aim to realize in vitro bone models with robust vascularization using human umbilical vein endothelial cells [129–131]. However, no in vitro 3D bone model is currently available that reflects the complexity of the human bone.

4.3.4. 3D Multicomponent Approaches: Reconstructing the Joint Structure

Multicomponent in vitro 3D co-cultures systems combining 3D in vitro models of articular cartilage and bone (osteochondral unit) with in vitro 3D models of the synovial membrane are necessary to study the cartilage degradation and bone erosive processes during RA that are linked to the invasiveness of the hyperplastic synovium (pannus) [132]. Currently, multicomponent engineering approaches are widely used to simulate key features of osteoarthritis instead of RA or are used to develop suitable artificial matrices that can replace damaged regions and promote tissue regeneration. Thus, many promising in vitro approaches have been recently developed using (i) scaffold-based bone and scaffold-free cartilage [133], (ii) different scaffolds for both bone and cartilage, (iii) a heterogeneous (bi-layered) scaffold, or (iv) a homogenous scaffold for both bone and cartilage (as reviewed in Reference [134]). Notably, bi-layered systems are most often fixed by adhesives, such as fibrin, creating a barrier for cell–cell contact. To avoid this, Lin et al. encapsulated iPSCs-derived MSCs (iMPCs) in a photocrosslinkable gelatin scaffold. Using a dual-flow bioreactor, encapsulated iMPCs were chondrogenic (top) and osteogenic (bottom) differentiated to directly form a stable bridging zone between the both tissue models [135]. So far, no appropriate multicomponent in vitro model exists that is able to mimic the physiologically relevant environment of a healthy or an inflamed joint, including all signaling molecules, cells, and tissue types. Consequently, we developed a valid in vitro 3D model to simulate the immune-mediated pathogenesis of arthritis. The in vitro model relies on the three main components of the joint: (i) the osteogenic and (ii) chondrogenic parts, and (iii) the synovial membrane with the synovial fluid. All components are based on differentiated MSCs from a single donor and thus include most relevant cell types involved, enabling crucial cell–cell interactions [136]. We simulated the inflamed joint using the application of RA-related cytokines, as well as immune cells [132]. Finally, we confirmed the suitability of the multicomponent in vitro 3D model, which may serve as a preclinical tool for the evaluation of both new targets and potential drugs in a more translational setup [137].

5. Microfluidic Approaches: Prospectively Systemic

In recent years, perfused cultivation systems have become increasingly important due to the advantages they provide for the cultivation of functional tissues. They ensure the permanent supply of nutrients and the defined real-time monitoring of environmental conditions, such as pH, temperature, and oxygen concentrations. Multi-chamber bioreactors provide the opportunity to cultivate two or more cell/tissue types in a defined manner [138,139]. Generally, microfluidic approaches provide inherent flexibility in combinatory design, which enables relevant concentration gradients, cellular spatial configuration, and co-culture and shear force conditions [140]. To date, only a few different microfluidic culture approaches have been reported that at least partially reflect the physiology of the joint structure, mimicking either subchondral bone, articular cartilage, or both together, namely the osteochondral part, as well as the synovial membrane, including spatial topology and mechanical loading [141,142]. However, these do not yet cover all the possibilities offered by these microfluidic systems (Figure 3).

In detail, using equine chondrocytes in a microfluidic culture, 3D cartilage constructs were formed by establishing a physiologic nutrient diffusion gradient across a simulated matrix. Additionally, the geometric design constraints of the microchambers drive native cartilage-like cellular behavior [141]. Calvo et al. developed a synovium-on-a-chip system by culturing patient-derived primary FLSs in a Matrigel™-based 3D micro-mass mimicking TNF-α-driven structural changes and synovial remodeling [142]. As a result, the activation of FLSs by TNF-α leads to induction of the expression of pro-inflammatory cytokines, such as *IL6* and *IL8*, as well as matrix-degrading metalloproteinases and pannus formation, which is a typical feature of RA. Since the rea-out parameters in a perfused system are often limited to endpoint assessments, the chip system reported by Calvo et al. (2017) facilitates the online monitoring of cellular parameters by incorporating a simplified light scattering method that enables the non-invasive detection of cell motility, proliferation, invasion, and even matrix condensation processes within the 3D tissue [142].

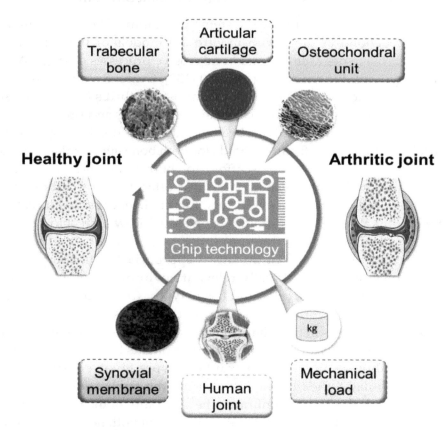

Figure 3. Overview of microfluidic approaches mimicking selected physiological interactions of the human joint tissues. Figure was modified from Servier Medical Art, licensed under a Creative Common Attribution 3.0 Generic License. http://smart.servier.com/.

Conclusively, the ultimate goals of microfluidic approaches are to (i) provide reliable information on the health and disease status of the integrated complex biological system, (ii) reproducibly promote the formation of the microphysiological tissue structure, and (iii) non-invasively and automatically monitor stimuli-driven tissue responses [143].

Additionally, a variety of organoids representing various tissues, such as liver, kidney, or heart, have been established and implemented in microfluidic systems as a single-tissue approach, namely organ-on-a-chip, or as multi-tissue approaches, such as multi-organ-on-a-chip or, if possible, human-on-a-chip (Figure 4). However, the human-joint-on-a-chip approach could be a promising in vitro tool to improve our understanding of the complex pathophysiological mechanisms in RA and to develop and verify new therapeutic strategies to further expand our repertoire in the battle against this potentially devastating disease. Future perspectives include human-joint-on-a-chip tailored to a single patient for use in a personalized medicine scenario to maintain human health.

Despite their advances and opportunities for translational studies and drug testing, microfluidic systems have still some limitations. So far, microfluidic systems are more challenging to operate and control than static systems, some organ functions, such as cognition on the brain and mechanical function in bone, cannot be readily modeled, and they are difficult to adapt to high-throughput screening and are difficult to standardize and scale up.

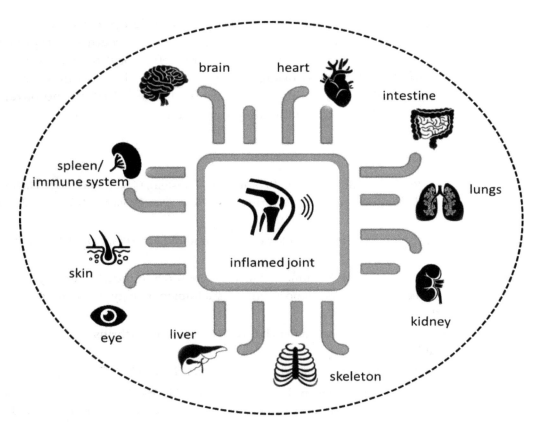

Figure 4. Next-generation preclinical in vitro approach based on microphysiological in vitro human-joint-on-a-chip systems in combination with pathophysiological-relevant human organs.

6. Conclusions and Outlook

Here, we comprehensively summarized key events in RA pathogenesis, which is the most common immune-mediated chronic inflammatory joint disease. Today's treatment goal of RA is to achieve remission or at least low disease activity. However, a strong unmet medical need remains, as by far not all patients reach sustained clinical remission and even about 25% still suffer from moderate or even high disease activity characterized by systemic inflammation, persistent synovitis, expansion of synovial cells (pannus formation) and progressive cartilage and bone destruction in late stages. In the last years, we have witnessed the failure of potential new therapies in clinical trials although their development was based on promising preclinical animal data, which can be attributed to the nature of these models. Animal models and simplified 2D cell cultures of arthritis have been useful to identify certain pathomechanisms underlying RA. However, they do not fully reflect human pathogenesis due to oversimplification of the pathophysiological processes or misleading in case of animal models which owe interspecies differences with regard to, e.g., chondrocyte biology, articular cartilage, and cartilage thickness [144–146].

Thus, we herewith suggest that shifting our traditional research approaches in biomedicine towards an improved human personalized patient-driven translation by using sophisticated in vitro models may enhance 'precision' in medicine. Finally, personalized in vitro models will provide guidance to replace today's inefficient standard treatment regimens (one fits it all) taking into account patient heterogeneity in terms of disease subtypes, endogenous circadian rhythms, autoantibodies, cytokine and infiltrating immune cell patterns, and the extent of pannus formation, ultimately preventing 'refractory' arthritis.

Along with the joints, RA can affect many of the body's organs [2,3]. Therefore, combining different 3D tissue models with state-of-the-art microfluidic devices must be the next generation in vitro approach to study the complex crosstalk between tissues/organs and the immune system, including the spreading of (auto)immune reactions across different organs, ultimately mimicking the systemic nature of rheumatic diseases.

Prospectively, the human-based approach will not only provide opportunities (i) to identify objective patient-related biomarkers to elucidate disease subtypes and treatment response but also (ii) enable strategies for the management of patients who are 'refractory' or resistant to available treatments. Thus, human-based cellular and tissue models will close the gaps in RA research and, finally, health care, increase clinical translatability, and contribute to the reduction and/or replacement of animal experiments used in basic and translational RA research.

References

1. Smolen, J.S.; Aletaha, D.; Barton, A.; Burmester, G.R.; Emery, P.; Firestein, G.S.; Kavanaugh, A.; McInnes, I.B.; Solomon, D.H.; Strand, V.; et al. Rheumatoid Arthritis. *Nat. Rev. Dis. Primers* **2018**, *4*, 18001. [CrossRef] [PubMed]

2. Burmester, G.R.; Pope, J.E. Novel treatment strategies in rheumatoid arthritis. *Lancet* **2017**, *389*, 2338–2348. [CrossRef]

3. Cassotta, M.; Pistollato, F.; Battino, M. Rheumatoid arthritis research in the 21st century: Limitations of traditional models, new technologies, and opportunities for a human biology-based approach. *Altex* **2019**, *37*, 223–242. [CrossRef]

4. Parish, L.C. An historical approach to the nomenclature of rheumatoid arthritis. *Arthritis Rheum.* **1963**, *6*, 138–158. [CrossRef] [PubMed]

5. Scott, D.L.; Wolfe, F.; Huizinga, T.W. Rheumatoid Arthritis. *Lancet* **2010**, *376*, 1094–1108. [CrossRef]

6. Deane, K.D.; Demoruelle, M.K.; Kelmenson, L.B.; Kuhn, K.A.; Norris, J.M.; Holers, V.M. Genetic and environmental risk factors for rheumatoid arthritis. *Best Pract. Res. Clin. Rheumatol.* **2017**, *31*, 3–18. [CrossRef] [PubMed]

7. Bustamante, M.F.; Garcia-Carbonell, R.; Whisenant, K.D.; Guma, M. Fibroblast-like synoviocyte metabolism in the pathogenesis of rheumatoid arthritis. *Arthritis Res.* **2017**, *19*, 1–12. [CrossRef] [PubMed]

8. Kiener, H.P.; Niederreiter, B.; Lee, D.M.; Jimenez-Boj, E.; Smolen, J.S.; Brenner, M.B. Cadherin 11 promotes invasive behavior of fibroblast-like synoviocytes. *Arthritis Rheum.* **2009**, *60*, 1305–1310. [CrossRef]

9. Pretzel, D.; Pohlers, D.; Weinert, S.; Kinne, R.W. In vitro model for the analysis of synovial fibroblast-mediated degradation of intact cartilage. *Arthritis Res. Ther.* **2009**, *11*, R25. [CrossRef]

10. Pollard, K.M.; Cauvi, D.M.; Toomey, C.B.; Morris, K.V.; Kono, D.H. Interferon-Gamma and Systemic Autoimmunity. *Discov. Med.* **2013**, *16*, 123–131.

11. Feldmann, M. Development of anti-TNF therapy for rheumatoid arthritis. *Nat. Rev. Immunol.* **2002**, *2*, 364–371. [CrossRef] [PubMed]

12. Yamada, H.; Haraguchi, A.; Sakuraba, K.; Okazaki, K.; Fukushi, J.-I.; Mizu-Uchi, H.; Akasaki, Y.; Esaki, Y.; Kamura, S.; Fujimura, K.; et al. Th1 is the predominant helper T cell subset that produces GM-CSF in the joint of rheumatoid arthritis. *RMD Open* **2017**, *3*, e000487. [CrossRef] [PubMed]

13. Gaffen, S.L.; Jain, R.; Garg, A.V.; Cua, D.J. The IL-23–IL-17 immune axis: From mechanisms to therapeutic testing. *Nat. Rev. Immunol.* **2014**, *14*, 585–600. [CrossRef]

14. Cascao, R.; Moura, R.A.; Perpetuo, I.; Canhao, H.; Vieira-Sousa, E.; Mourao, A.F.; Rodrigues, A.M.; Polido-Pereira, J.; Queiroz, M.V.; Rosario, H.S.; et al. Identification of a Cytokine Network Sustaining Neutrophil and Th17 Activation in Untreated Early Rheumatoid Arthritis. *Arthritis Res. Ther.* **2010**, *12*, R196. [CrossRef] [PubMed]

15. Pène, J.; Chevalier, S.; Preisser, L.; Vénéreau, E.; Guilleux, M.-H.; Ghannam, S.; Molès, J.-P.; Danger, Y.; Ravon, E.; Lesaux, S.; et al. Chronically Inflamed Human Tissues Are Infiltrated by Highly Differentiated Th17 Lymphocytes. *J. Immunol.* **2008**, *180*, 7423–7430. [CrossRef] [PubMed]

16. Manel, N.; Unutmaz, D.; Littman, D.R. The Differentiation of Human T(H)-17 Cells Requires Transforming Growth Factor-Beta and Induction of the Nuclear Receptor Rorgammat. *Nat. Immunol.* **2008**, *9*, 641–649. [CrossRef]

17.   Volpe, E.; Servant, N.; Zollinger, R.; Bogiatzi, S.I.; Hupé, P.; Barillot, E.; Soumelis, V. A critical function for transforming growth factor-β, interleukin 23 and proinflammatory cytokines in driving and modulating human TH-17 responses. *Nat. Immunol.* **2008**, *9*, 650–657. [CrossRef]

18.   van Hamburg, J.P.; Asmawidjaja, P.S.; Davelaar, N.; Mus, A.M.C.; Colin, E.M.; Hazes, J.M.W.; Dolhain, R.J.E.M.; Lubberts, E. Th17 cells, but not Th1 cells, from patients with early rheumatoid arthritis are potent inducers of matrix metalloproteinases and proinflammatory cytokines upon synovial fibroblast interaction, including autocrine interleukin-17A production. *Arthritis Rheum.* **2010**, *63*, 73–83. [CrossRef]

19.   Leipe, J.; Grunke, M.; de Chant, C.; Reindl, C.; Kerzendorf, U.; Schulze-Koops, H.; Skapenko, A. Role of Th17 cells in human autoimmune arthritis. *Arthritis Rheum.* **2010**, *62*, 2876–2885. [CrossRef]

20.   Shahrara, S.; Huang, Q.-Q.; Mandelin, A.M.; Pope, R.M. TH-17 cells in rheumatoid arthritis. *Arthritis Res. Ther.* **2008**, *10*, R93. [CrossRef]

21.   Yamada, H.; Nakashima, Y.; Okazaki, K.; Mawatari, T.; Fukushi, J.-I.; Kaibara, N.; Hori, A.; Iwamoto, Y.; Yoshikai, Y. Th1 but not Th17 cells predominate in the joints of patients with rheumatoid arthritis. *Ann. Rheum. Dis.* **2007**, *67*, 1299–1304. [CrossRef]

22.   Kaplan, M.J. Role of neutrophils in systemic autoimmune diseases. *Arthritis Res. Ther.* **2013**, *15*, 219. [CrossRef] [PubMed]

23.   Kotake, S.; Udagawa, N.; Takahashi, N.; Matsuzaki, K.; Itoh, K.; Ishiyama, S.; Saito, S.; Inoue, K.; Kamatani, N.; Gillespie, M.T.; et al. IL-17 in synovial fluids from patients with rheumatoid arthritis is a potent stimulator of osteoclastogenesis. *J. Clin. Investig.* **1999**, *103*, 1345–1352. [CrossRef] [PubMed]

24.   Orr, C.; Vieira-Sousa, E.; Boyle, D.L.; Buch, M.H.; Buckley, C.D.; Cañete, J.D.; Catrina, A.I.; Choy, E.H.S.; Emery, P.; Fearon, U.; et al. Synovial tissue research: A state-of-the-art review. *Nat. Rev. Rheumatol.* **2017**, *13*, 463–475. [CrossRef] [PubMed]

25.   Dennis, G.; Holweg, C.T.J.; Kummerfeld, S.K.; Choy, D.F.; Setiadi, A.F.; Hackney, J.A.; Haverty, P.M.; Gilbert, H.; Lin, W.Y.; Diehl, L.; et al. Synovial phenotypes in rheumatoid arthritis correlate with response to biologic therapeutics. *Arthritis Res. Ther.* **2014**, *16*, R90. [CrossRef]

26.   Smolen, J.S.; Landewé, R.B.M.; Bijlsma, J.W.J.; Burmester, G.R.; Dougados, M.; Kerschbaumer, A.; McInnes, I.B.; Sepriano, A.; van Vollenhoven, R.F.; de Wit, M.; et al. EULAR recommendations for the management of rheumatoid arthritis with synthetic and biological disease-modifying antirheumatic drugs: 2019 update. *Ann. Rheum. Dis.* **2020**. [CrossRef]

27.   Rao, D.A.; Gurish, M.F.; Marshall, J.L.; Slowikowski, K.; Fonseka, K.S.C.Y.; Liu, Y.; Donlin, L.T.; Henderson, L.A.; Wei, K.; Mizoguchi, F.; et al. Pathologically expanded peripheral T helper cell subset drives B cells in rheumatoid arthritis. *Nat. Cell Biol.* **2017**, *542*, 110–114. [CrossRef]

28.   Zhang, F.; Wei, K.; Slowikowski, K.; Fonseka, C.Y.; Rao, D.A.; Kelly, S.; Goodman, S.M.; Tabechian, D.; Hughes, L.B. Defining inflammatory cell states in rheumatoid arthritis joint synovial tissues by integrating single-cell transcriptomics and mass cytometry. *Nat. Immunol.* **2019**, *20*, 928–942. [CrossRef]

29.   O'Neil, L.J.; Kaplan, M.J. Neutrophils in Rheumatoid Arthritis: Breaking Immune Tolerance and Fueling Disease. *Trends Mol. Med.* **2019**, *25*, 215–227. [CrossRef]

30.   Rivellese, F.; Mauro, D.; Nerviani, A.; Pagani, S.; Fossati-Jimack, L.; Messemaker, T.; Kurreeman, F.A.S.; Toes, R.E.M.; Ramming, A.; Rauber, S.; et al. Mast cells in early rheumatoid arthritis associate with disease severity and support B cell autoantibody production. *Ann. Rheum. Dis.* **2018**, *77*, 1773–1781. [CrossRef]

31.   Schubert, N.; Dudeck, J.; Liu, P.; Karutz, A.; Speier, S.; Maurer, M.; Tuckermann, J.P.; Dudeck, A. Mast Cell Promotion of T Cell-Driven Antigen-Induced Arthritis Despite Being Dispensable for Antibody-Induced Arthritis in Which T Cells Are Bypassed. *Arthritis Rheumatol.* **2015**, *67*, 903–913. [CrossRef]

32.   Corrado, A.; Maruotti, N.; Cantatore, F.P. Osteoblast Role in Rheumatic Diseases. *Int. J. Mol. Sci.* **2017**, *18*, 1272. [CrossRef]

33.   Firestein, G.S. Rheumatoid arthritis in a mouse? *Nat. Clin. Pract. Rheumatol.* **2009**, *5*, 1. [CrossRef] [PubMed]

34.   Choudhary, N.; Bhatt, L.K.; Prabhavalkar, K.S. Experimental animal models for rheumatoid arthritis. *Immunopharmacol. Immunotoxicol.* **2018**, *40*, 193–200. [CrossRef]

35.   Bevaart, L.; Vervoordeldonk, M.J.; Tak, P.P. Evaluation of therapeutic targets in animal models of arthritis: How does it relate to rheumatoid arthritis? *Arthritis Rheum.* **2010**, *62*, 2192–2205. [CrossRef]

36. Asquith, D.L.; Miller, A.M.; McInnes, I.B.; Liew, F.Y. Animal models of rheumatoid arthritis. *Eur. J. Immunol.* **2009**, *39*, 2040–2044. [CrossRef] [PubMed]

37. Breschi, A.; Gingeras, T.R.; Guigó, A.B.R. Comparative transcriptomics in human and mouse. *Nat. Rev. Genet.* **2017**, *18*, 425–440. [CrossRef] [PubMed]

38. Schinnerling, K.; Rosas, C.; Soto, L.; Thomas, R.; Aguillón, J.C. Humanized Mouse Models of Rheumatoid Arthritis for Studies on Immunopathogenesis and Preclinical Testing of Cell-Based Therapies. *Front. Immunol.* **2019**, *10*, 10. [CrossRef]

39. Chatzidionysiou, K.; Emamikia, S.; Nam, J.; Ramiro, S.; Smolen, J.; van der Heijde, D.; Dougados, M.; Bijlsma, J.; Burmester, G.; Scholte, M.; et al. Efficacy of glucocorticoids, conventional and targeted synthetic disease-modifying antirheumatic drugs: A systematic literature review informing the 2016 update of the EULAR recommendations for the management of rheumatoid arthritis. *Ann. Rheum. Dis.* **2017**, *76*, 1102–1107. [CrossRef]

40. Winthrop, K.L.; Weinblatt, M.E.; Bathon, J.; Burmester, G.R.; Mease, P.J.; Crofford, L.; Bykerk, V.; Dougados, M.; Rosenbaum, J.T.; Mariette, X.; et al. Unmet need in rheumatology: Reports from the Targeted Therapies meeting 2019. *Ann. Rheum. Dis.* **2019**, *79*, 88–93. [CrossRef]

41. Holmdahl, R.; Jansson, L.; Larsson, E.; Rubin, K.; Klareskog, L. Homologous type II collagen induces chronic and progressive arthritis in mice. *Arthritis Rheum.* **1986**, *29*, 106–113. [CrossRef] [PubMed]

42. Courtenay, J.S.; Dallman, M.J.; Dayan, A.D.; Martin, A.; Mosedale, B. Immunisation against heterologous type II collagen induces arthritis in mice. *Nat. Cell Biol.* **1980**, *283*, 666–668. [CrossRef]

43. Trentham, D.E.; Townes, A.S.; Kang, A.H. Autoimmunity to type II collagen an experimental model of arthritis. *J. Exp. Med.* **1977**, *146*, 857–868. [CrossRef]

44. Nandakumar, K.S.; Holmdahl, R. Efficient promotion of collagen antibody induced arthritis (CAIA) using four monoclonal antibodies specific for the major epitopes recognized in both collagen induced arthritis and rheumatoid arthritis. *J. Immunol. Methods* **2005**, *304*, 126–136. [CrossRef]

45. Holmdahl, R.; Rubin, K.; Klareskog, L.; Larsson, E.; Wigzell, H. Characterization of the antibody response in mice with type II collagen–induced arthritis, using monoclonal anti–type II collagen antibodies. *Arthritis Rheum.* **1986**, *29*, 400–410. [CrossRef] [PubMed]

46. Kim, E.; Moudgil, K.D. The determinants of susceptibility/resistance to adjuvant arthritis in rats. *Arthritis Res. Ther.* **2009**, *11*, 239. [CrossRef] [PubMed]

47. Holmdahl, R.; Lorentzen, J.C.; Lu, S.; Olofsson, P.; Wester, L.; Holmberg, J.; Pettersson, U. Arthritis induced in rats with non-immunogenic adjuvants as models for rheumatoid arthritis. *Immunol. Rev.* **2001**, *184*, 184–202. [CrossRef]

48. Keystone, E.C.; Schorlemmer, H.U.; Pope, C.; Allison, A.C. Zymosan—Induced Arthritis. *Arthritis Rheum.* **1977**, *20*, 1396–1401. [CrossRef]

49. Frasnelli, M.E.; Tarussio, D.; Chobaz-Péclat, V.; Busso, N.; So, A. TLR2 modulates inflammation in zymosan-induced arthritis in mice. *Arthritis Res. Ther.* **2005**, *7*, R370–R379. [CrossRef]

50. Wilder, R.L. Streptococcal Cell Wall Arthritis. *Curr. Protoc. Immunol.* **1998**, *26*. [CrossRef]

51. Joosten, L.A.; Abdollahi-Roodsaz, S.; Heuvelmans-Jacobs, M.; Helsen, M.M.; van den Bersselaar, L.A.; Oppers-Walgreen, B.; Koenders, M.I.; van den Berg, W.B. T Cell Dependence of Chronic Destructive Murine Arthritis Induced by Repeated Local Activation of Toll-Like Receptor-Driven Pathways: Crucial Role of Both Interleukin-1beta and Interleukin-17. *Arthritis Rheum.* **2008**, *58*, 98–108. [CrossRef]

52. Carlsen, S.; Nandakumar, K.S.; Bäcklund, J.; Holmberg, J.; Hultqvist, M.; Vestberg, M.; Holmdahl, R. Cartilage oligomeric matrix protein induction of chronic arthritis in mice. *Arthritis Rheum.* **2008**, *58*, 2000–2011. [CrossRef]

53. Carlsén, S.; Hansson, A.-S.; Olsson, H.; Heinegård, D.; Holmdahl, R. Cartilage oligomeric matrix protein (COMP)-induced arthritis in rats. *Clin. Exp. Immunol.* **1998**, *114*, 477–484. [CrossRef]

54. Vingsbo, C.; Sahlstrand, P.; Brun, J.G.; Jonsson, R.; Saxne, T.; Holmdahl, R. Pristane-induced arthritis in rats: A new model for rheumatoid arthritis with a chronic disease course influenced by both major histocompatibility complex and non-major histocompatibility complex genes. *Am. J. Pathol.* **1996**, *149*, 1675–1683. [PubMed]

55. Wooley, P.H.; Seibold, J.R.; Whalen, J.D.; Chapdelaine, J.M. Pristane-induced arthritis. the immunologic and genetic features of an experimental murine model of autoimmune disease. *Arthritis Rheum.* **1989**, *32*, 1022–1030. [CrossRef] [PubMed]

56.   Brackertz, D.; Mitchell, G.F.; Vadas, M.A.; Mackay, I.R.; Miller, J.F. Studies on antigen-induced arthritis in mice. II. Immunologic correlates of arthritis susceptibility in mice. *J. Immunol.* **1977**, *118*, 1639–1644.

57.   Brackertz, D.; Mitchell, G.F.; Mackay, I.R. Antigen-Induced Arthritis in Mice. I. Induction of Arthritis in Various Strains of Mice. *Arthritis Rheum.* **1977**, *20*, 841–850. [CrossRef]

58.   Matsumoto, I.; Lee, D.M.; Goldbach-Mansky, R.; Sumida, T.; Hitchon, C.A.; Schur, P.H.; Anderson, R.J.; Coblyn, J.S.; Weinblatt, M.E.; Brenner, M.; et al. Low prevalence of antibodies to glucose-6-phosphate isomerase in patients with rheumatoid arthritis and a spectrum of other chronic autoimmune disorders. *Arthritis Rheum.* **2003**, *48*, 944–954. [CrossRef] [PubMed]

59.   van Gaalen, F.A.; Toes, R.E.M.; Ditzel, H.J.; Schaller, M.; Breedveld, F.C.; Verweij, C.L.; Huizinga, T.W.J. Association of autoantibodies to glucose-6-phosphate isomerase with extraarticular complications in rheumatoid arthritis. *Arthritis Rheum.* **2004**, *50*, 395–399. [CrossRef] [PubMed]

60.   Kouskoff, V.; Korganow, A.-S.; Duchatelle, V.; Degott, C.; Benoist, C.; Mathis, D. Organ-Specific Disease Provoked by Systemic Autoimmunity. *Cell* **1996**, *87*, 811–822. [CrossRef]

61.   Sakaguchi, N.; Takahashi, T.; Hata, H.; Nomura, T.; Tagami, T.; Yamazaki, S.; Sakihama, T.; Matsutani, T.; Negishi, I.; Nakatsuru, S.; et al. Altered thymic T-cell selection due to a mutation of the ZAP-70 gene causes autoimmune arthritis in mice. *Nat. Cell Biol.* **2003**, *426*, 454–460. [CrossRef] [PubMed]

62.   Butler, D.M.; Malfait, A.M.; Mason, L.J.; Warden, P.J.; Kollias, G.; Maini, R.N.; Feldmann, M.; Brennan, F.M. DBA/1 mice expressing the human TNF-alpha transgene develop a severe, erosive arthritis: Characterization of the cytokine cascade and cellular composition. *J. Immunol.* **1997**, *159*, 2867–2876. [PubMed]

63.   Keffer, J.; Probert, L.; Cazlaris, H.; Georgopoulos, S.; Kaslaris, E.; Kioussis, D.; Kollias, G. Transgenic mice expressing human tumour necrosis factor: A predictive genetic model of arthritis. *EMBO J.* **1991**, *10*, 4025–4031. [CrossRef]

64.   Marino, S.; Staines, K.A.; Brown, G.; Howard-Jones, R.A.; Adamczyk, M. Models of ex vivo explant cultures: Applications in bone research. *BoneKEy Rep.* **2016**, *5*, 818. [CrossRef]

65.   Gilbert, S.; Singhrao, S.K.; Khan, I.M.; Gonzalez, L.G.; Thomson, B.M.; Burdon, D.; Duance, V.C.; Archer, C. Enhanced Tissue Integration During Cartilage Repair In Vitro Can Be Achieved by Inhibiting Chondrocyte Death at the Wound Edge. *Tissue Eng. Part A* **2009**, *15*, 1739–1749. [CrossRef]

66.   Nozaki, T.; Takahashi, K.; Ishii, O.; Endo, S.; Hioki, K.; Mori, T.; Kikukawa, T.; Boumpas, D.T.; Ozaki, S.; Yamada, H. Development of an ex vivo cellular model of rheumatoid arthritis: Critical role of cd14-positive monocyte/macrophages in the development of pannus tissue. *Arthritis Rheum.* **2007**, *56*, 2875–2885. [CrossRef]

67.   Andersen, M.; Boesen, M.; Ellegaard, K.; Christensen, R.; Söderström, K.; Søe, N.H.; Spee, P.; Mørch, U.G.; Torp-Pedersen, S.; Bartels, E.; et al. Synovial explant inflammatory mediator production corresponds to rheumatoid arthritis imaging hallmarks: A cross-sectional study. *Arthritis Res. Ther.* **2014**, *16*, R107. [CrossRef] [PubMed]

68.   Chevrel, G.; Garnero, P.; Miossec, P. Addition of interleukin 1 (IL1) and IL17 soluble receptors to a tumour necrosis factor alpha soluble receptor more effectively reduces the production of IL6 and macrophage inhibitory protein-3alpha and increases that of collagen in an in vitro model of rheumatoid synoviocyte activation. *Ann. Rheum. Dis.* **2002**, *61*, 730–733. [CrossRef] [PubMed]

69.   Chabaud, M.; Miossec, P. The Combination of Tumor Necrosis Factor Alpha Blockade with Interleukin-1 and Interleukin-17 Blockade Is More Effective for Controlling Synovial Inflammation and Bone Resorption in an Ex Vivo Model. *Arthritis Rheum* **2001**, *44*, 1293–1303. [CrossRef]

70.   Hosaka, K.; Ryu, J.; Saitoh, S.; Ishii, T.; Kuroda, K.; Shimizu, K. The Combined Effects of Anti-Tnfalpha Antibody and Il-1 Receptor Antagonist in Human Rheumatoid Arthritis Synovial Membrane. *Cytokine* **2005**, *32*, 263–269. [CrossRef]

71.   Wu, J.; Li, Q.; Jin, L.; Qu, Y.; Liang, B.-B.; Zhu, X.-T.; Du, H.-Y.; Jie, L.-G.; Yu, Q.-H. Kirenol Inhibits the Function and Inflammation of Fibroblast-like Synoviocytes in Rheumatoid Arthritis in vitro and in vivo. *Front. Immunol.* **2019**, *10*, 1304. [CrossRef]

72.   Gotoh, H.; Kawaguchi, Y.; Harigai, M.; Hara, M.; Saito, S.; Yamaguchi, T.; Shimada, K.; Kawamoto, M.; Tomatsu, T.; Kamatani, N. Increased CD40 expression on articular chondrocytes from patients with rheumatoid arthritis: Contribution to production of cytokines and matrix metalloproteinases. *J. Rheumatol.* **2004**, *31*, 1506–1512. [PubMed]

73. Schultz, O.; Keyszer, G.; Zacher, J.; Sittinger, M.; Burmester, G.R. Development of in vitro model systems for destructive joint diseases: Novel strategies for establishing inflammatory pannus. *Arthritis Rheum.* **1997**, *40*, 1420–1428. [CrossRef]

74. Croft, A.P.; Naylor, A.J.; Marshall, J.L.; Hardie, D.L.; Zimmermann, B.; Turner, J.; de Santi, G.; Adams, H.; Yemm, A.I.; Müller-Ladner, U.; et al. Rheumatoid synovial fibroblasts differentiate into distinct subsets in the presence of cytokines and cartilage. *Arthritis Res.* **2016**, *18*, 1–11. [CrossRef] [PubMed]

75. Chen, H.-J.; Yim, A.Y.F.L.; Griffith, G.R.; de Jonge, W.J.; Mannens, M.M.A.M.; Ferrero, E.; Henneman, P.; de Winther, M.P.J. Meta-Analysis of in vitro-Differentiated Macrophages Identifies Transcriptomic Signatures That Classify Disease Macrophages in vivo. *Front. Immunol.* **2019**, *10*, 2887. [CrossRef]

76. Lewis, M.J.; Barnes, M.R.; Blighe, K.; Goldmann, K.; Rana, S.; Hackney, J.A.; Ramamoorthi, N.; John, C.R.; Watson, D.S.; Kummerfeld, S.K.; et al. Molecular Portraits of Early Rheumatoid Arthritis Identify Clinical and Treatment Response Phenotypes. *Cell Rep.* **2019**, *28*, 2455–2470.e5. [CrossRef]

77. Turner, R.; Counts, G.; Mashburn, H.; Treadway, W.; de Chatelet, L. Drug and rheumatoid factor effects on the uptake of immunoglobulin G aggregates by neurtrophil monolayers. *Inflammation* **1980**, *4*, 55–64. [CrossRef] [PubMed]

78. von der Mark, K.; Gauss, V.; von der Mark, H.; Müller, P. Relationship between cell shape and type of collagen synthesised as chondrocytes lose their cartilage phenotype in culture. *Nat. Cell Biol.* **1977**, *267*, 531–532. [CrossRef] [PubMed]

79. Benya, P.D.; Padilla, S.R.; Nimni, M.E. Independent regulation of collagen types by chondrocytes during the loss of differentiated function in culture. *Cell* **1978**, *15*, 1313–1321. [CrossRef]

80. Murphy, G.; Lee, M.H. What Are the Roles of Metalloproteinases in Cartilage and Bone Damage? *Ann. Rheum. Dis.* **2005**, *64*, 44–47. [CrossRef]

81. Schuerwegh, A.; Dombrecht, E.; Stevens, W.; van Offel, J.; Bridts, C.; de Clerck, L. Influence of pro-inflammatory (IL-1α, IL-6, TNF-α, IFN-γ) and anti-inflammatory (IL-4) cytokines on chondrocyte function. *Osteoarthr. Cartil.* **2003**, *11*, 681–687. [CrossRef]

82. Kim, H.A.; Song, Y.W. Apoptotic chondrocyte death in rheumatoid arthritis. *Arthritis Rheum.* **1999**, *42*, 1528–1537. [CrossRef]

83. Saito, S.; Murakoshi, K.; Kotake, S.; Kamatani, N.; Tomatsu, T. Granzyme B induces apoptosis of chondrocytes with natural killer cell-like cytotoxicity in rheumatoid arthritis. *J. Rheumatol.* **2008**, *35*, 1932–1943. [PubMed]

84. Tetlow, L.C.; Woolley, D.E. Comparative immunolocalization studies of collagenase 1 and collagenase 3 production in the rheumatoid lesion, and by human chondrocytes and synoviocytes in vitro. *Br. J. Rheumatol.* **1998**, *37*, 64–70. [CrossRef]

85. Goldring, M.; Berenbaum, F. Human chondrocyte culture models for studying cyclooxygenase expression and prostaglandin regulation of collagen gene expression. *Osteoarthr. Cartil.* **1999**, *7*, 386–388. [CrossRef] [PubMed]

86. Donlin, L.T.; Jayatilleke, A.; Giannopoulou, E.G.; Kalliolias, G.D.; Ivashkiv, L.B. Modulation of TNF-Induced Macrophage Polarization by Synovial Fibroblasts. *J. Immunol.* **2014**, *193*, 2373–2383. [CrossRef] [PubMed]

87. Pagani, S.; Torricelli, P.; Veronesi, F.; Salamanna, F.; Cepollaro, S.; Fini, M. An advanced tri-culture model to evaluate the dynamic interplay among osteoblasts, osteoclasts, and endothelial cells. *J. Cell. Physiol.* **2017**, *233*, 291–301. [CrossRef]

88. Kim, K.; Bou-Ghannam, S.; Thorp, H.; Grainger, D.W.; Okano, T. Human mesenchymal stem cell sheets in xeno-free media for possible allogenic applications. *Sci. Rep.* **2019**, *9*, 1–12. [CrossRef]

89. Weber, M.-C.; Fischer, L.; Damerau, A.; Ponomarev, I.; Pfeiffenberger, M.; Gaber, T.; Götschel, S.; Lang, J.; Röblitz, S.; Buttgereit, F.; et al. Macroscale mesenchymal condensation to study cytokine-driven cellular and matrix-related changes during cartilage degradation. *Biofabrication* **2020**, *12*, 045016. [CrossRef]

90. Dhivya, S.; Saravanan, S.; Sastry, T.P.; Selvamurugan, N. Nanohydroxyapatite-reinforced chitosan composite hydrogel for bone tissue repair in vitro and in vivo. *J. Nanobiotechnol.* **2015**, *13*, 1–13. [CrossRef]

91. Scheinpflug, J.; Pfeiffenberger, M.; Damerau, A.; Schwarz, F.; Textor, M.; Lang, A.; Schulze, F. Journey into Bone Models: A Review. *Genes* **2018**, *9*, 247. [CrossRef] [PubMed]

92. Smith, M.D. The Normal Synovium. *Open Rheumatol. J.* **2011**, *5*, 100–106. [CrossRef]

93. Kiener, H.P.; Watts, G.F.M.; Cui, Y.; Wright, J.; Thornhill, T.S.; Sköld, M.; Behar, S.M.; Niederreiter, B.; Lu, J.; Cernadas, M.; et al. Synovial fibroblasts self-direct multicellular lining architecture and synthetic function in three-dimensional organ culture. *Arthritis Rheum.* **2010**, *62*, 742–752. [CrossRef]

94. Karonitsch, T.; Beckmann, D.; Dalwigk, K.; Niederreiter, B.; Studenic, P.; A Byrne, R.; Holinka, J.; Sevelda, F.; Korb-Pap, A.; Steiner, G.; et al. Targeted inhibition of Janus kinases abates interfon gamma-induced invasive behaviour of fibroblast-like synoviocytes. *Rheumatology* **2017**, *57*, 572–577. [CrossRef]

95. Bonelli, M.; Dalwigk, K.; Platzer, A.; Calvo, I.O.; Hayer, S.; Niederreiter, B.; Holinka, J.; Sevelda, F.; Pap, T.; Steiner, G.; et al. IRF1 is critical for the TNF-driven interferon response in rheumatoid fibroblast-like synoviocytes. *Exp. Mol. Med.* **2019**, *51*, 1–11. [CrossRef] [PubMed]

96. Broeren, M.G.A.; Waterborg, C.E.J.; Wiegertjes, R.; Thurlings, R.M.; Koenders, M.I.; van Lent, P.; van der Kraan, P.M.; van de Loo, F.A.J. A Three-Dimensional Model to Study Human Synovial Pathology. *Altex* **2019**, *36*, 18–28. [CrossRef]

97. Kasperkovitz, P.V.; Timmer, T.C.G.; Smeets, T.J.; Verbeet, N.L.; Tak, P.P.; van Baarsen, L.G.M.; Baltus, B.; Huizinga, T.W.J.; Pieterman, E.; Fero, M.; et al. Fibroblast-like synoviocytes derived from patients with rheumatoid arthritis show the imprint of synovial tissue heterogeneity: Evidence of a link between an increased myofibroblast-like phenotype and high-inflammation synovitis. *Arthritis Rheum.* **2005**, *52*, 430–441. [CrossRef]

98. Denu, R.A.; Nemcek, S.; Bloom, D.D.; Goodrich, A.D.; Kim, J.; Mosher, D.F.; Hematti, P. Fibroblasts and Mesenchymal Stromal/Stem Cells Are Phenotypically Indistinguishable. *Acta Haematol.* **2016**, *136*, 85–97. [CrossRef]

99. Carballo, C.B.; Nakagawa, Y.; Sekiya, I.; Rodeo, S.A. Basic Science of Articular Cartilage. *Clin. Sports Med.* **2017**, *36*, 413–425. [CrossRef]

100. Adan, N.; Guzmán-Morales, J.; Ledesma-Colunga, M.G.; Perales-Canales, S.I.; Quintanar-Stephano, A.; López-Barrera, F.; Mendez, I.; Moreno-Carranza, B.; Triebel, J.; Binart, N.; et al. Prolactin promotes cartilage survival and attenuates inflammation in inflammatory arthritis. *J. Clin. Investig.* **2013**, *123*, 3902–3913. [CrossRef] [PubMed]

101. Goldring, M.B.; Otero, M.; Favero, M.; Dragomir, C.; el Hachem, K.; Hashimoto, K.; Plumb, D.A. Human chondrocyte cultures as models of cartilage-specific gene regulation. *Methods Mol. Med.* **2005**, *107*, 69–95. [CrossRef] [PubMed]

102. Zhang, W.; Chen, J.; Tao, J.; Jiang, Y.; Hu, C.; Huang, L.; Ji, J.; Ouyang, H.W. The use of type 1 collagen scaffold containing stromal cell-derived factor-1 to create a matrix environment conducive to partial-thickness cartilage defects repair. *Biomaterials* **2013**, *34*, 713–723. [CrossRef] [PubMed]

103. Peck, Y.; Leom, L.T.; Low, P.F.P.; Wang, D.-A. Establishment of an in vitro three-dimensional model for cartilage damage in rheumatoid arthritis. *J. Tissue Eng. Regen. Med.* **2017**, *12*, e237–e249. [CrossRef]

104. Andreas, K.; Lübke, C.; Häupl, T.; Dehne, T.; Morawietz, L.; Ringe, J.; Kaps, C.; Sittinger, M. Key regulatory molecules of cartilage destruction in rheumatoid arthritis: An in vitro study. *Arthritis Res. Ther.* **2008**, *10*, R9. [CrossRef]

105. Park, H.; Choi, B.; Hu, J.; Lee, M. Injectable chitosan hyaluronic acid hydrogels for cartilage tissue engineering. *Acta Biomater.* **2013**, *9*, 4779–4786. [CrossRef] [PubMed]

106. Andreas, K.; Häupl, T.; Lübke, C.; Ringe, J.; Morawietz, L.; Wachtel, A.; Sittinger, M.; Kaps, C. Antirheumatic drug response signatures in human chondrocytes: Potential molecular targets to stimulate cartilage regeneration. *Arthritis Res. Ther.* **2009**, *11*, R15. [CrossRef]

107. Ibold, Y.; Frauenschuh, S.; Kaps, C.; Sittinger, M.; Ringe, J.; Goetz, P.M. Development of a High-Throughput Screening Assay Based on the 3-Dimensional Pannus Model for Rheumatoid Arthritis. *J. Biomol. Screen.* **2007**, *12*, 956–965. [CrossRef]

108. Karimi, T.; Barati, D.; Karaman, O.; Moeinzadeh, S.; Jabbari, E. A developmentally inspired combined mechanical and biochemical signaling approach on zonal lineage commitment of mesenchymal stem cells in articular cartilage regeneration. *Integr. Biol.* **2014**, *7*, 112–127. [CrossRef]

109. Sato, M.; Yamato, M.; Hamahashi, K.; Okano, T.; Mochida, J. Articular Cartilage Regeneration Using Cell Sheet Technology. *Anat. Rec. Adv. Integr. Anat. Evol. Biol.* **2013**, *297*, 36–43. [CrossRef]

110. Furukawa, K.S.; Suenaga, H.; Toita, K.; Numata, A.; Tanaka, J.; Ushida, T.; Sakai, Y.; Tateishi, T. Rapid and large-scale formation of chondrocyte aggregates by rotational culture. *Cell Transplant.* **2003**, *12*, 475–479. [CrossRef] [PubMed]

111. Nakagawa, Y.; Muneta, T.; Otabe, K.; Ozeki, N.; Mizuno, M.; Udo, M.; Saito, R.; Yanagisawa, K.; Ichinose, S.; Koga, H.; et al. Cartilage Derived from Bone Marrow Mesenchymal Stem Cells Expresses Lubricin In Vitro and In Vivo. *PLoS ONE* **2016**, *11*, e0148777. [CrossRef]

112. Zhang, S.; Ba, K.; Wu, L.; Lee, S.; Peault, B.; Petrigliano, F.A.; McAllister, D.R.; Adams, J.S.; Evseenko, D.; Lin, Y. Adventitial Cells and Perictyes Support Chondrogenesis Through Different Mechanisms in 3-Dimensional Cultures With or Without Nanoscaffolds. *J. Biomed. Nanotechnol.* **2015**, *11*, 1799–1807. [CrossRef]

113. Penick, K.J.; Solchaga, L.A.; Welter, J.F. High-throughput aggregate culture system to assess the chondrogenic potential of mesenchymal stem cells. *Biotechniques* **2005**, *39*, 687–691. [CrossRef]

114. Teixeira, L.S.M.; Leijten, J.C.H.; Sobral, J.; Jin, R.; van Apeldoorn, A.A.; Feijen, J.; van Blitterswijk, C.; Dijkstra, P.J.; Karperien, M. High throughput generated micro-aggregates of chondrocytes stimulate cartilage formation in vitro and in vivo. *Eur. Cells Mater.* **2012**, *23*, 387–399. [CrossRef]

115. Sanjurjo-Rodríguez, C.; Castro-Viñuelas, R.; Hermida-Gómez, T.; Fuentes-Boquete, I.M.; de Toro, F.J.; Blanco, F.J.; Díaz-Prado, S.M. Human Cartilage Engineering in an In Vitro Repair Model Using Collagen Scaffolds and Mesenchymal Stromal Cells. *Int. J. Med. Sci.* **2017**, *14*, 1257–1262. [CrossRef]

116. Ando, W.; Tateishi, K.; Katakai, D.; Hart, D.; Higuchi, C.; Nakata, K.; Hashimoto, J.; Fujie, H.; Shino, K.; Yoshikawa, H.; et al. In Vitro Generation of a Scaffold-Free Tissue-Engineered Construct (TEC) Derived from Human Synovial Mesenchymal Stem Cells: Biological and Mechanical Properties and Further Chondrogenic Potential. *Tissue Eng. Part A* **2008**, *14*, 2041–2049. [CrossRef] [PubMed]

117. Karmakar, S.; Kay, J.; Gravallese, E.M. Bone Damage in Rheumatoid Arthritis: Mechanistic Insights and Approaches to Prevention. *Rheum. Dis. Clin. N. Am.* **2010**, *36*, 385–404. [CrossRef]

118. Clarke, B. Normal Bone Anatomy and Physiology. *Clin. J. Am. Soc. Nephrol.* **2008**, *3*, S131–S139. [CrossRef] [PubMed]

119. Wang, C.; Cao, X.; Zhang, Y. A Novel Bioactive Osteogenesis Scaffold Delivers Ascorbic Acid, Beta-Glycerophosphate, and Dexamethasone in Vivo to Promote Bone Regeneration. *Oncotarget* **2017**, *8*, 31612–31625. [CrossRef]

120. Ghassemi, T.; Shahroodi, A.; Ebrahimzadeh, M.H.; Mousavian, A.; Movaffagh, J.; Moradi, A. Current Concepts in Scaffolding for Bone Tissue Engineering. *Arch. Bone Jt. Surg.* **2018**, *6*, 90–99.

121. Vaccaro, A.R. The role of the osteoconductive scaffold in synthetic bone graft. *Orthopedics* **2002**, *25*, 571–578.

122. Duan, W.; Haque, M.; Kearney, M.T.; Lopez, M.J. Collagen and Hydroxyapatite Scaffolds Activate Distinct Osteogenesis Signaling Pathways in Adult Adipose-Derived Multipotent Stromal Cells. *Tissue Eng. Part C Methods* **2017**, *23*, 592–603. [CrossRef]

123. Bendtsen, S.T.; Quinnell, S.P.; Wei, M. Development of a novel alginate-polyvinyl alcohol-hydroxyapatite hydrogel for 3D bioprinting bone tissue engineered scaffolds. *J. Biomed. Mater. Res. Part A* **2017**, *105*, 1457–1468. [CrossRef] [PubMed]

124. Agarwal, T.; Kabiraj, P.; Narayana, G.H.; Kulanthaivel, S.; Kasiviswanathan, U.; Pal, K.; Giri, S.; Maiti, T.K.; Banerjee, I. Alginate Bead Based Hexagonal Close Packed 3D Implant for Bone Tissue Engineering. *ACS Appl. Mater. Interfaces* **2016**, *8*, 32132–32145. [CrossRef] [PubMed]

125. de Barros, A.P.D.N.; Takiya, C.M.; Garzoni, L.R.; Leal-Ferreira, M.L.; Dutra, H.S.; Chiarini, L.B.; Meirelles, M.N.; Borojevic, R.; Rossi, M.I.D. Osteoblasts and Bone Marrow Mesenchymal Stromal Cells Control Hematopoietic Stem Cell Migration and Proliferation in 3D In Vitro Model. *PLoS ONE* **2010**, *5*, e9093. [CrossRef]

126. de Witte, T.-M.; Fratila-Apachitei, L.E.; Zadpoor, A.; Peppas, N. Bone tissue engineering via growth factor delivery: From scaffolds to complex matrices. *Regen. Biomater.* **2018**, *5*, 197–211. [CrossRef]

127. Fernandes, G.; Wang, C.; Yuan, X.; Liu, Z.; Dziak, R.; Yang, S. Combination of Controlled Release Platelet-Rich Plasma Alginate Beads and Bone Morphogenetic Protein-2 Genetically Modified Mesenchymal Stem Cells for Bone Regeneration. *J. Periodontol.* **2016**, *87*, 470–480. [CrossRef]

128. Bouet, G.; Cruel, M.; Laurent, C.; Vico, L.; Malaval, L.; Marchat, D. Validation of an in vitro 3D bone culture model with perfused and mechanically stressed ceramic scaffold. *Eur. Cells Mater.* **2015**, *29*, 250–267. [CrossRef]

129. Yan, Y.; Chen, H.; Zhang, H.; Guo, C.; Yang, K.; Chen, K.; Cheng, R.; Qian, N.; Sandler, N.; Zhang, Y.S.; et al. Vascularized 3D printed scaffolds for promoting bone regeneration. *Biomaterilas* **2019**, 97–110. [CrossRef]

130. Chiesa, I.; de Maria, C.; Lapomarda, A.; Fortunato, G.M.; Montemurro, F.; di Gesù, R.; Tuan, R.S.; Vozzi, G.; Gottardi, R. Endothelial cells support osteogenesis in an in vitro vascularized bone model developed by 3D bioprinting. *Biofabrication* **2020**, *12*, 025013. [CrossRef]

131. Deng, Y.; Jiang, C.; Li, C.; Li, T.; Peng, M.; Wang, J.; Dai, K. 3d Printed Scaffolds of Calcium Silicate-Doped Beta-Tcp Synergize with Co-Cultured Endothelial and Stromal Cells to Promote Vascularization and Bone Formation. *Sci. Rep.* **2017**, *7*, 5588. [CrossRef]

132. Damerau, A.; Lang, A.; Pfeiffenberger, M.; Gaber, T.; Buttgereit, F. Fri0507 the Human-Based in Vitro 3d Arthritic Joint Model. *Ann. Rheum. Dis.* **2019**, *78*, 948–949. [CrossRef]

133. Wang, F.; Hu, Y.; He, D.; Zhou, G.; Ellis, E. Scaffold-free cartilage cell sheet combined with bone-phase BMSCs-scaffold regenerate osteochondral construct in mini-pig model. *Am. J. Transl. Res.* **2018**, *10*, 2997–3010.

134. Ng, J.; Bernhard, J.; Vunjak-Novakovic, G. Mesenchymal Stem Cells for Osteochondral Tissue Engineering. *Breast Cancer* **2016**, *1416*, 35–54. [CrossRef]

135. Lin, Z.; Li, Z.; Li, E.N.; Li, X.; del Duke, C.J.; Shen, H.; Hao, T.; O'Donnell, B.; Bunnell, B.A.; Goodman, S.B.; et al. Osteochondral Tissue Chip Derived From iPSCs: Modeling OA Pathologies and Testing Drugs. *Front. Bioeng. Biotechnol.* **2019**, *7*, 411. [CrossRef] [PubMed]

136. Damerau, A.; Lang, A.; Pfeiffenberger, M.; Buttgereit, F.; Gaber, T. FRI0002 Development of an in vitro multi-component 3d joint model to simulate the pathogenesis of arthritis. *Poster Presentat.* **2017**, *76*, 480. [CrossRef]

137. Damerau, A.; Pfeiffenberger, M.; Lang, A.; Gaber, T.; Buttgereit, F. Thu0069 Mimicking Arthritis in Vitro to Test Different Treatment Approaches. *Ann. Rheum. Dis.* **2020**, *79*, 247.

138. Lin, H.; Lozito, T.P.; Alexander, P.G.; Gottardi, R.; Tuan, R.S. Stem Cell-Based Microphysiological Osteochondral System to Model Tissue Response to Interleukin-1beta. *Mol. Pharm.* **2014**, *11*, 2203–2212. [CrossRef]

139. Steinhagen, J.; Bruns, J.; Niggemeyer, O.; Fuerst, M.; Ruther, W.; Schünke, M.; Kurz, B. Perfusion culture system: Synovial fibroblasts modulate articular chondrocyte matrix synthesis in vitro. *Tissue Cell* **2010**, *42*, 151–157. [CrossRef] [PubMed]

140. Mestres, G.; Perez, R.A.; D'Elía, N.L.; Barbe, L. Advantages of microfluidic systems for studying cell-biomaterial interactions—focus on bone regeneration applications. *Biomed. Phys. Eng. Express* **2019**, *5*, 032001. [CrossRef]

141. Rosser, J.; Bachmann, B.; Jordan, C.; Ribitsch, I.; Haltmayer, E.; Gueltekin, S.; Junttila, S.; Galik, B.; Gyenesei, A.; Haddadi, B.; et al. Microfluidic nutrient gradient-based three-dimensional chondrocyte culture-on-a-chip as an in vitro equine arthritis model. *Mater. Today Bio* **2019**, *4*, 100023. [CrossRef]

142. Rothbauer, M.; Hoell, G.; Eilenberger, C.; Kratz, S.R.A.; Farooq, B.; Schuller, P.; Calvo, I.O.; Byrne, R.A.; Meyer, B.; Niederreiter, B.; et al. Monitoring tissue-level remodelling during inflammatory arthritis using a three-dimensional synovium-on-a-chip with non-invasive light scattering biosensing. *Lab Chip* **2020**, *20*, 1461–1471. [CrossRef] [PubMed]

143. Giusti, S.; Mazzei, D.; Cacopardo, L.; Mattei, G.; Domenici, C.; Ahluwalia, A. Environmental Control in Flow Bioreactors. *Processes* **2017**, *5*, 16. [CrossRef]

144. Schulze-Tanzil, G.; Müller, R.D.; Kohl, B.; Schneider, N.; Ertel, W.; Ipaktchi, K.; Hünigen, H.; Gemeinhardt, O.; Stark, R.; John, T. Differing in vitro biology of equine, ovine, porcine and human articular chondrocytes derived from the knee joint: An immunomorphological study. *Histochem. Cell Biol.* **2008**, *131*, 219–229. [CrossRef] [PubMed]

145. Athanasiou, K.A.; Rosenwasser, M.P.; Buckwalter, J.A.; Malinin, T.I.; Mow, V.C. Interspecies comparisons of in situ intrinsic mechanical properties of distal femoral cartilage. *J. Orthop. Res.* **1991**, *9*, 330–340. [CrossRef] [PubMed]

146. McLure, S.W.D.; Fisher, J.; Conaghan, P.G.; Williams, S. Regional cartilage properties of three quadruped tibiofemoral joints used in musculoskeletal research studies. *Proc. Inst. Mech. Eng. H.* **2012**, *226*, 652–656. [CrossRef] [PubMed]

10

A Tissue-Engineered Human Psoriatic Skin Model to Investigate the Implication of cAMP in Psoriasis: Differential Impacts of Cholera Toxin and Isoproterenol on cAMP Levels of the Epidermis

Mélissa Simard [1,2,†], Sophie Morin [1,2,†], Geneviève Rioux [1,2], Rachelle Séguin [1,2], Estelle Loing [3] and Roxane Pouliot [1,2,*]

[1]   Centre de Recherche en Organogénèse Expérimentale de l'Université Laval/LOEX, Axe Médecine Régénératrice, Centre de Recherche du CHU de Québec—Université Laval, Québec, QC G1J 1Z4, Canada; melissa.simard.6@ulaval.ca (M.S.); sophie.morin.7@ulaval.ca (S.M.); genevieve.rioux.9@ulaval.ca (G.R.); rachelle.seguin.1@gmail.com (R.S.)

[2]   Faculté de Pharmacie, Université Laval, Québec, QC G1V 0A6, Canada

[3]   IFF-Lucas Meyer Cosmetics, Québec, QC G1V 4M6, Canada; estelle.loing@lucasmeyercosmetics.com

*   Correspondence: roxane.pouliot@pha.ulaval.ca

†   These authors contributed equally to this work.

Abstract: Pathological and healthy skin models were reconstructed using similar culture conditions according to well-known tissue engineering protocols. For both models, cyclic nucleotide enhancers were used as additives to promote keratinocytes' proliferation. Cholera toxin (CT) and isoproterenol (ISO), a beta-adrenergic agonist, are the most common cAMP stimulators recommended for cell culture. The aim of this study was to evaluate the impact of either CT or ISO on the pathological characteristics of the dermatosis while producing a psoriatic skin model. Healthy and psoriatic skin substitutes were produced according to the self-assembly method of tissue engineering, using culture media supplemented with either CT (10^{-10} M) or ISO (10^{-6} M). Psoriatic substitutes produced with CT exhibited a more pronounced psoriatic phenotype than those produced with ISO. Indeed, the psoriatic substitutes produced with CT had the thickest epidermis, as well as contained the most proliferating cells and the most altered expression of involucrin, filaggrin, and keratin 10. Of the four conditions under study, psoriatic substitutes produced with CT had the highest levels of cAMP and enhanced expression of adenylate cyclase 9. Taken together, these results suggest that high levels of cAMP are linked to a stronger psoriatic phenotype.

Keywords: psoriasis; cyclic adenosine monophosphate; cholera toxin; isoproterenol; tissue engineering

1. Introduction

Psoriasis is an autoimmune skin disease affecting 3% of the population worldwide and for which no cure currently exists [1]. Clinical manifestation of psoriasis is defined by the apparition of red plaques with white scales, which have detrimental consequences on patient's quality of life [2,3]. Psoriasis patches can range from a few spots of dandruff-like scaling to major eruptions that cover large areas of the body. The histological hallmarks of psoriasis are a marked thickening of the epidermis, due to keratinocyte hyperproliferation, abnormal epidermal differentiation, and immune keratinocytes activation accompanied with immune cell infiltrate [4]. Consequently, the granular layer of the epidermis is reduced in thickness and the horny layer contains some undifferentiated keratinocytes, which still contain cell nuclei [5]. Furthermore, altered keratinocyte differentiation in psoriasis results at a molecular level in the deregulation of the epidermal differentiation marker proteins, such as involucrin (up-regulated), filaggrin (down-regulated), and keratin 10 (down-regulated) [6–8]. It is now well documented that the complex etiology of psoriasis involves interactions between environmental factors and complex genetic background [9]. However, the exact cause of psoriasis is still unknown, making it difficult to develop an effective treatment for the pathology [1].

Since the development of the first tissue-engineered skin model in 1978, by Green H. and colleagues, great deal of research has been conducted in this field, leading to the generation of highly reproducible and sophisticated skin models. In the last decade, the limits of tissue engineering have been constantly pushed back by the development of new emerging techniques, such as bio-printing [10,11]. However, the reconstruction of pathological skin models, such as psoriatic skin substitutes, remains challenging, especially since the exact causes of the pathology are still unknown. However, many psoriatic skin models were reconstructed in vitro using different tissue-engineered techniques [12]. According to Niehues et al., the self-assembly method is the one which leads to the reconstruction of substitutes faithfully mimicking psoriatic characteristics and which, consequently, offers an effective tool for the screening of new molecules [12–14]. Indeed, reconstructed psoriatic skin substitutes produced using cells from patients with psoriasis were shown to closely mimic the pathology, as they displayed enhanced epidermal thickness, hyperproliferative keratinocytes and disturbed epidermal differentiation. The morphology of reconstructed skin models is directly affected by culture conditions. While developing the first healthy tissue engineered model of skin in 1978, Green H. reported the importance of adding cAMP stimulator, such as cholera toxin (CT) or isoproterenol (ISO), to induce the proliferation of cells in culture [15]. Therefore, skin models are produced using culture media supplemented with a stimulator of the AC system, which is crucial for stimulating the colony growth of human keratinocytes and thus essential to obtain a fully differentiated epidermis [15–20].

The cAMP signaling pathway is composed of the first messenger, the G protein-coupled receptor (GPCR), the adenylate cyclase enzyme (AC) and the cAMP-degrading enzyme (Figure 1a). The epidermis contains four independent receptor AC systems: the β-adrenergic, prostaglandin E, adenosine, and histamine receptors [21–23]. The cAMP mainly acts as a second messenger, by stimulating different proteins, such as protein kinase A (PKA) and exchange protein directly activated by cAMP (EPAC). PKA and EPAC are then able to modify cell activity by phosphorylating diverse proteins [18]. The action of cAMP ends upon its hydrolysis by phosphodiesterases [24–26]. CT and ISO induce the production of cAMP following a different pathway. ISO is an analogue to epinephrine and binds exclusively to β-adrenergic receptors, which then stimulates cAMP production following the classical pathway (Figure 1a) [27,28]. CT binds to ganglioside GM1 receptors on epithelial cells, which triggers endocytosis, thus transporting the receptors to the endoplasmic reticulum (ER). From the lumen of the ER, the A1 peptide of the CT is transported into the cytoplasm, where it prevents the G protein from cleaving GTP to GDP, leading to a tremendous increase in cAMP levels (Figure 1b) [16,29,30].

(a) Classical pathway (and isoproterenol)

(b) Cholera toxin pathway

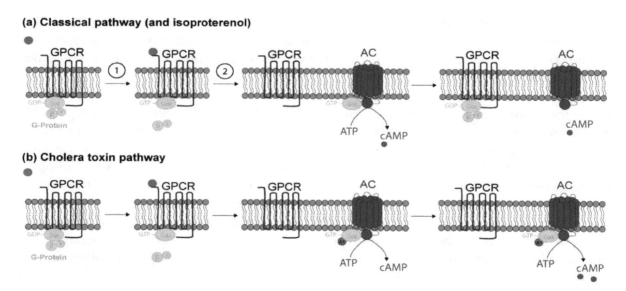

Figure 1. The cAMP signaling pathways. (**a**) 1: The first messenger (red) binds to the G protein-coupled receptor (GPCR). The receptor then changes conformation, leading to the replacement of GDP by GTP on the α subunit of the G protein and inducing the subsequent release of the α subunit from the β and γ subunits of the G protein [21,31]. 2: The α subunit binds to the catalytic domain of adenylate cyclase (AC). The α subunit can have inhibitory (α_i) or stimulating (α_s) properties, which will lead to the inhibition or stimulation of AC [16,32]. The activated AC will convert ATP into cAMP. Isoproterenol induces the conversion of ATP to cAMP following the classical pathway (**b**) The A1 peptide of cholera toxin (CT) (purple) binds to the complex composed of the GTP-α subunit of the G protein and AC and prevents the G protein from cleaving GTP to GDP, leading to a tremendous increase in cAMP levels [30].

The impact of different cAMP inducers (CT and ISO) on the morphology of psoriatic skin models has not yet been investigated, whereas it should be chosen with much consideration, especially since controversial results have been found regarding the involvement of cAMP in psoriasis. Indeed, Voorhes et al. suggested in the seventies that an alteration in cAMP levels could be involved in psoriasis [33]. Moreover, a few studies have reported lower levels of cAMP in psoriatic skin, suggesting that low levels of cAMP were linked with enhanced cell proliferation and thus contrasting with finding of Green H [33–37]. The aim of the present study was to therefore compare the use of both cAMP enhancers (CT or ISO) in the reconstruction of psoriatic skin substitutes to establish which one would lead to the better psoriatic phenotype, with traits, such as cell hyperproliferation and disturbed cell differentiation. The cAMP signaling pathway was then investigated in detail to demystify cAMP levels, as well as the activity of various agents of the AC system, in the pathology of psoriasis. Therefore, the current study brings new insights to the long-standing debate as to whether cAMP is increased or decreased in psoriasis.

2. Results

2.1. Differential Impact of CT and ISO on the Psoriatic Skin Substitute Phenotype: Epidermal Hyperproliferation

According to the macroscopic aspect of the skin substitutes, the epidermis of psoriatic substitutes reconstructed using culture media supplemented with CT (PS+CT) and the epidermis of psoriatic substitutes reconstructed using ISO (PS+ISO) were more disorganized since they displayed a less opaque and uniform surface than those of healthy substitutes (HS+CT) and (HS+ISO) (Figure 2c,d vs. Figure 2a,b). Moreover, both PS+CT and PS+ISO had a significantly thicker epidermis than their respective counterparts, showing higher proliferation of the keratinocytes (Figure 2g,h vs. Figure 2e,f). Hematoxylin and eosin staining of the skin substitutes are presented in Figure S1. This enhanced proliferation was confirmed with Ki67 immunofluorescence showing more basal keratinocytes in cellular division in PS than in HS (Figure 2k,l vs. Figure 2i,j). However, keratinocytes from PS+ISO were not as hyperproliferative as those in PS+CT. In fact, the epidermis of PS+ISO was not as thick as that of PS+CT (Figure 2m). The Ki67 staining confirmed these results (Figure 2n). Taken together, these results showed that both PS+CT and PS+ISO displayed higher levels of epidermal proliferation than their healthy counterparts HS+CT and HS+ISO, respectively. Moreover, the hyperproliferation was greater in the epidermis of PS+CT than PS+ISO.

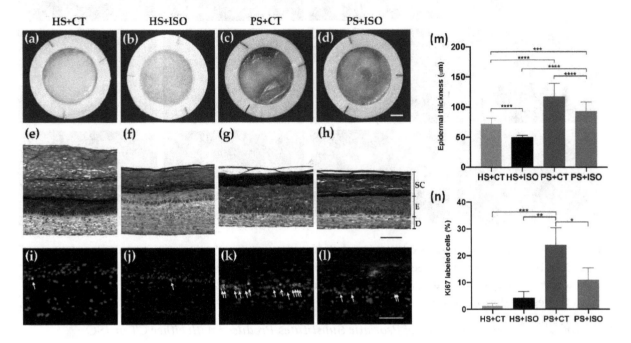

Figure 2. Morphology and epidermal proliferation of healthy substitutes (HS) and psoriatic substitutes (PS) produced with cholera toxin (+CT) or isoproterenol (+ISO). (**a–d**) Skin substitute macroscopic aspects; (**e–h**) Masson's trichrome staining of skin substitute histological sections (D: Dermis, E: Epidermis living layers, SC: *Stratum corneum*); (**i–l**) Ki67 immunofluorescence (green) detecting keratinocytes in proliferation. White arrows indicate Ki67-positive cells. Nuclei were stained with Hoechst (blue); (**m**) quantification of the thickness of the epidermal living layer (designated as E in panels (**e–h**) measured from Masson's trichrome staining using the ImageJ software; (**n**) ratio of Ki67 positive cells to the number of total keratinocytes in the basal layer. Scale bars: (**a–d**) 5 mm; (**e–h**) 100 μm. Data presented are the means +SD ($N = 2$ donors per condition, $n = 3$ skin substitutes per donor). The statistical significance was determined using one-way ANOVA followed by a Tukey's post-hoc test. (* p-value < 0.05; ** p-value < 0.01; *** p-value < 0.001; **** p-value < 0.0001).

2.2. Differential Impact of CT and ISO on the Psoriatic Skin Substitute Phenotype: Disturbed Epidermal Differentiation

The expression of differentiation markers was strongly altered in PS+CT compared with HS+CT, thus showing a psoriatic phenotype such as expected. Indeed, the late differentiation markers filaggrin and keratin 10 were both down-regulated while the expression of the early differentiation marker involucrin was up-regulated (Figure 3). On the other hand, the effects of ISO on the expression of the differentiation markers were not as conclusive as for CT. In fact, the expression of filaggrin and keratin 10 for PS+ISO was down-regulated but not as much as for PS+CT compared with their respective counterparts. The involucrin staining was even less conclusive, since its expression in PS+ISO appeared less intense than the other conditions. According to these results, PS produced with ISO displayed a complete epidermal differentiation similar to what was observed for HS, since late differentiation markers are expressed. Therefore, epidermal differentiation in PS+ISO did not properly mimic the disturbed epidermal differentiation characteristics of psoriasis, which further entail that ISO is not a suitable cAMP stimulator to produce psoriatic skin substitutes.

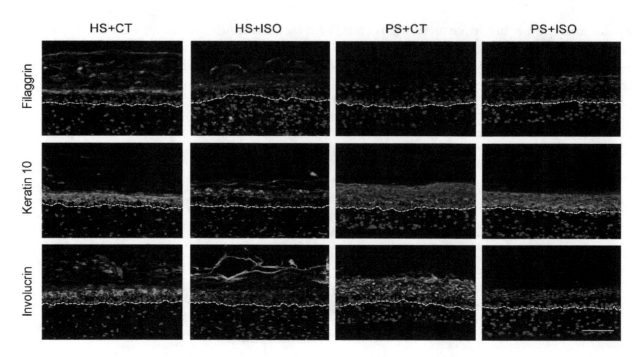

Figure 3. Expression of epidermal differentiation markers (green) in healthy substitutes (HS) and psoriatic substitutes (PS) produced with cholera toxin (+CT) or isoproterenol (+ISO) ($N = 2$ donors per condition, $n = 3$ skin substitutes per donor). The nuclei were stained with Hoechst (blue). The dotted line indicates the dermo-epidermal junction. Scale bar: 100 μm.

2.3. Levels of cAMP in the Epidermis of Psoriatic Substitutes Produced with either CT or ISO

The levels of cAMP in the epidermis of HS+CT and HS+ISO were approximately the same (Figure 4.) The two cAMP enhancers, therefore, have similar effects on the AC activity of HS. The level of cAMP for PS+CT was significantly higher than for the other three conditions, revealing a greater capacity of CT to stimulate the production of cAMP.

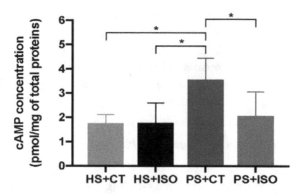

Figure 4. cAMP levels in the epidermis of healthy substitutes (HS) and psoriatic substitutes (PS) produced with either cholera toxin (+CT) or isoproterenol (+ISO). Data presented are the means +/− SD ($N = 2$ donors per condition, $n = 2$ skin substitutes per donor). The statistical significance was determined using one-way ANOVA followed by a Tukey's post-hoc test. (* p-value < 0.05).

2.4. Identification of Isoforms of cAMP-Related Protein Found in the Skin

Gene profiling on microarray was exploited to examine which AC were expressed in the skin substitutes (Table 1). Among the 10 AC isoform genes (*ADCY1-10*), only *ADCY3*, *ADCY7*, and *ADCY9* displayed a linear signal over 100 and were therefore identified as expressed genes in the HS+CT and PS+CT. Interestingly, the levels of expression of the three *ADCY* genes were higher (2-fold) in PS+CT than in HS+CT. Furthermore, expression of the *ADRB2* gene encoding for the beta-2 adrenergic receptor was also detected in both HS+CT and PS+CT. The expression of *ADRB2* was decreased (0.5-fold) in PS+CT.

Table 1. Linear signals and fold change for *ADCY1-10* and *ADRB2* genes between healthy and psoriatic substitutes produced with CT.

Gene Symbol	Gene Name	Linear Signal HS+CT	Linear Signal PS+CT	Fold Change PS/HS	*p*-Value
ADCY1	Adenylate cyclase type 1	6.688	11.806	1.765	0.0628
ADCY2	Adenylate cyclase type 2	2.953	3.933	1.332	0.0942
ADCY3	Adenylate cyclase type 3	1714.868	3504.636	2.044	0.0359
ADCY4	Adenylate cyclase type 4	59.773	43.994	0.736	0.1634
ADCY5	Adenylate cyclase type 5	7.595	5.103	0.672	0.2416
ADCY6	Adenylate cyclase type 6	55.333	127.404	2.302	0.2256
ADCY7	Adenylate cyclase type 7	275.730	614.908	2.230	0.0446
ADCY8	Adenylate cyclase type 8	5.560	4.526	0.814	0.5801
ADCY9	Adenylate cyclase type 9	228.869	549.184	2.400	0.0471
ADCY10	Adenylate cyclase type 10	7.348	5.591	1.286	0.1556
ADRB2	Beta-2 adrenergic receptor	338.779	160.034	0.472	0.0487

2.5. Levels of cAMP-Related Proteins in the Epidermis of Psoriatic Substitutes Produced with Either CT or ISO

Immunofluorescence staining was performed to validate the epidermal levels of AC9 and beta-2 adrenergic receptor, which are encoded by *ADCY3, ADCY7, ADCY9,* and *ADRB2* genes, respectively, as well as to compare the impact of CT and ISO on those protein levels (Figure 5a). Based on the immunofluorescence staining, the AC9 was found in the cells of both the dermis and the epidermis of the skin substitutes. Moreover, AC9 levels seemed up-regulated in psoriatic substitutes (PS+CT and PS+ISO) as compared with healthy ones. On the other hand, the β2-adrenergic receptor was found predominantly in the epidermis. Interestingly, high levels of the β2-adrenergic receptor were detected in HS+CT, while low levels were found in HS+ISO and it was not detected in PS+ISO. These results therefore support the conclusion that the levels of β2-adrenergic receptors are decreased in psoriatic skin.

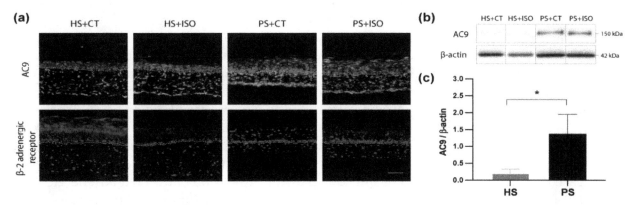

Figure 5. (a) Expression of adenylate cyclase 9 (AC9) (green) and β2-adrenergic receptor (red) in healthy substitutes (HS) and psoriatic substitutes (PS) produced with cholera toxin (+CT) or isoproterenol (+ISO). The nuclei were stained with Hoechst (blue). The dotted line indicates the dermo-epidermal junction. Scale bar: 100 μm. (b) Ten micrograms of total protein from skin substitute epidermis were analyzed by immunoblot for the presence of adenylate cyclase 9 (AC9). β-actin was used to control equal loading. (*N* = 2 donors per condition; *n* = 2 skin substitutes per donor). One representative immunoblot is shown. (c) Densitometric analyses of the immunoblot from panel (b) (*N* = 2 donors per condition; *n* = 2 skin substitutes per donor). Data from healthy or psoriatic substitutes were combined irrespective of treatment regime. The statistical significance was determined using an unpaired *t*-test (* *p*-value < 0.05).

Western blot analyses were next conducted to validate the levels of AC9 in the epidermis of the skin substitutes, since immunofluorescence staining of these proteins was used for qualitative purposes. AC9 was easily detected in the epidermis of all conditions, with a higher prevalence in PS+CT and PS+ISO (Figure 5b,c). AC3 and AC7 were not detected in the epidermis of either healthy nor psoriatic skin substitutes under our experimental conditions.

3. Discussion

Culture media used to generate tissue-engineered skin substitutes are supplemented with a cAMP stimulator, which ensures epithelial cell proliferation [15,17]. In the present study, it was shown that even if the psoriatic skin models reconstructed with either CT or ISO displayed pathological features of psoriasis, CT remains more effective for stimulating the cells towards a psoriatic phenotype with hyperproliferative epidermal cells. Psoriatic substitutes produced with CT also displayed higher cAMP levels, as well as higher AC9 expression, than other conditions. These results show that higher cAMP levels would be associated with a stronger psoriatic phenotype, such as epidermal hyperproliferation and altered epidermal differentiation, therefore supporting studies in which cAMP levels are found to be increased in psoriatic skin.

Our data suggest that, in the presence of healthy keratinocytes, both cAMP enhancers (CT and ISO) stimulate the production of cAMP in the same manner, although the concentration of CT used in the experiment was lower than the concentration of ISO. Therefore, these results are in accordance with previous studies, which have reported that even if CT and ISO are both cAMP stimulators of epidermal cells, CT is better than ISO for the improvement of keratinocyte growth since it may increase the overall growth rate of epidermal cells by reducing the doubling time of cells [15]. In vitro studies reported by Green et al. have shown that using 10^{-6} M of isoproterenol produced an increase in keratinocyte colony size but had less effect on cAMP levels than 10^{-6} M of CT. Indeed, CT was the strongest agent in affecting cAMP concentration among the four different cAMP stimulators tested [15]. Moreover, these results support the use of both cAMP inducers, at appropriate concentrations, when producing healthy reconstructed skin models, and are in accord with results reported by Cortez Ghio et al. [38]. Isoproterenol (ISO) is therefore a good candidate for replacing CT since it presents several advantages regarding safety and regulations [38]. Indeed, CT is known to be potentially toxic and therefore requires more precautions in its manipulation [39,40].

In psoriatic substitutes, CT induced a stronger psoriatic phenotype, as well as higher levels of cAMP, than ISO when using the same concentration of each inducer as for the healthy models. This suggests that CT would be a better cAMP stimulator for lesional cells and that unlike for healthy substitutes, ISO is not recommended to stimulate cAMP in lesional cells. This incapacity of ISO to stimulate cAMP production in psoriatic substitutes could be attributed to a beta-adrenergic defect in psoriatic skin. Although Archer et al. once reported that psoriasis is not associated with impaired beta-adrenergic reactivity [41], most studies have since reported a decrease in responsiveness to beta-adrenergic stimuli, such as isoproterenol, in the stimulation of cAMP production in psoriatic tissues [16,30,42–44]. This decreased responsiveness would be linked to a lower expression of the beta-adrenergic receptor in psoriatic skin, which was confirmed in the present study with immunofluorescence staining. Different beta-2-adrenergic receptor polymorphisms may also contribute to the pathogenesis [30,45]. This defect of the beta-2-adrenergic receptor in psoriatic skin does not affect cAMP stimulation by CT, since the CT mechanism of action is based on direct activation of AC9. Finally, in 1978, Das et al. reported that the topical application of 0.1% isoproterenol sulphate mixed with white Vaseline and applied to the psoriatic plaques of twelve patients induced a significant decrease in the scaliness and cell turnover of the treated psoriatic skin [46]. Even if no other literature was found on a possible commercial development of this formulation, it appears that isoproterenol could be a potential topical or systemic treatment for psoriatic patients.

In the present study, a stronger psoriatic phenotype and higher levels of cAMP were found in psoriatic substitutes as compared with healthy substitutes, especially with CT, for which the difference was significant. These results therefore imply that higher levels of cAMP would be found in psoriatic skin, which is in contrast to the levels of cAMP monitored in psoriatic skin in the late seventies [33]. However, these results are in accordance with studies on the effects of CT on psoriasis in vivo, in which it was revealed that the psoriatic epidermis accumulates much more cAMP than uninvolved body regions or normal human epidermis [18,21]. According to the present study, the higher levels of cAMP found in psoriatic epidermis could be attributed to enhanced AC9 expression. AC9 is mostly expressed in the brain, spinal cord, liver, heart, lung, kidney, muscle, and adrenal gland, but little is known about its localization in the skin, particularly in the epidermis [24,47,48]. Finally, enhanced levels of cAMP in psoriatic skin are more consistent with the levels of downstream mediators identified in the literature during the past decade [49–51]. Indeed, it is relatively well established that psoriatic skin displays increased levels of phosphorylated CREB, which controls cellular functions, such as the regulation of gene expression [52].

In summary, our study showed that, although both cAMP inducers stimulate the production of psoriatic skin substitutes displaying hyperproliferating epidermis and disturbed epidermal differentiation, CT induced a stronger psoriatic phenotype than ISO. Moreover, enhanced cAMP levels were found in the epidermis of our psoriatic substitutes, which could be attributed to enhanced expression of AC9 in psoriatic skin. Therefore, our study answers a highly controversial question, suggesting that cAMP levels are increased in psoriatic skin. Finally, cAMP levels were higher in the CT psoriatic substitutes than in the ISO psoriatic substitutes, confirming a defect of the beta2-adrenergic receptors in psoriatic skin. This study is of particular interest since to our knowledge it is one of the first to determine the concentrations of cAMP in a psoriatic reconstructed skin model produced by tissue engineering.

4. Materials and Methods

4.1. Cell Culture

This study was conducted in agreement with the Helsinki Declaration and performed under the guidelines of the Research Ethics Committee of the Centre de recherche du CHU de Québec – Université Laval. All patients were given adequate information for providing written consent. Two different patients with psoriasis aged 46 and 49 years old, respectively, were recruited. Six-millimeter punch biopsies were taken from psoriatic skin. As for the healthy skin substitutes, biopsies were obtained from healthy donors during breast reduction surgeries. Healthy donors were Caucasian females aged 46 and 49 years old. Cells were extracted according to the method based on thermolysin, trypsin, and collagenase digestions described elsewhere [53].

4.2. Skin Substitute Production

All skin substitutes were produced according to the self-assembly method of tissue engineering [13]. Briefly, fibroblasts at passage 6 were seeded in 6-well plates at 0.12×10^6 cells per well. Fibroblasts were cultured in Dulbecco's modified Eagle's medium (DME) (Gibco, Life Technologies, New York, NY, USA) supplemented with 10% Fetal Calf premium Serum (FCS) (Wiscent, Inc., St-Bruno, QC, Canada), 50 µg/mL ascorbate acid (Sigma, Oakville, ON, Canada) and antibiotics; 60 µg/mL penicillin G (Sigma, Oakville, ON, Canada) and 25 µg/mL gentamicin (Schering, Pointe-Claire, QC, Canada). After 28 days, dermal cells formed sheets that were superimposed and cultured for another 2 d. After that, keratinocytes at passage 2 were seeded at 1.2×10^6 cells upon each tissue sheet to form the epidermal layer. Another 7 days of culture allowed the epidermal cells to grow, and then each substitute was raised to the air-liquid interface and cultured for a total of 21 days. Keratinocytes were cultured in DME mixed with Ham's F12 medium (3:1) (DME-HAM) (Gibco, Life Technologies, New York, NY, USA) supplemented with 5% FetalClone II serum (Hyclone, Logan, UT, US), 5 µg/mL insulin (Sigma, Oakville, ON, Canada), 0.4 µg/mL hydrocortisone (Galenova, St-Hyacinthe, QC, Canada), 10 ng/mL human epidermal growth factor (EGF) (Ango Inc, San Ramon, CA, USA), 60 µg/mL penicillin, and 25 µg/mL gentamicin. Keratinocyte culture media were also supplemented with either 10^{-10} M cholera toxin (MP Biomedicals, Montreal, QC, Canada) or 10^{-6} M isoproterenol (rISO; Sigma Aldrich, St. Louis, MO, USA).

4.3. Histological Analyses

Biopsies of skin substitutes were analyzed by histological and immunohistochemical methods after 21 days of culture at the air-liquid interface. For living epidermal thickness analyses, two biopsies of each condition were fixed in Histochoice® (AMRESCO, Inc., Solon, OH, USA) and embedded in paraffin wax. After that, deparaffinized 5 µm tissue sections were cut and stained with Masson's Trichrome. The thickness of the living epidermis was obtained by measures made with Image J software (National Institutes of Health, USA, http://imagej.nih.gov/ij). Ten measurements in three different parts of each skin substitute were used to compare the living epidermis thickness between skin substitutes.

4.4. Indirect Immunofluorescence on Frozen Tissues

For immunofluorescence staining, tissues were embedded in Tissue-Tek O.C.T Compound (Sakura Finetek, CA, USA) and 6 µm thick sections were fixed in acetone at −20 °C before staining. The following antibodies were used and incubated in a dark room at room temperature for 45 min: rabbit anti-involucrin (Abcam, Cambridge, MA, USA), rabbit anti-filaggrin (Abcam, Cambridge, MA, USA), rabbit anti-keratin 10 (Abcam, Cambridge, MA, USA), mouse anti-Ki67 IgG1 (BD Biosciences, CA, USA), rabbit anti-AC9 (ab191423, Abcam, Cambridge, MA, USA) and rabbit anti-beta 2 adrenergic receptor (ab61778, Abcam, Cambridge, MA, USA). Tissues were then incubated with Alexa Fluor 488 donkey anti-rabbit IgG or Alexa Fluor 594 goat anti-mouse IgG (1:1600, Thermofisher Scientific, CA, USA) for 30 min also at room temperature. Nuclear counter staining using DAPI (SouthernBiotech, AL, USA) was then effected on different samples. Each tissue was observed using a Zeiss Axio Imager M2 microscope with an AxioCam ICc1 camera. The quantification of immunofluorescence staining was performed by densitometry using ImageJ software (from Wayne Rasband, National Institute of Health (NIH), USA).

4.5. Cyclic AMP Competitive ELISA Kit on Frozen Tissues

After 63 days of cell culture to prepare the reconstructed skin substitutes, the epidermis was mechanically separated from the dermis using scalpel and forceps, and samples were quick-frozen in liquid nitrogen to preserve the tissue integrity. Tissues were crushed in a Safe-Lock 2.0 mL Eppendorf tube (ATS Scientific, Inc., Burlington, ON, Canada) with two 6 mm stainless ball using a Cryomill MM400 (Retsch®, Newtown, PA, USA). The levels of cAMP were assayed using a Cyclic AMP Competitive ELISA Kit (ThermoFisher Scientific, Vienna, Austria). The non-acetylated version of the ELISA Kit was used following the protocol provided by the manufacturer, and 400 µg of cell lysate was used in the test. Total protein concentrations were determined using a BCA Protein Assay Kit according to the manufacturer's instructions (Thermo Scientific, Rockford, IL, USA).

4.6. Western Blots

Western blots were conducted using the following primary antibodies: rabbit anti-AC3 (ab14778, Abcam, Cambridge, MA, USA), rabbit anti-AC7 (ab14782, Abcam, Cambridge, MA, USA), and rabbit anti-AC9 (ab191423, Abcam, Cambridge, MA, USA). The secondary antibodies used were: goat anti-rabbit HRP labeled (1:60,000, Jackson ImmunoResearch Laboratories, Inc., West Grove, PA, USA); and goat anti-mouse HRP labeled (1:60,000, Jackson ImmunoResearch Laboratories, Inc., West Grove, PA, USA). Proteins of interest were detected using ECL Prime Western Blotting Detection Reagent (GE Healthcare, Little Chalfont, UK) and Fusion Fx7 imager (MBI Lab Equipment, Kirkland, QC, Canada). Quantification of immunoblots was performed by densitometry using ImageJ software (from Wayne Rasband, National Institute of Health (NIH), USA).

4.7. Gene Expression Profiling

Total RNA was isolated from skin substitutes using the RNeasy Mini Kit (QIAGEN, Toronto, ON, Canada) and its quality determined (2100 Bioanalyzer, Agilent Technologies, Mississauga, ON, Canada) as described [54]. Labeling of Cyanine 3-CTP-labeled targets, their hybridization on a G4851A SurePrint G3 Human GE 8 × 60K array slide (Agilent Technologies, Santa-Clara, CA, USA), and data acquisition and analyses were all performed as previously reported [54].

Author Contributions: Conceptualization, M.S., S.M., R.S. and R.P.; methodology, M.S., S.M., G.R. and R.P.; software, M.S.; validation, M.S., S.M., G.R., R.S. and R.P.; formal analysis, M.S., S.M., G.R. and R.P.; investigation, M.S., S.M., G.R., R.S. and R.P.; resources, R.P.; data curation, M.S., S.M., G.R. and R.P.; writing—original draft preparation, M.S., S.M., G.R. and R.P.; writing—review and editing, M.S., S.M., G.R., R.S., E.L., and R.P.; visualization, M.S., S.M., G.R., R.S. and R.P.; supervision, E.L. and R.P.; project administration, R.P.; funding acquisition, R.P. All authors have read and agreed to the published version of the manuscript.

Acknowledgments: Authors would like to acknowledge the support provided by Israel Martel in cell culture. We thank Jacques Soucy for kindly providing psoriatic biopsies for the in vitro study on psoriatic cells. Mélissa Simard received studentships from the Fonds de recherche Québec-Santé (FRQS) and from the Fonds d'Enseignement et de Recherche (FER) of the Faculté de Pharmacie of Université Laval. Sophie Morin received studentships from the FER, from the Fondation du CHU de Québec—Université Laval and from the FRQS. Geneviève Rioux received studentships from the FER and the Centre de recherche en organogénèse experimentale de l'Université Laval/LOEX. Rachelle Séguin received studentships from the FRQS, the FER and from the Centre de recherche en organogénèse experimentale de l'Université Laval/LOEX. Roxane Pouliot is a career award scholar from FRQS. We would also like to acknowledge the Réseau de thérapie cellulaire, tissulaire et génique du Québec-ThéCell (a thematic network supported by the FRQS).

Abbreviations

CT	Cholera toxin
ISO	Isoproterenol
AC	Adenylate cyclase enzyme
GPCR	G protein-coupled receptor
PKA	Protein kinase A
ER	Endoplasmic reticulum
HS	Healthy substitutes
PS	Psoriatic substitutes

References

1. Boehncke, W.-H.; Schön, M.P. Psoriasis. *Lancet* **2015**, *386*, 983–994. [CrossRef]
2. Lowes, M.A.; Bowcock, A.M.; Krueger, J.G. Pathogenesis and therapy of psoriasis. *Nature* **2007**, *445*, 866–873. [CrossRef] [PubMed]
3. Krueger, G.; Koo, J.; Lebwohl, M.; Menter, A.; Stern, R.S.; Rolstad, T. The impact of psoriasis on quality of life: Results of a 1998 National Psoriasis Foundation patient-membership survey. *Arch. Dermatol.* **2001**, *137*, 280–284. [PubMed]
4. Mak, R.K.; Hundhausen, C.; Nestle, F.O. Progress in understanding the immunopathogenesis of psoriasis. *Actas Dermo Sifiliogr.* **2009**, *100*, 2–13. [CrossRef]
5. Schön, M.P.; Boehncke, W.-H. Psoriasis. *N. Engl. J. Med.* **2005**, *352*, 1899–1912. [CrossRef]
6. Chen, J.-Q.; Man, X.-Y.; Li, W.; Zhou, J.; Landeck, L.; Cai, S.-Q.; Zheng, M. Regulation of Involucrin in Psoriatic Epidermal Keratinocytes: The Roles of ERK1/2 and GSK-3β. *Cell Biochem. Biophys.* **2013**, *66*, 523–528. [CrossRef]
7. Wolf, R.; Orion, E.; Ruocco, E.; Ruocco, V. Abnormal epidermal barrier in the pathogenesis of psoriasis. *Clin. Dermatol.* **2012**, *30*, 323–328. [CrossRef]
8. Danso, M.; Boiten, W.; Van Drongelen, V.; Meijling, K.G.; Gooris, G.; El Ghalbzouri, A.; Absalah, S.; Vreeken, R.J.; Kezic, S.; Van Smeden, J.; et al. Altered expression of epidermal lipid bio-synthesis enzymes in atopic dermatitis skin is accompanied by changes in stratum corneum lipid composition. *J. Dermatol. Sci.* **2017**, *88*, 57–66. [CrossRef]

9.   Benhadou, F.; Mintoff, D.; Del Marmol, V. Psoriasis: Keratinocytes or Immune Cells—Which Is the Trigger? *Dermatology* **2019**, *235*, 91–100. [CrossRef]

10.   Shi, Y.; Xing, T.L.; Zhang, H.B.; Yin, R.X.; Yang, S.M.; Wei, J.; Zhang, W.J. Tyrosinase-doped bioink for 3D bioprinting of living skin constructs. *Biomed. Mater.* **2018**, *13*, 035008. [CrossRef]

11.   Lee, V.; Singh, G.; Trasatti, J.P.; Bjornsson, C.; Xu, X.; Tran, T.N.; Yoo, S.-S.; Dai, G.; Karande, P. Design and Fabrication of Human Skin by Three-Dimensional Bioprinting. *Tissue Eng. Part C Methods* **2014**, *20*, 473–484. [CrossRef] [PubMed]

12.   Desmet, E.; Ramadhas, A.; Lambert, J.; Van Gele, M. In vitro psoriasis models with focus on reconstructed skin models as promising tools in psoriasis research. *Exp. Biol. Med.* **2017**, *242*, 1158–1169. [CrossRef] [PubMed]

13.   Jean, J.; Lapointe, M.; Soucy, J.; Pouliot, R. Development of an in vitro psoriatic skin model by tissue engineering. *J. Dermatol. Sci.* **2009**, *53*, 19–25. [CrossRef] [PubMed]

14.   Niehues, H.; van den Bogaard, E.H. Past, present and future of in vitro 3D reconstructed inflammatory skin models to study psoriasis. *Exp. Dermatol.* **2018**, *27*, 512–519. [CrossRef]

15.   Green, H. Cyclic AMP in relation to proliferation of the Epidermal cell: A new view. *Cell* **1978**, *15*, 801–811. [CrossRef]

16.   Tamura, T.; Takahashi, H.; Ishida-Yamamoto, A.; Hashimoto, Y.; Iizuka, H. Functional alteration of guanine nucleotide binding proteins (Gs and Gi) in psoriatic epidermis. *J. Dermatol. Sci.* **1998**, *17*, 61–66. [CrossRef]

17.   Okada, N.; Kitano, Y.; Ichihara, K. Effects of Cholera Toxin on Proliferation of Cultured Human Keratinocytes in Relation to Intracellular Cyclic AMP Levels. *J. Investig. Dermatol.* **1982**, *79*, 42–47. [CrossRef]

18.   Takahashi, H.; Honma, M.; Miyauchi, Y.; Nakamura, S.; Ishida-Yamamoto, A.; Iizuka, H. Cyclic AMP differentially regulates cell proliferation of normal human keratinocytes through ERK activation depending on the expression pattern of B-Raf. *Arch. Dermatol. Res.* **2004**, *296*, 74–82. [CrossRef]

19.   Duque-Fernandez, A.; Gauthier, L.; Simard, M.; Jean, J.; Gendreau, I.; Morin, A.; Soucy, J.; Auger, M.; Pouliot, R. A 3D-psoriatic skin model for dermatological testing: The impact of culture conditions. *Biochem. Biophys. Rep.* **2016**, *8*, 268–276. [CrossRef]

20.   Löwa, A.; Vogt, A.; Kaessmeyer, S.; Hedtrich, S. Generation of full-thickness skin equivalents using hair follicle-derived primary human keratinocytes and fibroblasts. *J. Tissue Eng. Regen. Med.* **2018**, *12*, e2134–e2146. [CrossRef]

21.   Iizuka, H.; Matsuo, S.; Tamura, T.; Ohkuma, N. Increased Cholera Toxin-, and Forskolin-induced Cyclic AMP Accumulations in Psoriatic Involved Versus Uninvolved or Normal Human Epidermis. *J. Investig. Dermatol.* **1988**, *91*, 154–157. [CrossRef] [PubMed]

22.   Lee, T.P.; Busse, W.W.; Reed, C.E. Epidermal adenyl cyclase of human and mouse. A study of the atopic state. *J. Allergy Clin. Immunol.* **1974**, *53*, 283–287. [CrossRef]

23.   Andrés, R.M.; Terencio, M.C.; Arasa, J.; Payá, M.; Valcuende-Cavero, F.; Navalón, P.; Montesinos, M.C. Adenosine A2A and A2B Receptors Differentially Modulate Keratinocyte Proliferation: Possible Deregulation in Psoriatic Epidermis. *J. Investig. Dermatol.* **2017**, *137*, 123–131. [CrossRef] [PubMed]

24.   Adachi, K.; Iizuka, H.; Halprin, K.M.; Levine, V. Specific refractoriness of adenylate cyclase in skin to epinephrine, prostaglandin E, histamine and AMP. *Biochim. Biophys. Acta (BBA) Gen. Subj.* **1977**, *497*, 428–436. [CrossRef]

25.   Stratakis, C.A. Cyclic AMP, Protein Kinase A, and Phosphodiesterases: Proceedings of an International Workshop. *Horm. Metab. Res.* **2012**, *44*, 713–715. [CrossRef]

26.   Sakkas, L.I.; Mavropoulos, A.; Bogdanos, D.P. Phosphodiesterase 4 Inhibitors in Immune-mediated Diseases: Mode of Action, Clinical Applications, Current and Future Perspectives. *Curr. Med. Chem.* **2017**, *24*, 3054–3067. [CrossRef]

27.   Warne, A.; Moukhametzianov, R.; Baker, J.G.; Nehmé, R.; Edwards, P.C.; Leslie, A.G.W.; Schertler, G.F.X.; Tate, C.G. The structural basis for agonist and partial agonist action on a β1-adrenergic receptor. *Nature* **2011**, *469*, 241–244. [CrossRef]

28. Ji, Y.; Chen, S.-Y.; Li, K.; Xiao, X.; Zheng, S.; Xu, T. The role of β-adrenergic receptor signaling in the proliferation of hemangioma-derived endothelial cells. *Cell Div.* **2013**, *8*, 1. [CrossRef]

29. Androutsellis-Theotokis, A.; Walbridge, S.; Park, D.M.; Lonser, R.R.; McKay, R.D.G. Cholera Toxin Regulates a Signaling Pathway Critical for the Expansion of Neural Stem Cell Cultures from the Fetal and Adult Rodent Brains. *PLoS ONE* **2010**, *5*, e10841. [CrossRef]

30. Sivamani, R.K.; Lam, S.T.; Isseroff, R.R. Beta Adrenergic Receptors in Keratinocytes. *Dermatol. Clin.* **2007**, *25*, 643–653. [CrossRef]

31. Choi, E.J.; Toscano, W.A., Jr. Modulation of adenylate cyclase in human keratinocytes by protein kinase C. *J. Biol. Chem.* **1988**, *263*, 17167–17172. [PubMed]

32. Cumbay, M.G.; Watts, V.J. Novel Regulatory Properties of Human Type 9 Adenylate Cyclase. *J. Pharmacol. Exp. Ther.* **2004**, *310*, 108–115. [CrossRef] [PubMed]

33. Voorhees, J.J.; Duell, E.A. Psoriasis as a possible defect of the adenyl cyclase-cyclic AMP cascade. A defective chalone mechanism? *Arch. Dermatol.* **1971**, *104*, 352–358. [CrossRef]

34. Bass, L.J.; Powell, J.A.; Voorhees, J.J.; Duell, E.A.; Harrell, E.R. The Cyclic Amp System in Normal and Psoriatic Epidermis. *J. Investig. Dermatol.* **1972**, *59*, 114–120. [CrossRef] [PubMed]

35. Marcelo, C.L.; Tomich, J. Cyclic AMP, Glucocorticoid, and Retinoid Modulation of in Vitro Keratinocyte Growth. *J. Investig. Dermatol.* **1983**, *81*, S64–S68. [CrossRef]

36. Halprin, K.M.; Adachi, K.; Yoshikawa, K.; Levine, V.; Mui, M.M.; Hsia, S.L. Cyclic Amp And Psoriasis. *J. Investig. Dermatol.* **1975**, *65*, 170–178. [CrossRef]

37. Billi, A.C.; Gudjonsson, J.E.; Voorhees, J.J. Psoriasis: Past, Present, and Future. *J. Investig. Dermatol.* **2019**, *139*, e133–e142. [CrossRef]

38. Cortez Ghio, S.; Cantin-Warren, L.; Guignard, R.; Larouche, D.; Germain, L. Are the Effects of the Cholera Toxin and Isoproterenol on Human Keratinocytes' Proliferative Potential Dependent on Whether They Are Co-Cultured with Human or Murine Fibroblast Feeder Layers? *Int. J. Mol. Sci.* **2018**, *19*, 2174. [CrossRef]

39. Bullock, A.J.; Higham, M.C.; MacNeil, S. Use of Human Fibroblasts in the Development of a Xenobiotic-Free Culture and Delivery System for Human Keratinocytes. *Tissue Eng.* **2006**, *12*, 245–255. [CrossRef]

40. Bharati, K.; Ganguly, N.K. Cholera toxin: A paradigm of a multifunctional protein. *Indian J. Med. Res.* **2011**, *133*, 179–187.

41. Archer, C.B.; Hanson, J.M.; Morley, J.; Macdonald, D.M. Mononuclear Leukocyte Cyclic Adenosine Monophosphate Responses in Psoriasis Are Normal. *J. Investig. Dermatol.* **1984**, *82*, 316–317. [CrossRef]

42. Iizuka, H.; Adachi, K.; Halprin, K.M.; Levine, V. Cyclic Amp Accumulation in Psoriatic Skin: Differential Responses to Histamine, Amp, and Epinephrine by the Uninvolved and Involved Epidermis. *J. Investig. Dermatol.* **1978**, *70*, 250–253. [CrossRef] [PubMed]

43. Yoshikawa, K.; Adachi, K.; Halprin, K.M.; Levine, V. On the lack of response to catecholamine stimulation by the adenyl cyclase system in psoriatic lesions*. *Br. J. Dermatol.* **1975**, *92*, 619–624. [CrossRef] [PubMed]

44. Lanna, C.; Cesaroni, G.M.; Mazzilli, S.; Bianchi, L.; Campione, E. Small Molecules, Big Promises: Improvement of Psoriasis Severity and Glucidic Markers with Apremilast: A Case Report. *Diabetes Metab. Syndr. Obes. Targets Ther.* **2019**, *12*, 2685–2688. [CrossRef] [PubMed]

45. Steinkraus, V.; Steinfath, M.; Stove, L.; Korner, C.; Abeck, D.; Mensing, H. β-Adrenergic receptors in psoriasis: Evidence for down-regulation in lesional skin. *Arch. Dermatol. Res.* **1993**, *285*, 300–304. [CrossRef] [PubMed]

46. Das, N.S.; Chowdary, T.N.; Sobhanadri, C.; Rao, K.V. The effect of topical isoprenaline on psoriatic skin. *Br. J. Dermatol.* **1978**, *99*, 197–200. [CrossRef] [PubMed]

47. Pierre, S.; Eschenhagen, T.; Geisslinger, G.; Scholich, K. Capturing adenylyl cyclases as potential drug targets. *Nat. Rev. Drug Discov.* **2009**, *8*, 321–335. [CrossRef]

48. Pálvölgyi, A.; Simpson, J.; Bodnár, I.; Bíró, J.; Palkovits, M.; Radovits, T.; Skehel, P.; Antoni, F.A. Auto-inhibition of adenylyl cyclase 9 (AC9) by an isoform-specific motif in the carboxyl-terminal region. *Cell. Signal.* **2018**, *51*, 266–275. [CrossRef]

49. Tsuge, K.; Inazumi, T.; Shimamoto, A.; Sugimoto, Y. Molecular mechanisms underlying prostaglandin E2-exacerbated inflammation and immune diseases. *Int. Immunol.* **2019**, *31*, 597–606. [CrossRef]

50. Shi, Q.; Yin, Z.; Zhao, B.; Sun, F.; Yu, H.; Yin, X.; Zhang, L.; Wang, S. PGE2 Elevates IL-23 Production in Human Dendritic Cells via a cAMP Dependent Pathway. *Mediat. Inflamm.* **2015**, *2015*, 984690. [CrossRef]

51. Litvinov, I.V.; Bizet, A.A.; Binamer, Y.M.; Sasseville, D.; Jones, D.A.; Philip, A. CD109 release from the cell surface in human keratinocytes regulates TGF-β receptor expression, TGF-β signalling and STAT3 activation: Relevance to psoriasis. *Exp. Dermatol.* **2011**, *20*, 627–632. [CrossRef] [PubMed]

52. Funding, A.T.; Johansen, C.; Kragballe, K.; Iversen, L. Mitogen- and Stress-Activated Protein Kinase 2 and Cyclic AMP Response Element Binding Protein are Activated in Lesional Psoriatic Epidermis. *J. Investig. Dermatol.* **2007**, *127*, 2012–2019. [CrossRef] [PubMed]

53. Germain, L.; Rouabhia, M.; Guignard, R.; Carrier, L.; Bouvard, V.; Auger, F.A. Improvement of human keratinocyte isolation and culture using thermolysin. *Burns* **1993**, *19*, 99–104. [CrossRef]

54. Rioux, G.; Pouliot-Bérubé, C.; Simard, M.; Benhassine, M.; Soucy, J.; Guérin, S.L.; Pouliot, R. The Tissue-Engineered Human Psoriatic Skin Substitute: A Valuable In Vitro Model to Identify Genes with Altered Expression in Lesional Psoriasis. *Int. J. Mol. Sci.* **2018**, *19*, 2923. [CrossRef]

11

Patterned Piezoelectric Scaffolds for Osteogenic Differentiation

Teresa Marques-Almeida [1,2], **Vanessa F. Cardoso** [1,3], **Miguel Gama** [2],
Senentxu Lanceros-Mendez [4,5,*] **and Clarisse Ribeiro** [1,2,*]

[1] CF-UM-UP, Centro de Física das Universidades do Minho e Porto, Campus de Gualtar, Universidade do Minho, 4710-057 Braga, Portugal; talmeida@fisica.uminho.pt (T.M.-A.); vanessa@dei.uminho.pt (V.F.C.)
[2] CEB, Centro de Engenharia Biológica, Campus de Gualtar, Universidade do Minho, 4710-057 Braga, Portugal; fmgama@deb.uminho.pt
[3] CMEMS-UMinho, Campus de Azurém, Universidade do Minho, 4800-058 Guimarães, Portugal
[4] BCMaterials, Basque Center for Materials, Applications and Nanostructures, UPV/EHU Science Park, 48940 Leioa, Spain
[5] IKERBASQUE, Basque Foundation for Science, 48009 Bilbao, Spain
* Correspondence: senentxu.lanceros@bcmaterials.net (S.L.-M.); cribeiro@fisica.uminho.pt (C.R.)

Abstract: The morphological clues of scaffolds can determine cell behavior and, therefore, the patterning of electroactive polymers can be a suitable strategy for bone tissue engineering. In this way, this work reports on the influence of poly(vinylidene fluoride-co-trifluoroethylene) (P(VDF-TrFE)) electroactive micropatterned scaffolds on the proliferation and differentiation of bone cells. For that, micropatterned P(VDF-TrFE) scaffolds were produced by lithography in the form of arrays of lines and hexagons and then tested for cell proliferation and differentiation of pre-osteoblast cell line. Results show that more anisotropic surface microstructures promote bone differentiation without the need of further biochemical stimulation. Thus, the combination of specific patterns with the inherent electroactivity of materials provides a promising platform for bone regeneration.

Keywords: piezoelectric; electroactive; patterning; cell differentiation; bone tissue engineering

1. Introduction

Bone tissue regeneration represents one of the major challenges of biomedicine. As in other areas of biomedicine, efforts are being conducted on replacing conventional approaches with more biomimetic ones. In this scope, tissue-specific active scaffolds are being developed, combining stem or precursor cells and physic/chemical cues, that synergistically stimulate the repairing process, eventually being replaced by the patient's own tissue [1].

Bone can be differentiated, according to the macrostructure, in trabecular (porous) and cortical (compact). At the cellular and molecular levels, bone is composed of cells (osteoblasts, osteoclasts, osteocytes, and bone lining cells) merged in a non-oriented collagen type I matrix, mineralized by hydroxyapatite (HA) that is responsible for toughening the bone [2]. When an injury takes place, a defective microenvironment compromises the normal resorption and regrowth of bone tissue, and consequently its regeneration.

Investigations are being developed based on different strategic cues, such as electromechanical [3,4], chemical [5,6], and morphological [7,8], in an attempt to recreate tissue-specific microenvironments and thus trigger their natural recovery. Morphological cues have been demonstrated to effectively influence cellular proliferation and differentiation, the cell–scaffold interaction triggering a series of physical-chemical reactions. Cells sense the site they are attached to and mechanically transduce that information (hardness, curvature, and shape) into morphological changes [9]. When favorable topographical signals are presented at the surface of the scaffold, they can trigger the initiation of mechanosensitive cell cascades and thus a cell's differentiation signaling pathways [10–12]. However, the effective mechanism by which morphological cues regulate cell fate, in terms of orientation, morphology, proliferation, and differentiation, is still barely understood. Given the topographic complexity of its natural microenvironment, bone cells are adaptable to different scaffolds' architectures, although it is known that their phenotype is not favored in aligned morphologies, unlike for instance in the case of myoblasts or neurons. Different structures have been developed to grow bone tissue, but only a few trials with micropatterned scaffolds have been reported so far. Micropatterned scaffolds based on polycaprolactone (PCL)/polylactic-co-glycolic acid (PLGA) have been applied for periodontal tissue regeneration [13], demonstrating that micropatterning can effectively enhance tissue responses. HA ceramics with surface micropatterning have been demonstrated to promote the osteogenic differentiation [14]. In addition to the influence of the morphology, studies have demonstrated that piezoelectric biomaterials, capable of providing mechano-electrical stimuli, can enhance bone cell differentiation and regeneration [15,16], as those electro active stimuli effectively mimic the natural cell's microenvironment.

As previously shown [7], pre-osteoblast cells maintain their phenotype when adhered to a scaffold with isotropic hexagonal surface topography, unlike what happens in the linear topography. The influence of both morphologies, hexagonal and linear, on pre-osteoblasts proliferation and differentiation is here studied in an attempt to determine whether it is possible to physically induce the differentiation of bone precursor cells, avoiding the use of biochemical differentiation factors. In addition, it is known that bone tissue presents inherent piezoelectricity, and therefore, morphological features were patterned on a piezoelectric polymer, using the non-biodegradable polymer of poly(vinylidene fluoride-co-trifluoroethylene) (P(VDF-TrFE) once it presented the highest piezoelectric coefficient among all the polymers [17]. This was done to allow the development of electroactive platforms for bone tissue engineering that combines the morphology and the possibility of further mechano-electrical stimuli to the cells.

2. Results and Discussion

2.1. Cellular Proliferation

The preference of MC3T3-E1 cells for adhesion on surfaces with specific topographies has been previously reported [7] and was further assessed in the present study. Further, it is to notice that it is essential, from a molecular point of view, to obtain the materials in the electroactive phase, i.e., in the all-trans b-phase chain conformation, to provide electromechanical cues to the cells at the nano- [18] or microscale [15], depending on the poling state of the material [19]. It is confirmed that the scaffolds are obtained in the β-phase, identified by the vibration modes at ~510, 840, 1287, and 1400 cm^{-1} [20], representing, respectively, the CF2 bending; CF2 symmetric stretching; CF2 and CC symmetric stretching and CCC bending; and CH2 wagging and CC antisymmetric stretching (see supplementary information Figure S1).

Proliferation assays were performed using control scaffolds (non-patterned) and patterned ones featuring topographies with 25, 75, and 150 μm wide lines or hexagons.

For both linear and hexagonal topographies, the smaller lines and hexagons (25 μm) show higher cell viability than the larger ones (Figure 1a). Checking the immunofluorescence images (Figure 1b), it is observed that cells show, contrary to the non-patterned control samples (see supplementary information Figure S4), the preferential orientation of the microstructure of the patterned P(VDF-TrFE) scaffolds, being slightly lower with smaller features (25 μm). Thus, the perception of more cellular viability is provided by the fact that cells have more contact area to proliferate, since they adhere to all the scaffolds' surfaces. Bone cells are quite resilient to different surroundings, since they are naturally present in different microenvironments. Therefore, MC3T3-E1 proliferate quite well in both topography types and dimensions, although significantly more on the isotropic hexagonal topography. Immunofluorescent images disclose the compromised cellular phenotype over linear topography, being elongated and not round as is common. Their elongated phenotype suggests that the normal fate of cells is being negatively influenced by this physical stimulus, contrary to the hexagonal topography. This may indicate that, in a differentiation phase, this stimulus may cause cells to be unable to differentiate, or to differentiate into an unwanted cell type.

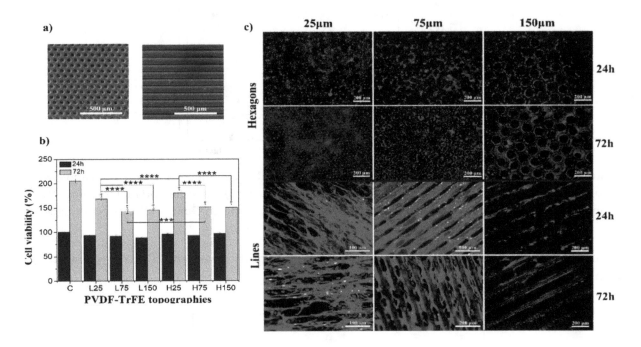

Figure 1. (**a**) SEM images of patterned poly(vinylidene fluoride-co-trifluoroethylene) (P(VDF-TrFE)) scaffolds of hexagons and lines microstructures with 75 μm dimensional features; (**b**) cellular viability obtained after MTS assay of MC3T3-E1 cells in contact with control; (**c**) representative images of pre-osteoblast culture on linear topographies with 25, 75, and 150 μm (L25, L75, and L150, respectively) and hexagonal topographies with 25, 75, and 150 μm (H25, H75 and H150, respectively), at 24 h and 72 h timepoints (nucleus stained with DAPI-blue and cytoskeleton stained with tetramethylrhodamine (TRITC)-red). $^{\tau}$ $p < 0.0005$ vs. control; **** $p < 0.0001$, *** $p < 0.001$ (two-way ANOVA); (**c**) MC3T3-E1 adhesion on patterned P(VDF-TrFE) scaffolds with 25, 75, and 150 μm dimensions, in 24 and 72 h of contact.

2.2. Differentiation Assays

This study was performed in order to investigate the effect of the surface topography on the osteoblasts differentiation, using alkaline phosphatase (ALP) quantification assay as early marker, since it plays a critical role in bone formation [21,22], and alizarin red staining as late marker of osteoblast differentiation.

For the differentiation assays, control scaffolds (non-patterned) and patterned ones with 75 μm features (linear or hexagonal topographies) were tested. The differentiation in the two conditions was assessed in the absence (GM) and in the presence (DM) of the biochemical inducer for 21 days. The goal is to understand the physical effect, namely, the topography, of the patterned scaffolds on the cells' differentiation fate, with or without the cells being exposed to biochemical inducers.

After 7 days, ALP was evaluated in the different samples. Alkaline phosphatase is an enzyme mostly found in liver, kidney, and bones, and the measurement of its activity has been found to be suitable for monitoring changes in bone formation and thus in bone cells differentiation [23]. It was found that ALP activity levels are significantly higher for cells under DM, compared to GM (Figure 2). The physical influence of hexagons topographies on the improvement of differentiation can be confirmed by cells' ALP activity in 7 days of culture with GM, compared to cells over the control and lines patterned scaffolds. These results demonstrating the positive influence of the hexagonal topography are further supported by the alizarin red staining results.

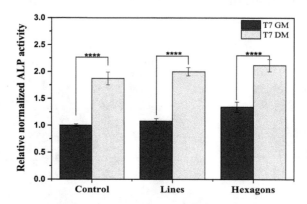

Figure 2. Osteogenic differentiation determined by relative alkaline phosphatase quantification assay expression after 7 days of culture (T7), using growth medium (GM) and differentiation medium (DM). The ALP expression was normalized against DNA content using CyQuant cell proliferation assay. **** $p < 0.0001$ (two-way ANOVA).

After 14 and 21 days of cell differentiation, the cell viability assay and alizarin red staining were performed (Figure 3).

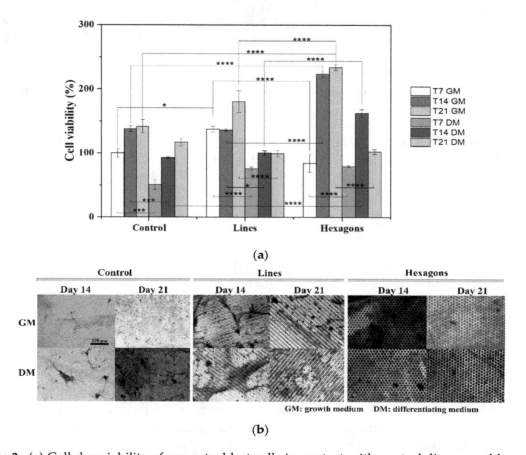

Figure 3. (a) Cellular viability of pre-osteoblast cells in contact with control, linear, and hexagonal scaffold topographies, in a differentiation essay of 7, 14, and 21 days, with GM and DM. * $p < 0.05$, *** $p < 0.001$, **** $p < 0.0001$ (two-way ANOVA); (b) Alizarin red staining for mineral deposition during osteogenesis induction, at day 14 and 21, with and without DM. Calcified areas are presented with pink color. The scale bar 500 μm is valid for all the images.

Cell viability results (Figure 3a) were compared to a cell mineralization assay (Figure 3b). The presence of DM induces differentiating pathways on cells, providing chemical stimuli to stop proliferating and begin differentiation processes. On the contrary, basal GM provides cells with all the nutrients necessary to proliferate continuously until reaching the confluency. In DM conditions, cells present lower viability for all topographies and dimensions, compared to GM (Figure 3a). Complementing the viability results with the alizarin red images, in order to evaluate the mineralization of bone cells, a lower osteogenic differentiation of cells over linear topographies can be seen, which was predictable from the proliferation assays. On the other hand, the degree of mineralization on isotropic hexagon patterned scaffolds is very similar both with GM and DM after both 14 and 21 days (Figure 3b). These results indicate that the provided physical effect is able to regulate cell fate and may activate differentiation signaling pathways, with no need of biochemical inducers.

In this way, it is concluded that pre-osteoblast cells can be differentiated into osteoblast by specific patterns that also support matrix mineralization.

3. Materials and Methods

3.1. Materials

Poly(vinylidenefluoride-co-trifluoroethylene); P(VDF-TrFE), with 70 mol% vinylidene fluoride and 30 mol% trifluoroethylene, from Solvay (Póvoa de Santa Iria, Portugal); and N,N-dimethylformamide, DMF, from Merck (Sintra, Portugal), were used as received. An 8% (w/w) polymer solution in DMF was prepared under magnetic stirring at room temperature.

3.2. Samples Processing and Main Characteristics

Patterned scaffolds were obtained through photolithography and replica molding techniques, as described in [7,24]. In short, lines and hexagonal patterns with different dimensions (25, 75, and 150 µm) were designed in AutoCAD 2018 software and printed in photolithographic masks by Microlitho (see supplementary information—Figures S2 and S3). The photolithographic masks were used for the fabrication of SU-8 patterned molds, through photolithography, which were replicated in flexible and reusable polydimethylsiloxane (PDMS) molds, through replica molding. The P(VDF-TrFE) in DMF polymer solution was deposited on both patterned and non-patterned PDMS molds, previously treated with an oxygen plasma for 10 min, and evaporated at 100 °C [25]. The obtained samples present hydrophobic behavior (>100°) and a crystallinity degree of ≈ 32%, as shown previously [7].

3.3. Samples Sterilization

Circular samples with 6 mm diameter were cut, exposed to ultraviolet (UV) light on each side for 1 h, and then placed in standard 48-well cell culture plates. All samples were washed five times (5 min each time) with sterile phosphate buffer saline (PBS) 1× solution.

3.4. Cell Culture

MC3T3-E1 pre-osteoblasts (Riken bank, Tsukuba, Japan) were grown in a 75 cm^2 cell-culture flask with modified Eagle's medium (DMEM, Biochrom, Berlin, Germany) containing 1 g·L^{-1} glucose, 1% penicillin/streptomycin (P/S, Biochrom, Berlin, Germany) and 10% Fetal Bovine Serum (FBS, Biochrom, Berlin, Germany). The flask was placed in a 37 °C incubator under 95% humidified air and 5% CO$_2$ conditions. Culture medium was changed every two days and, at a 60–70% confluence, cells were trypsinized with 0.05% trypsin-EDTA (Biochrom, Berlin, Germany). A 25 µL drop of cell suspension was added over each P(VDF-TrFE) sample with a density of 10×10^4 cells.mL^{-1} for the proliferation assays (cell viability) and at a density of 50×10^4 cells·mL^{-1} for the differentiation studies (cell viability, ALP and alizarin red). After that, the plates were incubated during 30 min for cell adhesion. The well volume was then completed with the growth medium (GM) and incubated once more for 24 h.

For proliferation assessment, cells were cultivated for 3 days, with 24 h and 72 h as timepoints. For the differentiation assays, the medium was exchanged by osteogenic differentiating medium (DM) after 24 h and cells were maintained up to 21 days, with 7, 14, and 21 days as timepoints. DM was composed of the GM supplemented with 0.1 μM dexamethasone (Sigma-Aldrich, Sintra, Portugal), 50 μg.mL^{-1} of ascorbic acid (Sigma-Aldrich, city, state, country), and 10 mM of b-glycerophosphate (Sigma-Aldrich, Sintra, Portugal). Culture media were changed every two days.

3.5. Cell Viability

Viable cells in proliferation and differentiation assays were quantified at the different timepoints using a CellTiter 96® AQueous One Solution Cell Proliferation Assay (MTS, Promega, Madison, WI, USA). For that, the samples were incubated in an MTS:GM (1:5) solution for 3 h at 37 °C, under 95% humidified air and 5% CO_2 conditions, and the absorbance at 490 nm was recorded with a microplate reader (Biotech Synergy HT, Winooski, VT, USA,). The quantitative results of proliferative cells will be presented as mean ± standard deviation (SD) and the results of the cells differentiation will be presented as mean ± standard error of mean (SEM), both of quadruplicated samples.

3.6. Immunofluorescence Staining

At the proliferation assay timepoints (24 and 72 h), two replicates of each sample were fixed with 4% formaldehyde (Panreac AppliChem, Barcelona, Spain) and subjected to immunofluorescence staining, to analyze their behavior in the different culture patterns. Cell's cytoskeleton was stained with 1 μg.mL^{-1} of phalloidin tetramethylrhodamine (TRITC, Sigma Aldrich, Sintra, Portugal) solution for 45 min at room temperature, and cells' nucleus with 1 μg.mL^{-1} of a 4,6-diamidino-2-phenylindole (DAPI, Sigma Aldrich, Sintra, Portugal) solution for 5 min. Samples were washed with PBS 1× before, after, and during the steps. Finally, the samples were visualized with fluorescence microscopy (Olympus BX51 Microscope, Lisboa, Portugal).

3.7. Quantification of DNA and Alkaline Phosphatase Activity

Cells' DNA content was measured using a CyQUANT® Cell Proliferation Assay Kit (Life Technologies, Porto, Portugal) and the osteogenic capacity was determined through an ALP (Sigma Aldrich, Sintra, Portugal), both of which are described in [26]. Briefly, after 7 days of cell culture on both GM and DM, cells were lysed with Triton 0.1% buffer and frozen at 70 °C. After thawing, 50 μL p-nitrophenyl phosphate and 50 μL 2-amino-2-methyl-1-propanol were added, according to manufacturer's protocol. Using a microplate reader, the amount of the produced p-NP (p-nitrophenol) was measured, reading the absorbance at 405 nm. To proceed to the ALP activity normalization, cells' DNA content was quantified from the cell lysate using the CyQUANT® Cell Proliferation Assay Kit, according to the manufacturer's protocol, measuring its fluorescence by exciting the sample at 480 nm and measuring the emission at 520 nm, using the same microplate reader. The results will be presented as mean ± SEM of quadriplicated samples.

3.8. Mineralization Assay

At the differentiation timepoints (7, 14, and 21 days), the samples were marked with 10% alizarin red solution in acetic acid, allowing the qualitative detection of calcium deposits, characteristic of osteogenic differentiation.

3.9. Statistical Analysis

Statistical analysis was carried out by ANOVA using Tukey test (Graphpad Prism 8, San Diego, CA, USA), with p values < 0.05 considered to be statistically significant.

4. Conclusions

This study reports on the influence of different patterned piezoelectric P(VDF-TrFE) scaffolds in the preosteoblasts' proliferation and differentiation. It was observed that MC3T3-E1 cells differentiation can be induced solely by a physical stimulus, specifically by an isotropic hexagonal surface topography, with no need of using a chemical inducer. Finding that bone cells differentiation can be positively influenced by this type of geometry, it is concluded that novel scaffolds based on electromechanically active materials with specific geometries can provide the necessary geometrical and electroactive components of the bone microenvironment to support bone regeneration.

Author Contributions: Conceptualization, V.F.C., S.L.-M. and C.R.; Data curation, T.M.-A.; Formal analysis, T.M.-A. and C.R.; Funding acquisition, S.L.-M. and C.R.; Investigation, T.M.-A., V.F.C., M.G., S.L.-M. and C.R.; Methodology, C.R.; Resources, M.G.; Supervision, S.L.-M. and C.R.; Writing—original draft, T.M.-A.; Writing—review & editing, V.F.C., M.G., S.L.-M. and C.R. All authors have read and agreed to the published version of the manuscript.

References

1. Langer, R.; Vacanti, J.P. Tissue engineering. *Science* **1993**, *260*, 920–926. [CrossRef] [PubMed]
2. Setiawati, R.; Rahardjo, P. Bone Development and Growth. In *Osteogenesis and Bone Regeneration*; Yang, H., Ed.; IntechOpen: London, UK, 2018; pp. 1–20.
3. Ribeiro, C.; Moreira, S.; Correia, V.; Sencadas, V.; Rocha, J.G.; Gama, F.M.; Ribelles, J.G.; Lanceros-Méndez, S. Enhanced proliferation of pre-osteoblastic cells by dynamic piezoelectric stimulation. *RSC Adv.* **2012**, *2*, 11504–11509. [CrossRef]
4. Ribeiro, S.; Gomes, A.C.; Etxebarria, I.; Lanceros-Méndez, S.; Ribeiro, C. Electroactive biomaterial surface engineering effects on muscle cells differentiation. *Mater. Sci. Eng. C* **2018**, *92*, 868–874. [CrossRef] [PubMed]
5. Wu, A.T.H.; Aoki, T.; Sakoda, M.; Ohta, S.; Ichimura, S.; Ito, T.; Ushida, T.; Furukawa, K.S. Enhancing osteogenic differentiation of MC3T3-E1 cells by immobilizing inorganic polyphosphate onto hyaluronic acid hydrogel. *Biomacromolecules* **2015**, *16*, 166–173. [CrossRef] [PubMed]
6. Budiraharjo, R.; Neoh, K.G.; Kang, E.T. Enhancing bioactivity of chitosan film for osteogenesis and wound healing by covalent immobilization of BMP-2 or FGF-2. *J. Biomater. Sci. Polym. Ed.* **2013**, *24*, 645–662. [CrossRef] [PubMed]
7. Marques-Almeida, T.; Cardoso, V.F.; Ribeiro, S.; Gama, F.M.; Ribeiro, C.; Lanceros-Méndez, S. Tuning Myoblast and Pre-osteoblast Cell Adhesion Site, Orientation and Elongation through Electroactive Micropatterned Scaffolds. *ACS Appl. Bio Mater.* **2019**, *2*, 1591–1602. [CrossRef]
8. Watari, S.; Hayashi, K.; Wood, J.A.; Russell, P.; Nealey, P.F.; Murphy, C.J.; Genetos, D.C. Modulation of osteogenic differentiation in hMSCs cells by submicron topographically-patterned ridges and grooves. *Biomaterials* **2012**, *33*, 128–136. [CrossRef]
9. Yin, Z.; Chen, X.; Chen, J.L.; Shen, W.L.; Hieu Nguyen, T.M.; Gao, L.; Ouyang, H.W. The regulation of tendon stem cell differentiation by the alignment of nanofibers. *Biomaterials* **2010**, *31*, 2163–2175. [CrossRef]
10. Yang, K.; Jung, H.; Lee, H.R.; Lee, J.S.; Kim, S.R.; Song, K.Y.; Cheong, E.; Bang, J.; Im, S.G.; Cho, S.W. Multiscale, hierarchically patterned topography for directing human neural stem cells into functional neurons. *ACS Nano* **2014**, *8*, 7809–7822. [CrossRef] [PubMed]
11. Pan, Z.; Yan, C.; Peng, R.; Zhao, Y.; He, Y.; Ding, J. Control of cell nucleus shapes via micropillar patterns. *Biomaterials* **2012**, *33*, 1730–1735. [CrossRef]
12. Kim, J.; Yang, K.; Jung, K.; Ko, E.; Kim, J.; Park, K.I.; Cho, S.W. Nanotopographical manipulation of focal adhesion formation for enhanced differentiation of human neural stem cells. *ACS Appl. Mater. Interfaces* **2013**, *5*, 10529–10540.
13. Pilipchuk, S.P.; Fretwurst, T.; Yu, N.; Larsson, L.; Kavanagh, N.M.; Asa'ad, F.; Cheng, K.C.K.; Lahann, J.; Giannobile, W.V. Micropatterned Scaffolds with Immobilized Growth Factor Genes Regenerate Bone and Periodontal Ligament-Like Tissues. *Adv. Healthc. Mater.* **2018**, *7*, 1800750. [CrossRef]

14. Zhao, C.; Xia, L.; Zhai, D.; Zhang, N.; Liu, J.; Fang, B.; Chang, J.; Lin, K. Designing ordered micropatterned hydroxyapatite bioceramics to promote the growth and osteogenic differentiation of bone marrow stromal cells. *J. Mater. Chem. B* **2015**, *3*, 968–976. [CrossRef] [PubMed]

15. Ribeiro, C.; Pärssinen, J.; Sencadas, V.; Correia, V.; Miettinen, S.; Hytönen, V.P.; Lanceros-Méndez, S. Dynamic piezoelectric stimulation enhances osteogenic differentiation of human adipose stem cells. *J. Biomed. Mater. Res. Part A* **2015**, *103*, 2172–2175. [CrossRef] [PubMed]

16. Ribeiro, C.; Correia, D.M.; Rodrigues, I.; Guardão, L.; Guimarães, S.; Soares, R.; Lanceros-Méndez, S. In vivo demonstration of the suitability of piezoelectric stimuli for bone reparation. *Mater. Lett.* **2017**, *209*, 118–121. [CrossRef]

17. Cardoso, V.F.; Correia, D.M.; Ribeiro, C.; Fernandes, M.M.; Lanceros-Méndez, S. Fluorinated polymers as smart materials for advanced biomedical applications. *Polymers* **2018**, *10*, 161. [CrossRef]

18. Zheng, T.; Yue, Z.; Wallace, G.G.; Du, Y.; Higgins, M.J. Nanoscale piezoelectric effect of biodegradable PLA-based composite fibers by piezoresponse force microscopy. *Nanotechnology* **2020**, *31*, 375708. [CrossRef]

19. Ribeiro, C.; Sencadas, V.; Correia, D.M.; Lanceros-Méndez, S. Piezoelectric polymers as biomaterials for tissue engineering applications. *Colloids Surf. B Biointerfaces* **2015**, *136*, 46–55. [CrossRef]

20. Martins, P.; Lopes, A.C.; Lanceros-Mendez, S. Electroactive phases of poly(vinylidene fluoride): Determination, processing and applications. *Prog. Polym. Sci.* **2014**, *39*, 683–706. [CrossRef]

21. Jafary, F.; Hanachi, P.; Gorjipour, K. Osteoblast differentiation on collagen scaffold with immobilized alkaline phosphatase. *Int. J. Organ Transplant. Med.* **2017**, *8*, 195–202.

22. Sobreiro-Almeida, R.; Tamaño-Machiavello, M.N.; Carvalho, E.O.; Cordón, L.; Doria, S.; Senent, L.; Correia, D.M.; Ribeiro, C.; Lanceros-Méndez, S.; Sabater, I.; et al. Human mesenchymal stem cells growth and osteogenic differentiation on piezoelectric poly(Vinylidene fluoride) microsphere substrates. *Int. J. Mol. Sci.* **2017**, *18*, 2391. [CrossRef]

23. Van Straalen, J.P.; Sanders, E.; Prummel, M.F.; Sanders, G.T.B. Bone-alkaline phosphatase as indicator of bone formation. *Clin. Chim. Acta* **1991**, *201*, 27–33. [CrossRef]

24. Pinto, V.C.; Sousa, P.J.; Cardoso, V.F.; Minas, G. Optimized SU-8 processing for low-cost microstructures fabrication without cleanroom facilities. *Micromachines* **2014**, *5*, 738–755. [CrossRef]

25. Ribeiro, C.; Costa, C.M.; Correia, D.M.; Nunes-Pereira, J.; Oliveira, J.; Martins, P.; Gonçalves, R.; Cardoso, V.F.; Lanceros-Méndez, S. Electroactive poly(vinylidene fluoride)-based structures for advanced applications. *Nat. Protoc.* **2018**, *13*, 681–704. [CrossRef] [PubMed]

26. Lindroos, B.; Mäenpää, K.; Ylikomi, T.; Oja, H.; Suuronen, R.; Miettinen, S. Characterisation of human dental stem cells and buccal mucosa fibroblasts. *Biochem. Biophys. Res. Commun.* **2008**, *368*, 329–335. [CrossRef]

12

Mitochondrial Dysfunction and Calcium Dysregulation in Leigh Syndrome Induced Pluripotent Stem Cell Derived Neurons

Teresa Galera-Monge [1,2,3,4], Francisco Zurita-Díaz [1,2,3,4], Isaac Canals [5],
Marita Grønning Hansen [5], Laura Rufián-Vázquez [3,4,6], Johannes K. Ehinger [7], Eskil Elmér [7],
Miguel A. Martin [3,4,6], Rafael Garesse [1,2,3], Henrik Ahlenius [5,*,†]
and M. Esther Gallardo [1,2,3,4,8,*,†]

[1]   Departamento de Bioquímica, Facultad de Medicina, Universidad Autónoma de Madrid, 28029 Madrid, Spain;
      tgaleramonge@gmail.com (T.G.-M.); frazurdia@gmail.com (F.Z.-D.); rafael.garesse@uam.es (R.G.)
[2]   Departamento de Modelos Experimentales de Enfermedades Humanas, Instituto de Investigaciones
      Biomédicas "Alberto Sols" UAM-CSIC, 28029 Madrid, Spain
[3]   Centro de Investigación Biomédica en Red (CIBERER), 28029 Madrid, Spain; laurarufian@gmail.com (L.R.-V.);
      mamcasanueva.imas12@h12o.es (M.A.M.)
[4]   Instituto de Investigación Sanitaria Hospital 12 de Octubre (i + 12), 28041 Madrid, Spain
[5]   Department of Clinical Sciences, Neurology, Lund Stem Cell Center, Lund University, 221 00 Lund, Sweden;
      isaac.canals@med.lu.se (I.C.); marita.gronning_hansen@med.lu.se (M.G.H.)
[6]   Laboratorio de enfermedades mitocondriales y Neurometabólicas, Hospital 12 de Octubre, 28041 Madrid, Spain
[7]   Mitochondrial Medicine, Department of Clinical Sciences Lund, Faculty of Medicine, Lund University,
      BMC A13, 221 84 Lund, Sweden; johannes.ehinger@med.lu.se (J.K.E.); eskil.elmer@med.lu.se (E.E.)
[8]   Grupo de Investigación Traslacional con células iPS. Instituto de Investigación Sanitaria Hospital 12 de
      Octubre (i + 12), 28041 Madrid, Spain
*   Correspondence: henrik.ahlenius@med.lu.se (H.A.); egallardo.imas12@h12o.es (M.E.G.)
†   These authors contributed equally to this work.

Abstract: Leigh syndrome (LS) is the most frequent infantile mitochondrial disorder (MD) and is characterized by neurodegeneration and astrogliosis in the basal ganglia or the brain stem. At present, there is no cure or treatment for this disease, partly due to scarcity of LS models. Current models generally fail to recapitulate important traits of the disease. Therefore, there is an urgent need to develop new human in vitro models. Establishment of induced pluripotent stem cells (iPSCs) followed by differentiation into neurons is a powerful tool to obtain an in vitro model for LS. Here, we describe the generation and characterization of iPSCs, neural stem cells (NSCs) and iPSC-derived neurons harboring the mtDNA mutation m.13513G>A in heteroplasmy. We have performed mitochondrial characterization, analysis of electrophysiological properties and calcium imaging of LS neurons. Here, we show a clearly compromised oxidative phosphorylation (OXPHOS) function in LS patient neurons. This is also the first report of electrophysiological studies performed on iPSC-derived neurons harboring an mtDNA mutation, which revealed that, in spite of having identical electrical properties, diseased neurons manifested mitochondrial dysfunction together with a diminished calcium buffering capacity. This could lead to an overload of cytoplasmic calcium concentration and the consequent cell death observed in patients. Importantly, our results highlight the importance of calcium homeostasis in LS pathology.

Keywords: Leigh syndrome; mitochondrial disorder; iPSC; NSC; neuron; disease modeling; mtDNA

1. Introduction

Leigh syndrome (LS) is the most frequent infantile mitochondrial disorder (MD) with a prevalence of 1 in 40,000 births [? ? ?]. Forty-five percent of LS patients die before reaching 20 years of age, most of them by respiratory failure. However, there are several complications that may further increase morbidity and mortality, such as refractory seizures and cardiovascular deterioration [?]. The common feature of these patients is the presence of bilateral symmetric necrotic areas in the basal ganglia or the brain stem, which correspond with regions of demyelination, neuronal death and astrogliosis [? ?]. However, LS is characterized by a prominent clinical and genetic variability. More than 75 genes have been associated to LS [?], all of them involved in mitochondrial energy production [?].

Mitochondria are cellular organelles considered to be the powerhouse of the cell because of their participation in cellular energy production through a process known as oxidative phosphorylation (OXPHOS) [?]. The OXPHOS process is carried out by five multiheteromeric complexes located in the inner mitochondrial membrane and collectively termed the respiratory chain (RC) [?]. Moreover, mitochondria are also involved in other pivotal processes such as reactive oxygen species (ROS) production, apoptosis and calcium homeostasis, whose role in pathology is being increasingly recognized. Given the crucial role of mitochondria for functionality of neuronal cells, it is not surprising that diseases affecting mitochondria result in neurological conditions such as LS [?].

Mitochondria possess their own DNA [?] and the human mitochondrial DNA (mtDNA) is a double-stranded circular molecule of 16.5 kb that encodes 13 subunits of the OXPHOS complexes as well as two rRNAs and 22 tRNAs [?]. Hundreds to thousands of copies of mtDNA are present per cell [? ?], allowing the possibility of coexistence of healthy and pathogenic mtDNA molecules, a phenomenon called heteroplasmy. Genetically, LS is highly heterogeneous; to date, a broad variety of causative mutations have been described in nuclear- and mitochondrial-encoded genes involved in energy metabolism. mtDNA mutations are responsible for 10–20% of LS cases and, more specifically, mutations in genes affecting complex I of the respiratory chain have been a well-recognized cause of LS. Among them, the m.13513G>A mutation located in the *MT-ND5* gene is a frequent cause of LS [?]. In spite of the advances in the molecular diagnosis of LS, the molecular pathogenesis of this disease remains poorly understood due, in part, to the lack of suitable disease models.

For that reason, generation of induced pluripotent stem cells (iPSCs) and differentiation into the affected tissue could be an interesting approach for modeling LS [?]. Several studies have used iPSC technology to generate in vitro models of LS harboring the mutation m.8993T>G in the *MT-ATP6* gene [? ?]. These models recapitulate the mitochondrial dysfunction in muscle [?] or in neurons, the principally affected cell type [?]. Moreover, this model recapitulated the typical neurodegeneration in brains of LS patients [?]. The m.13513G>A mutation in the *MT-ND5* gene is responsible for mitochondrial myopathy, encephalopathy, lactic acidosis and stroke (MELAS) and LS. Until now, one report has described the behavior of heteroplasmy during reprogramming and extended culture of iPSCs harboring this mutation in association with MELAS syndrome, but neuronal characterization is missing. Here, we generate iPSC-derived neurons from a described patient suffering LS caused by m.13513G>A mutation in heteroplasmy [?] and explore the mechanisms by which this mutation could cause the disease. Although LS iPSC-derived neurons were electrophysiologically normal, they manifested a decreased respiration and a diminished calcium buffering capacity. The slower removal of cytoplasmic calcium could lead to an overload and the consequent neuronal death observed in patients.

2. Results

2.1. Diminished Respiration in LS Fibroblasts Caused by a Decrease of Mitochondrial Mass

As a first step to test mitochondrial function in human LS cells, we subjected LS fibroblasts harboring the *MT-ND5* m.13513G>A mutation, with a mutant load of 55%, and control fibroblasts to high-resolution respirometry on an Oroboros Oxygraph-2k. Measurements revealed that basal respiration (Cr-ROX), maximal respiratory capacity (CrU-ROX) and complex I contribution to respiration, CrU-(CRot-ROX), were lower in patient as compared to control fibroblasts (Figure ??A). However, albeit lowered to a similar extent, no significant difference was detected in complex II contribution to respiration (CRot-ROX) (Figure ??A). To further understand if the respiratory deficiency was provoked by diminished mitochondrial mass or quantity of RC complexes, a Western blot against subunits of complexes I-V and citrate synthase (CS) as a marker of mitochondrial mass was performed. A diminished mitochondrial mass was indeed observed in Leigh fibroblasts, accounting for the defect in complexes I, III+V and IV (Figure ??B). A defect in complexes I, III+V and IV was detected when normalizing with GAPDH but not when normalizing with both CS and GAPDH (Figure ??B). Moreover, analysis of enzymatic activities of respiratory complexes highlighted a normal function of complexes II, III and IV (CS normalized) but a decreased activity of CS in LS fibroblasts (Figure ??C), supporting a decrease in mitochondrial mass as the main contribution to the lower levels of respiration. Finally, extracellular lactate measurements demonstrated that LS fibroblasts produced more than twice the lactate generated by control fibroblasts (Figure ??D), which is consistent with the decreased respiration of LS fibroblasts.

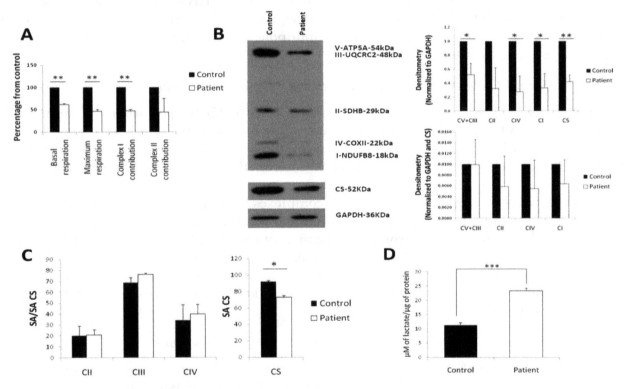

Figure 1. Decreased mitochondrial mass and respiration in Leigh syndrome (LS) fibroblasts. (**A**) Oxygen consumption measured in Oroboros Oxygraph-2k. All data are displayed as a percentage of control. (**B**) Western blot assay against mitoprofile, citrate synthase (CS) and GAPDH (left). Quantification of the Western blot, normalized with GAPDH as a marker of the total protein (right, top panel) or GAPDH and citrate synthase (CS) as a marker of mitochondrial mass (right, bottom panel). All data are displayed as a percentage of control. (**C**) Spectrophotometric measurements of the activity of electron transfer chain (ETC) complexes (left) and citrate synthase (CS) (right); SA: specific activity. (**D**) Extracellular lactate production normalized by total protein. (* p-value < 0.05; ** p-value < 0.01; *** p-value < 0.001)

2.2. Generation of the Control iPSc Line N44SV.1

Since LS models are scarce, generation of LS iPSCs and their differentiation into neural lineages would allow to shed light on the pathological mechanisms causing this disease.

To model LS, we used a previously obtained iPSC line, LND554SV.4, with the mutation m.13513G>A at a percentage of 45% [?]. This line was derived from the same patient fibroblasts where we detected diminished respiration. As a control, we generated the iPSC line (N44SV.1) from normal human dermal fibroblasts (NHDFs) using Sendai virus. We could confirm by sequencing that it, in contrast to the LS iPSC line, lacked the m.13513G>A mutation in the mtDNA (See Supplementary Figure S1A). Moreover, the N44SV.1 line displayed hES-like colonies positive for the pluripotency marker alkaline phosphatase (see Supplementary Figure S1B,C). N44SV.1 also showed high levels of mRNAs of pluripotency-related genes *OCT4*, *SOX2*, *CRIPTO*, *NANOG* and *REX1* in comparison with the original fibroblasts (see Supplementary Figure S1D). At the protein level, we detected by immunofluorescence the presence of pluripotent surface proteins SSEA3, SSEA4, Tra-1-81 and Tra-1-60 and the pluripotent transcription factors OCT4, SOX2 and NANOG (see Supplementary Figure S1E). The ability to differentiate into the three germ layers was tested using an embryoid body (EB)-based methodology. Endoderm (positive for AFP), mesoderm (positive for SMA) and neuroectoderm cells (positive for TUJ1) were observed (see Supplementary Figure S1F). Moreover, N44SV.1 presented a complete clearance of the Sendai viruses used for inducing reprogramming (see Supplementary Figure S1G) and a normal karyotype (see Supplementary Figure S1H). Finally, DNA fingerprinting analysis revealed genetic identity of N44SV.1 with control NHDFs (see Supplementary Figure S1I). Importantly, we did not detect any difference in the reprogramming process or in the pluripotency markers between LND554SV.4 [?] and the control iPSC line, N44SV.1.

2.3. LS iPSCs Manifest a Decreased Basal Respiration and a Combined RC Deficiency

Afterwards, we analyzed metabolic function of LS iPSCs by performing a mitochondrial characterization. Oxygen consumption measurements revealed that basal respiration (Cr-ROX) was diminished in LS iPSCs, while maximum respiration (CrU-ROX) and complex I contribution, CrU-(CRot-ROX), were not significantly decreased. Complex II contribution (CRot-ROX) was unaltered as compared to control iPSCs (Figure ??A). Except for a significant increase in CS, analysis of mitochondrial mass and protein content of RC complexes by Western blot demonstrated no major differences between the patient and the control (Figure ??B). However, activity measurements showed a prominent defect in the activity of complexes I and III of LS iPSC (Figure ??C). Similar to fibroblasts, lactate levels were increased in LS iPSCs as compared to control iPSCs (Figure ??D).

These findings show that hyperlactacidemia (a molecular marker of mitochondrial dysfunction commonly found in LS) can be detected as increased lactate levels in both fibroblasts and iPSCs derived from LS patient fibroblasts.

2.4. Similar Proliferation and Differentiation Capacity of LS iPSC-Derived NSCs

In order to assess pathology in cells more relevant to the disease, we derived neural stem cells (NSCs) from control and LS iPSCs. NSCs were efficiently derived from both patient and control iPSCs and stained positive for the NSC marker nestin (Figure ??A). The m.13513G>A mutation was retained in patient NSCs at a percentage of 19.26% while absent in control NSCs (Figure ??B). No differences were observed in the percentages of EdU+ cells between control and patient NSCs (Figure ??C,D), indicating a similar proliferative rate. Differentiation capacity was tested at 3 weeks of culture in differentiation media. At that time point, control NSCs had developed a network of neuronal cells with clustering of somas and interconnecting neurites (Figure ??E). The tendency to cluster was less strong in patient NSCs, which showed a less organized network. Immunocytochemistry for the neuronal marker MAP2 and the astrocytic marker GFAP revealed that most of the cells, both in control and patient, were MAP2+, and only a few of them were GFAP+ (Figure ??F). In order to study the subtype of neurons present in the culture, we allowed NSCs to differentiate for 6 weeks and analyzed by immunofluorescence the presence of the glutamatergic marker KGA and the GABAergic marker GAD65/67. Both control and patient NSCs generated MAP2+ or TUJ1+ cells, which co-stained with KGA or GAD65/67, without any obvious differences between groups (Figure ??G,H), proving the presence of both glutamatergic and GABAergic neurons in our cultures.

2.5. Cell death and Complex I deficiency in LS iPSC-Derived Neurons

In order to specifically analyze neuronal properties, we used a previously published protocol [?] to induce a pure population of neurons (iPSC-iNs) from control and LS iPSCs. To analyze appearance of neuronal networks, sparse cytoplasmic lentiviral GFP labeling was used. Initially, neuronal networks derived from control and patient iPSCs appeared similar (Figure ??A). However, after prolonged culture on mouse astrocytes, pronounced neuronal death was observed in cultures from patient iPSCs at day 21; this was further aggravated at day 42 (Figure ??B).

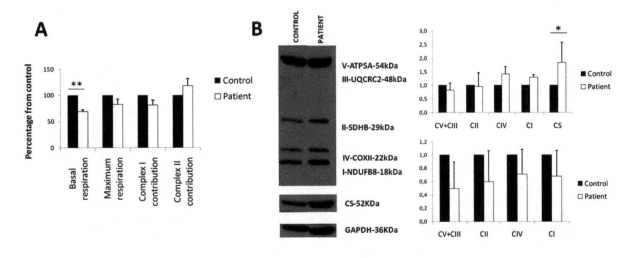

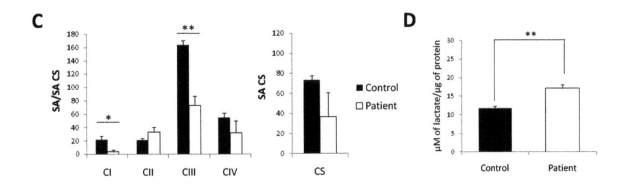

Figure 2. LS induced pluripotent stem cells (iPSCs) manifest a decreased basal respiration and a combined respiratory chain (RC) deficiency. (**A**) Oxygen consumption measured in Oroboros Oxygraph-2k. All data are displayed as a percentage of control. (**B**) Western blot assay against mitoprofile, citrate synthase (CS) and GAPDH (left). Quantification of the Western blot, normalized with GAPDH as a marker of the total protein (right, top panel) or GAPDH and citrate synthase (CS) as a marker of mitochondrial mass (right, bottom panel). All data are displayed as a percentage of control. (**C**) Spectrophotometric measurements of the activity of ETC complexes (left) and citrate synthase (CS) (right); SA: specific activity. (**D**) Extracellular lactate production normalized by total protein. (* p-value < 0.05; ** p-value < 0.01; *** p-value < 0.001)

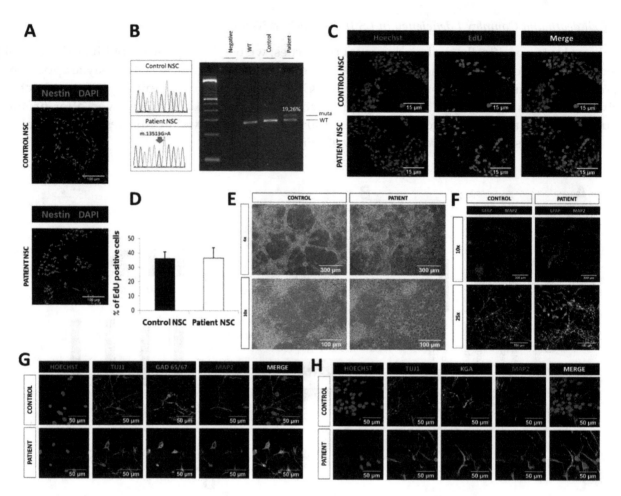

Figure 3. LS neural stem cells (NSCs) manifested similar proliferative and differentiation capacity. (**A**) Immunofluorescence analysis of the neural stem cell marker nestin, manifesting an efficient generation of NSCs from iPSCs; scale bar: 100 μm. (**B**) Electropherograms showing the mutation m.13513G>A in patient NSCs and its absence in control NSCs (left) and heteroplasmy levels of m.13513G>A mutation by RFLP followed by Agilent quantification. (**C**) Proliferation assay of NSCs with the thymidine analogue 5-ethynyl-2'-deoxyuridine (EdU). Scale bar: 15 μm. (**D**) Quantification of EdU (percentage of EdU+/Hoechst+). (**E**) Bright field images (4× and 10×) of neural populations obtained after differentiation of NSCs. (**F**) Immunofluorescence analysis of MAP2, a marker of mature neurons, and GFAP, a marker of astrocytes, in the neural populations obtained after differentiation of NSCs in N2B27 for 3 weeks (**G–H**) Immunofluorescence analysis of the GABAergic marker GAD 65/67 and glutamatergic marker KGA together with neuronal markers (Tuj1 and MAP2).

Measurement of oxygen consumption was performed on a Seahorse Analyzer directly on attached iNs without replating to avoid cell death. A complex I deficiency was detected in patient iNs evidenced by a decrease of the basal respiration (Cr-ROX), maximum respiratory capacity (CrU-ROX) and complex I contribution, CrU-(CRot-ROX), to maximal respiratory capacity (Figure ??C,D). Complex II contribution (CRot-ROX) was not different in patient iNs as compared to control (Figure ??C,D). Treatment with the cell-permeable succinate prodrug NV241 resulted in a similar increase in routine respiration in both patient and control iNs, however the response in patient iNs vehicle (DMSO) control also displayed increased respiration which rendered the NV241 response data in the patient iNs inconclusive (Figure ??C). Further, the treatment with the succinate prodrug increased complex II contribution to maximal uncoupled respiration in control iNs (Figure ??E). In patient iNs, the succinate prodrug induced a similar level of complex II contribution to maximal uncoupled respiration, but the difference to vehicle (DMSO) control did not reach significance (Figure ??E).

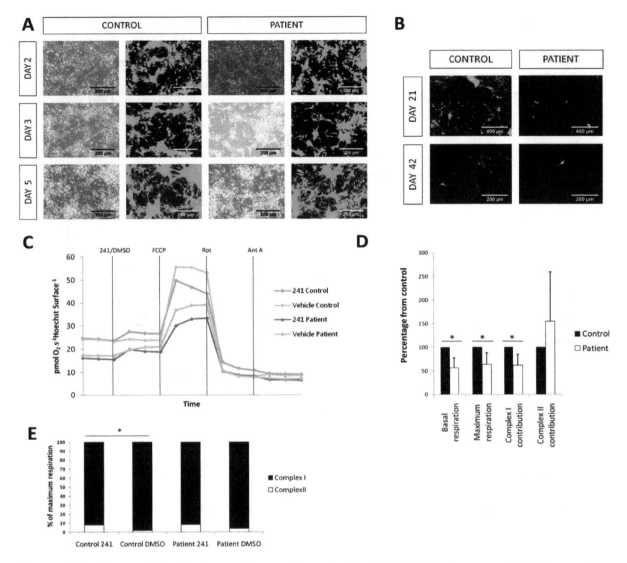

Figure 4. Respiratory defect and neurodegeneration of patient iPSC-derived neurons. (**A**) iN generation from iPSCs using lentiviral vectors for NgN2, rtTA and GFP showing no alterations in derivation of iNs from the patient. (**B**) iNs co-cultured with mouse astrocytes showing a marked neurodegeneration in the patient in comparison with the control both at days 21 and 42. (**C**) Oxygen consumption plots of the different treatments (Control/Patient and NV241/DMSO). (**D**) Quantification of oxygen consumption measured in a Seahorse XFe96 Analyzer. All data are normalized with the control. (**E**) Quantification of the contributions of complexes I and II to the maximum respiration, in percentages. (* p-value < 0.05)

2.6. LS iPSC-Derived Neurons are Functional

In order to test whether the alteration in mitochondrial function has an effect on neuronal function, electrophysiological properties of LS neurons obtained, after a differentiation period of six weeks, were analyzed using whole-cell patch clamp. Neurons differentiated from both control and patient iPSC-derived NSCs were able to fire action potentials (APs) upon current injection (Figure **??**A), and the number of elicited APs in each current step injected was not different between groups (Figure **??**B). Moreover, the maximal number of elicited APs was similar (Figure **??**C), and the application of ramps of currents triggered trains of APs in patient neurons in the same way as in controls (Figure **??**D). Furthermore, both control and patient neurons had depolarizing inward Na^+ currents blocked by TTX (Figure **??**E) and repolarizing outward K^+ currents blocked by TTX + TEA (Figure **??**F). In conclusion, no abnormalities were detected in intrinsic properties or AP characteristics of patient neurons as compared to control.

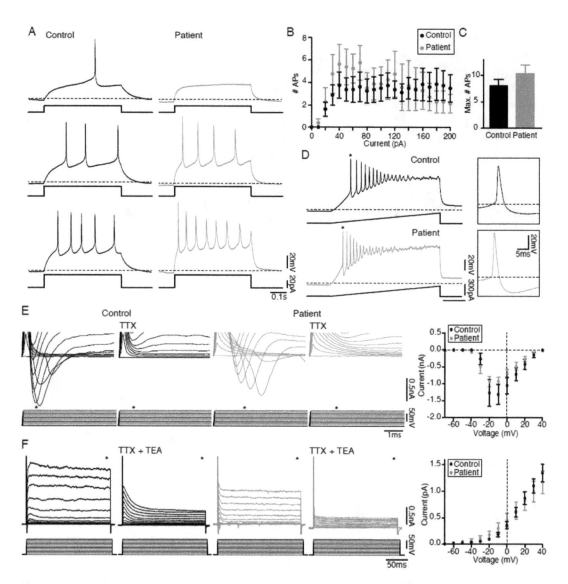

Figure 5. LS iPSC-derived neurons are electrophysiologically functional. (**A**) Neurons derived from control (black) and patient (orange) neural stem cells were able to generate action potentials (APs) upon current injection. (**B**) Graph showing the injected current versus the number of APs elicited ($n = 8$). (**C**) Bar diagram of the maximal number of APs induced ($n = 8$, n.s.). (**D**) Voltage traces show that current injection (ramp from 0–300 pA) induces trains of APs; * denotes the expanded AP17. (**E,F**) Left: Current traces of the fast inward current peak (**E**) and the sustained outward current (**F**) activated by step depolarizations from a holding potential of −70 mV in the absence and presence of 1 mM TTX (**E**) or 1 mM TTX and 10 mM TEA (**F**); * denotes the fast inward current peak (**E**) and the sustained outward current (**F**). Right: Voltage–current plot of the inward current peak (**E**) and sustained outward current (**F**).

2.7. Disturbed Calcium Regulation in LS iPSC-Derived Neurons

Given the role of the mitochondria in regulating intracellular calcium we wanted to assess calcium dynamics in LS neurons. Intracellular calcium concentrations were analyzed by live cell calcium imaging in NSC-derived neurons, after 6 weeks of differentiation. KCl was added in order to stimulate neuronal activity, and the number of evoked cells (responding to KCl) was drastically diminished in patient neurons as compared to controls (see Supplementary Video S1 (Control) and Supplementary Video S2 (Patient)). Moreover, patient evoked cells showed a very different response, compared to controls, with increase in the width of peaks (time from basal to basal level) and time between peaks (Figure ??A). Quantification confirmed the increase in the width of peaks (Figure ??B), which could be explained by a slower increase in cytoplasmic Ca^{2+} upon a depolarizing stimulus or a decrease in the calcium buffering capacity. In order to understand the specific step affected, peaks were measured as time to peak (from basal to maximum level) and time to decay (from maximum to basal level). Although both parameters were increased in patient cells, time to decay was more affected, indicating a Ca^{2+} buffering defect (Figure **??**B). Time between peaks was also assessed and showed an increase of the refractory period in patient neurons (Figure **??**C). All these results together indicate a dysregulation in calcium homeostasis caused by the m.13513G>A mutation.

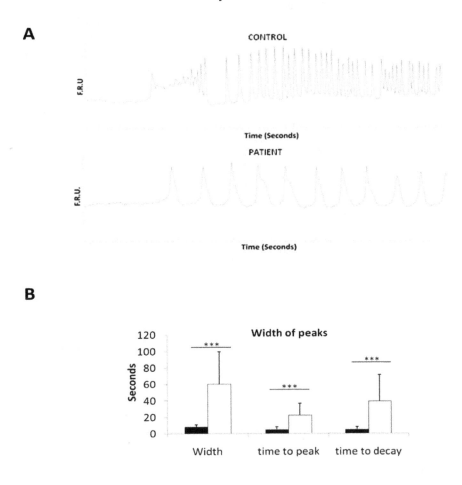

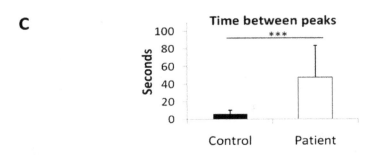

Figure 6. Calcium dysregulation in LS iPSC-derived neurons. (**A**) Representative plots of calcium imaging displaying a different response to KCl in LS iPSC-derived neurons. FRU: fluorescence relative units. (**B**) Quantification of the width of the peaks (from basal to basal), time to peak (from basal to peak) and time to decay (from peak to basal). (**C**) Quantification of the time between peaks. (*** *p*-value < 0.001)

3. Discussion

All mammalian cells produce ATP by glycolysis and OXPHOS. The balance between these processes is different in each cell type [?], and its disturbance can cause disease. Here, we show a clearly compromised OXPHOS function in LS patient fibroblasts and neurons; however only basal respiration was altered in patient iPSCs. This indicates that the defect in iPSC lines is less evident than in fibroblasts or neurons, probably due to the main dependence of iPSCs on glycolysis [? ? ?]. This conclusion is supported by the low levels of oxygen consumed by iPSCs: basal respiration in control iPSCs was approximately half and maximum respiration one-third of that of control fibroblasts.

Theoretically, a defect in complex I of the respiratory chain could be bypassed by an increase in the electron flow through complex II that could restore $\Delta\Psi$mit and ATP levels. In this regard, succinate fuels complex II but it cannot cross biological membranes and does not reach mitochondria. Recently, several cell-permeable succinate prodrugs have been described [?]. As LS patient neurons manifested a decreased complex I respiration, we attempted to rescue that phenotype using one of the succinate prodrugs, NV241, described by Ehinger et al. [?]. Administration of the succinate prodrug increased both routine (basal) respiration and the complex II contribution to maximal respiration to similar levels in patient and control iPSC-derived neurons. However, patient neurons responded with increased respiration, possibly uncoupling, to the drug vehicle (DMSO), which rendered the drug response data inconclusive for the patient iPSC-derived neurons. For the control neurons, the stimulating effects of complex II on respiration were significant. In the initial publication describing the cell-permeable succinate prodrugs, respiration and spare respiratory capacity were increased by prodrug administration in LS patient fibroblasts with a recessive *NDUFS2* mutation.

The question remains as to why iPSCs with their low dependence on OXPHOS manifest a combined RC defect with decreased activities of complexes I and III, while fibroblasts, which rely on mitochondrial OXPHOS more than iPSCs [?], do not show a diminished activity of complexes III and IV similar to muscle. Due to technical problems it was not possible to measure complex I deficiency. Therefore, we cannot discard the possibility of an isolated complex I deficiency in fibroblasts. Sequencing of mitochondrial complex I related genes, however, ruled out the possibility that control fibroblasts had mutations that were not previously detected. One possible explanation for normal specific activity of complexes III and IV in fibroblasts is that the alteration of the ETC function in patient fibroblasts could lead to a decreased $\Delta\Psi$mit, which in turn could induce an increase in the selective elimination of abnormal mitochondria through mitophagy and mask the underlying defect when enzymatic activities are measured. We do observe, as the main finding in patient fibroblasts, diminished mitochondrial mass; considering that mitophagy is the core mechanism of mitochondrial quality and quantity control [?], it is reasonable to think that mitophagy could account for the decrease in mitochondrial mass. This compensatory effect has already been described in other fibroblasts with OXPHOS mutations [? ? ? ?], and its protective role during pathogenesis is recognized [?]. However, in one of these studies, no increase of mitophagy in fibroblasts harboring the m.13513G>A mutation was observed [?]. We therefore cannot rule out the possibility that the observed reduction in protein levels and activity in CS could be the consequence of a decreased mitochondrial biogenesis or an increase in the random elimination of mitochondria by bulk macroautophagy. This alternative explanation, in which no selection against abnormal mitochondria occurs, would not, however, explain the normal activity found in complexes III and IV in the fibroblasts.

In contrast to what happen in fibroblasts, iPSCs, similar to cancer cells, are considered to mainly rely on glycolysis [? ? ?]. In theory, the low OXPHOS function in iPSCs would lead both in controls and patients to a decrease of $\Delta\Psi$mit and the consequent activation of mitophagy [? ?]. It has been demonstrated that iPSCs display higher $\Delta\Psi$mit than differentiated cells [? ?], even using hydrolase activity of ATP synthase to maintain $\Delta\Psi$mit [?]. The maintenance of normal $\Delta\Psi$mit would prevent mitophagy and the compensatory effect on patient iPSCs; as a consequence, no abnormal complexes would be removed, and a compromised activity of complexes I and III would be detected even when the defect is not evident at physiological conditions.

For disease modeling, it is important to recapitulate the principal pathological features of the disease. LS is characterized by the presence of bilateral symmetric necrotic areas in the basal ganglia or the brain stem which correspond with regions of demyelination, neuronal death and astrocytic gliosis [? ?]. Diseased neurons manifested a compromised respiratory capacity and evident neuronal death after being replated on mouse astrocytes. These results indicate that this in vitro model, at least in part, recapitulates the in vivo phenotype of LS. Moreover, increased lactate levels in blood or cerebrospinal fluid are common criteria in the diagnosis of MD. Here, both cell types analyzed (fibroblasts and iPSCs) manifest elevated production of lactate, independently of the main pathway used by the cell type to generate energy.

Patch clamp recordings revealed normal electrophysiological function of individual patient neurons, indicating that mtDNA mutation m.13513G>A does not impair Na^+ or K^+ currents or the ability of neurons to fire APs. This is important, because it indicates that electrically functional neurons can be derived from LS patient NSCs. However, those neurons manifested a marked dysregulation of calcium homeostasis.

Ca^{2+} is an intracellular signal responsible for controlling numerous cellular processes [?]. The Ca^{2+} signaling network requires first the ON mechanisms, by which there is a 10-fold rise of cytoplasmic Ca^{2+}, followed by the OFF mechanisms, which ensure that Ca^{2+} is recovered to basal levels [?]. The increase of cytoplasmic calcium is the consequence of the entry of external Ca^{2+} through different types of channels (VOCs, ROCs, SOCs) and the release of Ca^{2+} from internal stores [?]. At the same time, the OFF mechanisms include the extrusion of calcium to the outside, its return to internal stores (endoplasmic reticulum and mitochondrion) and its association with cytosolic buffers (parvalbumin, calbindin-D28k, calretinin) [? ?]. In the presynaptic membrane of neurons, the rise of cytoplasmic Ca^{2+} triggers the release of neurotransmitters. As important as the rise is the buffering of that calcium to ensure the possibility of a new activation. In fact, sustained high concentrations of cytoplasmic Ca^{2+} are associated with neuronal death [?] through apoptosis if there is ATP available or necrosis if there is ATP depletion [? ?]. It has been postulated that the addition of KCl induces a slow depolarization of membrane potential ($\Delta\Psi$), consequent activation of Ca^{2+} voltage-operated channels (VOCs) and Ca^{2+} influx into the cell [? ?]. Cells that respond to this KCl are called evoked cells [?]. The analysis of the response of evoked cells harboring the mtDNA mutation m.13513G>A manifested a slower buffering capacity and an increased refractory period. This is not surprising, because mitochondria play a significant role in shaping global Ca^{2+} signals [?] through direct or indirect participation [?]. The decreased buffering capacity that we observe can lead to higher amounts of calcium in the cytoplasm and consequent neuronal necrosis observed in LS patients [?]. The concrete mechanism by which this mtDNA mutation causes inappropriate buffering of calcium remains unknown. Recently, neural progenitor cells (NPCs) harboring the homoplasmic m.9185T>C mutation in the *MT-ATP6* causative of episodic paralysis with spinal neuropathy (NPC_ATP6) has been reported [?]. In contrast to our results, NPC_ATP6 mutated cells manifested a decreased calcium release (ON mechanisms) rather than the decrease in calcium buffering observed here (OFF mechanism). These differences could be the consequence of the $\Delta\Psi$mit, in which mutations that affect the ATP synthase increase $\Delta\Psi$mit and affect calcium release, while mutations in the ETC provoke a diminishment in $\Delta\Psi$mit and affect calcium buffering. In support of our findings, iPSC-derived neurons from other neurodegenerative diseases such as Parkinson's disease (PD) or frontotemporal lobar degeneration tauopathy (FTLD-Tau) have also shown an impaired calcium homeostasis as the underlying pathogenic mechanism [? ?]. It is also very interesting that the necrosis observed after myocardial ischemia or stroke [? ?] occurs in response to a lack of oxygen to the mitochondria, suggesting that MD, similarly to PD and TFLD-Tau and other conditions such as myocardial infarction or ischemic stroke, could occur through the same calcium overload process, which leads to necrosis. Although other studies have concluded that mutations in the *MT-ND5* gene in fibroblasts or cybrids can be associated to calcium handling defects [? ?], the phenotype of the neurons, the cell type in which the disease occurs, remained unexplored.

Here, we demonstrate defective OXPHOS and a clear dysregulation in calcium homeostasis in neurons harboring the m.13513G>A mutation, with profound consequences for the pathology.

4. Materials and Methods

4.1. Cell Culture, iPSC Generation and Characterization

This study was approved by the Institutional Ethical Committee of the Autonoma University of Madrid according to Spanish and European Union legislation and complies with the principles of the 1964 Helsinki declaration. Normal human dermal fibroblasts (NHDFs) were purchased from Promocell (C12300) and maintained in Dulbecco's modified Eagle medium (DMEM) supplemented with 10% FBS, 50 U/mL penicillin and 50 μg/mL streptomycin. Leigh fibroblasts from a described patient harboring the heteroplasmic mutation in the mtDNA m.13513G>A (p.D393N) were kindly provided by Dr. Francina Munell from the Hospital Universitario Vall d'Hebron (Barcelona, Spain), with prior informed consent for study participation. LS fibroblasts were maintained in the same growth medium as NHDF but supplemented with 50 μg/mL uridine. A previously described iPSC line, LND554SV.4 [?], and a control iPSC line reported here (N44SV.1) have been derived, maintained and characterized following the protocols described in [?].

4.2. Oxygen Consumption

Oxygen consumption measurements in fibroblasts and iPSCs were performed in an Oroboros Oxygraph-2k and analyzed using DatLab4 software (Oroboros Instruments). Approximately one million fibroblasts or two million iPSCs were used in each chamber. An intact cell measurement protocol has been used. The experimental regime started with routine respiration (Cr), which is defined as respiration in cell-culture medium without additional substrates. After reaching steady-state respiratory flux, ATP synthase was inhibited with oligomycin (2 μg/mL), followed by uncoupling of oxidative phosphorylation by titration of FCCP (carbonyl cyanide p-trifluoromethoxyphenylhydrazone) with 0.5 μM steps (CrU). Finally, respiration was inhibited by sequential addition of rotenone at 0.5 μM (to test for the effect of inhibiting complex I activity) (CRot) and antimycin A at 2.5 μM (for inhibiting complex III) (non-mitochondrial respiration, ROX). Values were normalized by number of cells, and ROX was subtracted. Basal respiration (Cr-ROX), maximum respiration (CrU-ROX), complex I contribution, CrU-(CRot-ROX), or complex II contribution (CRot-ROX) were calculated from a minimum of three independent experiments. Oxygen consumption measurements of induced neurons (iNs) were performed in a Seahorse XFe96 Analyzer and analyzed by Wave software (Agilent Technologies). Induction of iNs from iPSCs was performed in the seahorse microplate following a previously described protocol [?] and measured at day 7, after puromycin selection. Before the experiment, growth medium was replaced by XF-Base Medium (Agilent Technologies) supplemented with 2 mM L-glutamine, 5 mM sodium pyruvate and 10 mM glucose (pH 7.4), and cells were kept 1h at 37 °C at atmospheric O_2 and CO_2. Oxygen consumption was determined at basal conditions (Cr) as well as after addition of the following drugs: 500 μM NV241 or DMSO as vehicle, FCCP titration (0.125, 1 or 2 μM), 2 μM rotenone (CRot) and 1 μg/mL antimycin A (ROX). After the experiment, Hoechst 33342 was added to the wells, and pictures were acquired in a fluorescence microscope to quantify DAPI surface. Values were normalized by DAPI surface and non-mitochondrial respiration (ROX) was subtracted. Basal respiration (Cr-ROX), maximum respiration (CrU-ROX), complex I contribution, CrU-(CRot-ROX), or complex II contribution (CRot-ROX) and effect of NV241 were calculated from three independent experiments.

4.3. Protein Extraction and Western Blot

Protein was extracted using cold RIPA buffer (50 mM Tris-HCl, pH 7.4, 1% NP-40, 0.5% Na-deoxycholate, 0.1% SDS, 150 mM NaCl, 2 mM EDTA, 50 mM NaF) supplemented with protease inhibitor cocktail (Roche, 11873580001) and quantified using BCA Protein Assay Kit (Thermo

Fisher Scientific 23225). Fifty micrograms of protein was separated on a 12% SDS-PAGE gel and electrotransferred to Immobilon-P membranes (Millipore, IPVH00010). Primary antibodies used were Mitoprofile (Abcam ab110411; 1:1000), CS (Abcam ab96600; 1:10000) and GAPDH (Abcam ab8245; 1:6000). Appropriate secondary antibodies coupled to horseradish peroxidase were used, and peroxidase activity was tested using ECL (GE Healthcare, RPN2209). Quantification of WB was performed by densitometric analysis in Image J; values were normalized by total protein amount (GAPDH) and/or by mitochondrial mass (CS).

4.4. Lactic Acid Production

Extracellular lactic acid concentration in the cell culture medium was determined using a lactate-dehydrogenase activity assay. First, 500,000 fibroblasts were plated on P100 culture dishes. The next day, growth medium was changed by fresh medium; exactly 24 h after medium was changed, 100 μL was removed, deproteinized and adjusted to pH 6–8. Samples were kept at −80 °C until assay. For the assay, 15 μL of the sample was incubated with 30 μL of NAD+ 15 mM, 5 μL of LDH 1 mg/mL (Roche, 10 127 230 001), 150 μL of assay buffer (consisting of 0.5 M glycine, 0.2 M hydrazine and 3.4 mM EDTA; pH 9.5) and adjusted to a final volume of 300 μL with bidistilled water. This reaction was incubated for 105 min at 37 °C before measuring absorbance at 340 nm. The lactate present in the sample together with NAD^+ was transformed by lactate dehydrogenase in NADH, with concentration proportional to the increase in absorbance at 340 nm. A standard curve of lactate ranging from 4 to 0.25 mM was used to extrapolate the lactate concentration present in the sample. These values were normalized by the total amount of protein present in the culture measured with BCA Protein Assay Kit (Thermo Scientific, Waltham, MA, USA, Ref 23225). For iPSCs, a similar protocol was used, but starting cell density was 80% confluent. The choice of measurement at 24 h after changing the medium was determined using previous kinetics data. At least three independent experiments were performed.

4.5. Respiratory Chain Activity Determination by Spectrophotometry

Measurements of the specific activities (SA) of respiratory chain complexes were determined following the protocol established in [?]. Values were normalized with the specific activity (SA) of citrate synthase (CS). A minimum of three independent experiments was performed.

4.6. NSC Generation and Neuronal Differentiation

For generation of NSC, the commercially available PSC Neural Induction Medium (ThermoFisher, Waltham, MA, USA, A1647801) was used, following manufacturer's instructions. NSCs were routinely maintained in NEM medium (ThermoFisher, Waltham, MA, USA). For neuronal differentiation, 42,000 cells/cm^2 were seeded in GFR Matrigel-coated plates in NEM medium and maintained for 48 h. After this, growth medium was replaced with differentiation medium consisting of DMEMF12 (Thermo Fisher; Waltham, MA, USA, 11330-057) supplemented with 1x N2 (Thermo Fisher Scientific, Waltham, MA, USA, 1750200), 1x B27 (Thermo Fisher Scientific, Waltham, MA, USA, 17504044), 1x NEAA (Thermo Fisher Scientific, Waltham, MA, USA, 11140035) and 100 μM β-mercaptoethanol (Thermo Fisher Scientific, Waltham, MA, USA, 21985023). Culture medium was changed every other day for 6 weeks. For Patch Clamp and Calcium Imaging experiments, BrainPhys (Stemcell Technologies, Grenoble, France, 05792) was used instead of DMEMF12.

4.7. Mutation Analysis and Heteroplasmy Quantification of mtDNA m.13513G>A Mutation

For this purpose, the protocol described in Galera-Monge et al. [?] was followed. Briefly, total DNA was extracted using a standard phenol-chloroform protocol. For mutation analysis, amplification by PCR of a mtDNA region containing the m.13513G>A position was carried out using the following primers: mt-20F: 5′ ATCTGTACCCACGCCTTC 3′ and mt-20R: 5′ AGAGGGGTCAGGGTTGATTC 3′. Following PCR amplification, direct sequencing of amplicons was performed on both strands in an ABI 3730 sequencer (Applied Biosystems,

Foster City, CA, USA) using a dye terminator cycle sequencing kit (Applera, Rockville, MD). Heteroplasmy quantification of m.13513G>A mutation was studied by RFLP. PCR amplification with the following primers: 13513F: 5′GACTGACTGACTGACAAGTCAACTAGGACTCATAATA3′ and 13513R: 5′CAGGCGTTTGTGTATGATATGTTTGCGGTTTCGATGACGTGG3′ was followed by digestion with restriction enzyme PflFI (New England Biolabs. Reference: R0595S) and quantified with the Agilent DNA 1000 Kit (Agilent, Santa Clara, CA, USA, 5067-1504) in an Agilent 2100 Bioanalyzer.

4.8. Proliferation Assay

Here, 90,000 NSCs were plated onto matrigel-coated 13 mm coverslips in P24 multiwell plates with NEM medium. At 48 h, 10 μM EdU was added to the media and incubated for one hour. After that, cells were washed with PBS, fixed with 4% PFA and permeabilized with 0.025% Triton in PBS. The Click-iT EdU Kit (Invitrogen, Carlsbad, CA, USA, C10338) was used for the detection of EdU, following the manufacturer′s instructions. Hoechst 33342 was added to stain nuclei. Quantification was performed manually at 40× in a randomized and blinded manner (n = t5).

4.9. Immunofluorescence

Cells were fixed with 4% PFA, permeabilized and blocked with 0.3% Triton X-100 and 3% Donkey Serum in TBS. Samples were incubated with the following primary antibodies overnight at 4 °C: MAP2 (Sigma, St. Louis, MO, USA, M1406,1:250); GFAP (Dako, Carpinteria, CA, USA, Z0334, 1:500); GAD65/67 (Abcam, Branford, CT, USA, ab11070, 1:500); KGA (Abcam, Branford, CT, USA, ab156876, 1:200); TUJ1 (Sigma, St. Louis, MO, USA, T8660, 1:400); MAP2 Chicken (Abcam, Branford, CT, USA, ab5392, 1:10000). Alexa Fluor Dye secondary antibodies were applied (1:500) at room temperature for 30 min. Images were acquired with a spectral confocal microscope LSM710 (Zeiss, Matesalka, Hungary).

4.10. iN Morphology Analysis

iPSC-iNs were generated following the protocol described in [?]. Lentiviral GFP+ and unlabeled iNs were mixed in a proportion of 1:6 in order to assess morphology of individual neurons. Images were directly acquired using a BX61 fluorescence microscope (Olympus, Tokyo, Japan).

4.11. Electrophysiology

iPSC-derived neurons cultured on coverslips were transferred to the recording chamber and held down with a small piece of platinum wire [?]. The coverslip was constantly perfused with carbonated artificial cerebral spinal fluid (in mM: 119 NaCl, 2.5 KCl, 1.3 MgSO$_4$, 2.5 CaCl$_2$, 26 NaHCO$_3$, 1.25 NaH$_2$PO$_4$ and 11 glucose; pH ~7.4) at 34 °C. Recording pipettes were filled with intracellular solution (in mM: 122.5 potassium gluconate, 12.5 KCl, 10 HEPES, 0.2 EGTA, 2.0 MgATP, 0.3 Na$_2$-GTP, 8.0 NaCl; pH ~7.3) and had a resistance of 4–12 MΩ. Target cells were visualized using a water immersion objective (Olympus, 40×), and whole-cell patch clamp recordings were performed with a HEKA double patch clamp EPC10 amplifier using Patch Master for data acquisition. Voltage and current clamp recordings were used for the electrophysiological characterization. Sodium and potassium currents were evoked by a series of 200 ms long voltage steps (from −70 to +40 mV in 10 mV steps) and inhibited with 1 μM TTX and 10 mM TEA, respectively. Series of current steps (0–200 pA in 10 pA steps) and current ramps from 0–300 pA were performed to determine the cells' ability to generate action potentials. Data were analyzed offline with FitMaster and IgorPro. At least nine cells of each condition were analyzed.

4.12. Calcium Imaging

For calcium imaging, NSCs were allowed to differentiate for 6 weeks (see NSC differentiation section) on Glass-Bottom P35 culture dishes coated with 15 µg/mL Poly-L-Ornithine (Sigma, St. Louis, MO, USA, P3655) at RT for 1 h and 10 µg/mL laminin at 37 °C for 2 h (Thermo Fisher Scientific, Waltham, MA, USA, 23017-015) [?]. The day of the experiment, cells were loaded with 1 ng/µL of Fluo-4 AM (Molecular Probes, Eugene, OR, USA, F14201) diluted in growth medium without Phenol Red and incubated at 37 °C and 5% CO_2 for 30 min. Before imaging, medium was changed to fresh growth medium without Phenol Red to wash away excess Fluo-4 AM. Time-lapse recordings were acquired using a Zeiss confocal microscope at maximal speed during 30 min. At minute 6, 90 mM KCl was added to the culture plate. Four videos of each condition and independent culture plate were recorded for analysis. For quantification, 10 cells with response to KCl were chosen randomly, and the type of response was analyzed.

4.13. Statistical Analysis

Data are presented as mean ± SD. An unpaired two-tailed t-test or Mann–Whitney test was performed; p-values of less than 0.05 were considered statistically significant. (* *p*-value < 0.05; ** *p*-value < 0.01; *** *p*-value < 0.001).

Author Contributions: Conceptualization, R.G. and M.E.G.; methodology, T.G.-M., F.Z.D., I.C., L.R.-V., M.-G.H., J.K.-E, H.A. and M.E.G.-; validation, T.G.-M., F.Z.D., I.C., M.-G.H., J.K.-E., E.E., M.A.M., H.A. and M.E.-G.; formal analysis, T.G.-M., H.A. and M.E.G.; investigation, T.G.-M., F.Z.-D., I.C., M.-G.H., L.R.-V. and J.K.-E; project administration, M.E.G.; writing—original draft preparation, T.G.-M., H.A. and M.E.G.; writing—review and editing, E.E., H.A. and M.E.G.; supervision, H.A. and M.E.G.; funding acquisition, R.G., H.A. and M.E.G. All authors have read and agreed to the published version of the manuscript.

Acknowledgments: We are very grateful to Francina Munell for kindly providing us with the patient fibroblasts.

Abbreviations

LS	Leigh syndrome
MD	Mitochondrial disorder
iPSCs	Induced pluripotent stem cells
mtDNA	Mitochondrial DNA
NSCs	Neural stem cells
OXPHOS	Mitochondrial oxidative phosphorylation system
DMSO	Dimethyl sulfoxide
FCCP	Carbonyl cyanide p-trifluoromethoxyphenylhydrazone
RC	Respiratory chain
ROS	Reactive oxygen species
MELAS	Mitochondrial myopathy, encephalopathy, lactic acidosis and stroke
ATP	Adenosine triphosphate
ETC	Electron transfer chain
rRNA	Ribosomal ribonucleic acid
tRNA	Transfer RNA
CS	Citrate synthase
$\Delta\Psi$mit	Mitochondrial membrane potential
AP	Action potential
FRU	Fluorescence relative units

References

1. Johnson, S.C.; Yanos, M.E.; Kayser, E.B.; Quintana, A.; Sangesland, M.; Castanza, A.; Uhde, L.; Hui, J.; Wall, V.Z.; Gagnidze, A.; et al. mTOR inhibition alleviates mitochondrial disease in a mouse model of Leigh syndrome. *Science* **2013**, *342*, 1524–1528. [CrossRef] [PubMed]

2. Quintana, A.; Zanella, S.; Koch, H.; Kruse, S.E.; Lee, D.; Ramirez, J.M.; Palmiter, R.D. Fatal breathing dysfunction in a mouse model of Leigh syndrome. *J. Clin. Investig.* **2012**, *122*, 2359–2368. [CrossRef] [PubMed]

3. Lake, N.J.; Compton, A.G.; Rahman, S.; Thorburn, D.R. Leigh syndrome: One disorder, more than 75 monogenic causes. *Ann. Neurol.* **2016**, *79*, 190–203. [CrossRef]

4. Ernster, L.; Schatz, G. Mitochondria: A historical review. *J. Cell Biol.* **1981**, *91*, 227s–255s. [CrossRef]

5. Viscomi, C.; Bottani, E.; Zeviani, M. Emerging concepts in the therapy of mitochondrial disease. *Biochim. Biophys. Acta* **2015**, *1847*, 544–557. [CrossRef]

6. Mlody, B.; Lorenz, C.; Inak, G.; Prigione, A. Energy metabolism in neuronal/glial induction and in iPSC models of brain disorders. *Semin. Cell Dev. Biol.* **2016**, *52*, 102–109. [CrossRef]

7. Anderson, S.; Bankier, A.T.; Barrell, B.G.; de Bruijn, M.H.; Coulson, A.R.; Drouin, J.; Eperon, I.C.; Nierlich, D.P.; Roe, B.A.; Sanger, F.; et al. Sequence and organization of the human mitochondrial genome. *Nature* **1981**, *290*, 457–465. [CrossRef]

8. Stewart, J.B.; Chinnery, P.F. The dynamics of mitochondrial DNA heteroplasmy: Implications for human health and disease. *Nat. Rev. Genet.* **2015**, *16*, 530–542. [CrossRef]

9. Monlleo-Neila, L.; Toro, M.D.; Bornstein, B.; Garcia-Arumi, E.; Sarrias, A.; Roig-Quilis, M.; Munell, F. Leigh Syndrome and the Mitochondrial m.13513G>A Mutation: Expanding the Clinical Spectrum. *J. Child Neurol.* **2013**, *28*, 1531–1534. [CrossRef]

10. Galera, T.; Zurita-Diaz, F.; Garesse, R.; Gallardo, M.E. iPSCs, a Future Tool for Therapeutic Intervention in Mitochondrial Disorders: Pros and Cons. *J. Cell. Physiol.* **2016**, *231*, 2317–2318. [CrossRef]

11. Ma, H.; Folmes, C.D.; Wu, J.; Morey, R.; Mora-Castilla, S.; Ocampo, A.; Ma, L.; Poulton, J.; Wang, X.; Ahmed, R.; et al. Metabolic rescue in pluripotent cells from patients with mtDNA disease. *Nature* **2015**, *524*, 234–238. [CrossRef] [PubMed]

12. Zheng, X.; Boyer, L.; Jin, M.; Kim, Y.; Fan, W.; Bardy, C.; Berggren, T.; Evans, R.M.; Gage, F.H.; Hunter, T. Alleviation of neuronal energy deficiency by mTOR inhibition as a treatment for mitochondria-related neurodegeneration. *eLife* **2016**, *5*, e13378. [CrossRef] [PubMed]

13. Galera, T.; Zurita, F.; Gonzalez-Paramos, C.; Moreno-Izquierdo, A.; Fraga, M.F.; Fernandez, A.F.; Garesse, R.; Gallardo, M.E. Generation of a human iPSC line from a patient with Leigh syndrome. *Stem Cell Res.* **2016**, *16*, 63–66. [CrossRef] [PubMed]

14. Zhang, Y.; Pak, C.; Han, Y.; Ahlenius, H.; Zhang, Z.; Chanda, S.; Marro, S.; Patzke, C.; Acuna, C.; Covy, J.; et al. Rapid single-step induction of functional neurons from human pluripotent stem cells. *Neuron* **2013**, *78*, 785–798. [CrossRef]

15. Teslaa, T.; Teitell, M.A. Pluripotent stem cell energy metabolism: An update. *EMBO J.* **2015**, *34*, 138–153. [CrossRef]

16. Zhang, J.; Nuebel, E.; Wisidagama, D.R.; Setoguchi, K.; Hong, J.S.; Van Horn, C.M.; Imam, S.S.; Vergnes, L.; Malone, C.S.; Koehler, C.M.; et al. Measuring energy metabolism in cultured cells, including human pluripotent stem cells and differentiated cells. *Nat. Protoc.* **2012**, *7*, 1068–1085. [CrossRef]

17. Ehinger, J.K.; Piel, S.; Ford, R.; Karlsson, M.; Sjovall, F.; Frostner, E.A.; Morota, S.; Taylor, R.W.; Turnbull, D.M.; Cornell, C.; et al. Cell-permeable succinate prodrugs bypass mitochondrial complex I deficiency. *Nat. Commun.* **2016**, *7*, 12317. [CrossRef]

18. Wei, H.; Liu, L.; Chen, Q. Selective removal of mitochondria via mitophagy: Distinct pathways for different mitochondrial stresses. *Biochim. Biophys. Acta* **2015**, *1853*, 2784–2790. [CrossRef]

19. Rodriguez-Hernandez, A.; Cordero, M.D.; Salviati, L.; Artuch, R.; Pineda, M.; Briones, P.; Gomez Izquierdo, L.; Cotan, D.; Navas, P.; Sanchez-Alcazar, J.A. Coenzyme Q deficiency triggers mitochondria degradation by mitophagy. *Autophagy* **2009**, *5*, 19–32. [CrossRef]

20. Cotan, D.; Cordero, M.D.; Garrido-Maraver, J.; Oropesa-Avila, M.; Rodriguez-Hernandez, A.; Gomez Izquierdo, L.; De la Mata, M.; De Miguel, M.; Lorite, J.B.; Infante, E.R.; et al. Secondary coenzyme Q10 deficiency triggers mitochondria degradation by mitophagy in MELAS fibroblasts. *FASEB J. Off. Publ. Fed. Am. Soc. Exp. Biol.* **2011**, *25*, 2669–2687.

21. De la Mata, M.; Garrido-Maraver, J.; Cotan, D.; Cordero, M.D.; Oropesa-Avila, M.; Izquierdo, L.G.; De Miguel, M.; Lorite, J.B.; Infante, E.R.; Ybot, P.; et al. Recovery of MERRF fibroblasts and cybrids pathophysiology by coenzyme Q10. *Neurother. J. Am. Soc. Exp. NeuroTher.* **2012**, *9*, 446–463. [CrossRef] [PubMed]

22. Granatiero, V.; Giorgio, V.; Cali, T.; Patron, M.; Brini, M.; Bernardi, P.; Tiranti, V.; Zeviani, M.; Pallafacchina, G.; De Stefani, D.; et al. Reduced mitochondrial Ca(2+) transients stimulate autophagy in human fibroblasts carrying the 13514A>G mutation of the ND5 subunit of NADH dehydrogenase. *Cell Death Differ.* **2016**, *23*, 231–241. [CrossRef] [PubMed]

23. Moran, M.; Delmiro, A.; Blazquez, A.; Ugalde, C.; Arenas, J.; Martin, M.A. Bulk autophagy, but not mitophagy, is increased in cellular model of mitochondrial disease. *Biochim. Biophys. Acta* **2014**, *1842*, 1059–1070. [CrossRef] [PubMed]

24. Armstrong, L.; Tilgner, K.; Saretzki, G.; Atkinson, S.P.; Stojkovic, M.; Moreno, R.; Przyborski, S.; Lako, M. Human induced pluripotent stem cell lines show stress defense mechanisms and mitochondrial regulation similar to those of human embryonic stem cells. *Stem Cells* **2010**, *28*, 661–673. [CrossRef]

25. Prigione, A.; Hossini, A.M.; Lichtner, B.; Serin, A.; Fauler, B.; Megges, M.; Lurz, R.; Lehrach, H.; Makrantonaki, E.; Zouboulis, C.C.; et al. Mitochondrial-associated cell death mechanisms are reset to an embryonic-like state in aged donor-derived iPS cells harboring chromosomal aberrations. *PLoS ONE* **2011**, *6*, e27352. [CrossRef]

26. Zhang, J.; Khvorostov, I.; Hong, J.S.; Oktay, Y.; Vergnes, L.; Nuebel, E.; Wahjudi, P.N.; Setoguchi, K.; Wang, G.; Do, A.; et al. UCP2 regulates energy metabolism and differentiation potential of human pluripotent stem cells. *EMBO J.* **2011**, *30*, 4860–4873. [CrossRef]

27. Berridge, M.J.; Lipp, P.; Bootman, M.D. The versatility and universality of calcium signalling. *Nat. Rev. Mol. Cell Biol.* **2000**, *1*, 11–21. [CrossRef]

28. Grienberger, C.; Konnerth, A. Imaging calcium in neurons. *Neuron* **2012**, *73*, 862–885. [CrossRef]

29. Orrenius, S.; Zhivotovsky, B.; Nicotera, P. Regulation of cell death: The calcium-apoptosis link. *Nat. Rev. Mol. Cell Biol.* **2003**, *4*, 552–565. [CrossRef]

30. Eguchi, Y.; Shimizu, S.; Tsujimoto, Y. Intracellular ATP levels determine cell death fate by apoptosis or necrosis. *Cancer Res.* **1997**, *57*, 1835–1840.

31. Leist, M.; Single, B.; Castoldi, A.F.; Kuhnle, S.; Nicotera, P. Intracellular adenosine triphosphate (ATP) concentration: A switch in the decision between apoptosis and necrosis. *J. Exp. Med.* **1997**, *185*, 1481–1486. [CrossRef] [PubMed]

32. Cameron, M.; Kekesi, O.; Morley, J.W.; Tapson, J.; Breen, P.P.; van Schaik, A.; Buskila, Y. Calcium Imaging of AM Dyes Following Prolonged Incubation in Acute Neuronal Tissue. *PLoS ONE* **2016**, *11*, e0155468. [CrossRef]

33. Duchen, M.R. Contributions of mitochondria to animal physiology: From homeostatic sensor to calcium signalling and cell death. *J. Physiol.* **1999**, *516 Pt 1*, 1–17. [CrossRef]

34. Collins, T.J.; Lipp, P.; Berridge, M.J.; Li, W.; Bootman, M.D. Inositol 1,4,5-trisphosphate-induced Ca^{2+} release is inhibited by mitochondrial depolarization. *Biochem. J.* **2000**, *347*, 593–600. [CrossRef] [PubMed]

35. Lorenz, C.; Lesimple, P.; Bukowiecki, R.; Zink, A.; Inak, G.; Mlody, B.; Singh, M.; Semtner, M.; Mah, N.; Aure, K.; et al. Human iPSC-Derived Neural Progenitors Are an Effective Drug Discovery Model for Neurological mtDNA Disorders. *Cell Stem Cell* **2017**, *20*, 659–674. [CrossRef] [PubMed]

36. Schondorf, D.C.; Aureli, M.; McAllister, F.E.; Hindley, C.J.; Mayer, F.; Schmid, B.; Sardi, S.P.; Valsecchi, M.; Hoffmann, S.; Schwarz, L.K.; et al. iPSC-derived neurons from GBA1-associated Parkinson's disease patients show autophagic defects and impaired calcium homeostasis. *Nat. Commun.* **2014**, *5*, 4028. [CrossRef]

37. Imamura, K.; Sahara, N.; Kanaan, N.M.; Tsukita, K.; Kondo, T.; Kutoku, Y.; Ohsawa, Y.; Sunada, Y.; Kawakami, K.; Hotta, A.; et al. Calcium dysregulation contributes to neurodegeneration in FTLD patient iPSC-derived neurons. *Sci. Rep.* **2016**, *6*, 34904. [CrossRef]

38. Fleckenstein, A.; Janke, J.; Doring, H.J.; Leder, O. Myocardial fiber necrosis due to intracellular Ca overload-a new principle in cardiac pathophysiology. *Recent Adv. Stud. Card. Struct. Metab.* **1974**, *4*, 563–580.

39. Tuttolomondo, A.; Di Sciacca, R.; Di Raimondo, D.; Arnao, V.; Renda, C.; Pinto, A.; Licata, G. Neuron protection as a therapeutic target in acute ischemic stroke. *Curr. Top. Med. Chem.* **2009**, *9*, 1317–1334. [CrossRef]

40. McKenzie, M.; Duchen, M.R. Impaired Cellular Bioenergetics Causes Mitochondrial Calcium Handling Defects in MT-ND5 Mutant Cybrids. *PLoS ONE* **2016**, *11*, e0154371. [CrossRef]

41. Galera-Monge, T.; Zurita-Diaz, F.; Garesse, R.; Gallardo, M.E. The mutation m.13513G>A impairs cardiac function, favoring a neuroectoderm commitment, in a mutant-load dependent way. *J. Cell. Physiol.* **2019**, *234*, 19511–19522. [CrossRef] [PubMed]

42. Medja, F.; Allouche, S.; Frachon, P.; Jardel, C.; Malgat, M.; Mousson de Camaret, B.; Slama, A.; Lunardi, J.; Mazat, J.P.; Lombes, A. Development and implementation of standardized respiratory chain spectrophotometric assays for clinical diagnosis. *Mitochondrion* **2009**, *9*, 331–339. [CrossRef] [PubMed]

43. Hansen, M.G.; Tornero, D.; Canals, I.; Ahlenius, H.; Kokaia, Z. In Vitro Functional Characterization of Human Neurons and Astrocytes Using Calcium Imaging and Electrophysiology. *Methods Mol. Biol.* **2019**, *1919*, 73–88. [PubMed]

Permissions

List of Contributors

Katarzyna Haraźna, Tomasz Witko, Małgorzata Zimowska, Robert Socha, Małgorzata Witko and Maciej Guzik
Jerzy Haber Institute of Catalysis and Surface Chemistry Polish Academy of Sciences, Niezapominajek 8, 30-239 Kraków, Poland

Ewelina Cichoń, Szymon Skibiński and Aneta Zima
Faculty of Materials Science and Ceramics, AGH University of Science and Technology, 30 Mickiewicza Ave, 30-059 Kraków, Poland

Daria Solarz
Faculty of Physics, Astronomy and Applied Computer Science, Jagiellonian University, Lojasiewicza 11, 30-348 Kraków, Poland

Iwona Kwiecień
Department of Physical Chemistry and Technology of Polymers, Silesian University of Technology, M. Strzody 9, 44-100 Gliwice, Poland

Elena Marcello
School of Life Sciences, College of Liberal Arts and Sciences, University of Westminster, New Cavendish Street, London W1W 6UW, UK

Ewa Szefer, Konstantinos N. Raftopoulos and Krzysztof Pielichowski
Department of Chemistry and Technology of Polymers, Cracow University of Technology, Warszawska 24, 31-155 Kraków, Poland

Ipsita Roy
Department of Materials Science and Engineering, University of Sheffield , Broad Lane, Sheffield S3 7HQ, UK

Ashim Gupta
General Therapeutics, Cleveland Heights, OH 44118, USA
Future Biologics, Lawrenceville, GA 30043, USA
BioIntegrate, Lawrenceville, GA 30043, USA
South Texas Orthopaedic Research Institute, Laredo, TX 78045, USA
Veterans in Pain, Valencia, CA 91354, USA

Anne-Marie Fauser
Bohlander Stem Cell Research Laboratory, Department of Biology, Bradley University, Peoria, IL 61625, USA

Craig Cady
General Therapeutics, Cleveland Heights, OH 44118, USA
Bohlander Stem Cell Research Laboratory, Department of Biology, Bradley University, Peoria, IL 61625, USA

Hugo C. Rodriguez
Future Biologics, Lawrenceville, GA 30043, USA
South Texas Orthopaedic Research Institute, Laredo, TX 78045, USA
School of Osteopathic Medicine, University of the Incarnate Word, San Antonio, TX 78235, USA
Future Physicians of South Texas, San Antonio, TX 78235, USA

R. Justin Mistovich
General Therapeutics, Cleveland Heights, OH 44118, USA
Department of Orthopaedics, School of Medicine, Case Western Reserve University, Cleveland, OH 44106, USA

Anish G. R. Potty
General Therapeutics, Cleveland Heights, OH 44118, USA
South Texas Orthopaedic Research Institute, Laredo, TX 78045, USA
School of Osteopathic Medicine, University of the Incarnate Word, San Antonio, TX 78235, USA
Laredo Sports Medicine Clinic, Laredo, TX 78041, USA

Nicola Maffulli
Department of Musculoskeletal Disorders, School of Medicine and Surgery, University of Salerno, 84084 Fisciano, Italy
San Giovanni di Dio e Ruggi D'Aragona Hospital "Clinica Orthopedica" Department, Hospital of Salerno, 84124 Salerno, Italy
Barts and the London School of Medicine and Dentistry, Centre for Sports and Exercise Medicine, Queen Mary University of London, London E1 4DG, UK
School of Pharmacy and Bioengineering, Keele University School of Medicine, Stoke on Trent ST5 5BG, UK

Juliana Baranova
Department of Biochemistry, Institute of Chemistry, University of São Paulo, Avenida Professor Lineu Prestes 748, Vila Universitária, São Paulo 05508-000, Brazil

Dominik Buchner, Margit Schulze and Edda Tobiasch
Department of Natural Sciences, Bonn-Rhein-Sieg University of Applied Sciences, von-Liebig-Strase 20, 53359 Rheinbach, NRW, Germany

Werner Gotz
Oral Biology Laboratory, Department of Orthodontics, Dental Hospital of the University of Bonn, Welschnonnenstrase 17, 53111 Bonn, NRW, Germany

Raquel Ruiz-Hernandez and Sugoi Retegi-Carrion
Center for Cooperative Research in Biomaterials (CIC biomaGUNE), Basque Research and Technology Alliance (BRTA), 20014 Donostia-San Sebastian, Spain

Unai Mendibil
Center for Cooperative Research in Biomaterials (CIC biomaGUNE), Basque Research and Technology Alliance (BRTA), 20014 Donostia-San Sebastian, Spain
TECNALIA, Basque Research and Technology Alliance (BRTA), 20009 Donostia-San Sebastian, Spain

Nerea Garcia-Urquia and Beatriz Olalde-Graells
TECNALIA, Basque Research and Technology Alliance (BRTA), 20009 Donostia-San Sebastian, Spain

Ander Abarrategi
Center for Cooperative Research in Biomaterials (CIC biomaGUNE), Basque Research and Technology Alliance (BRTA), 20014 Donostia-San Sebastian, Spain
Ikerbasque, Basque Foundation for Science, 48013 Bilbao, Spain

Saltanat Smagul, Yevgeniy Kim, Aiganym Smagulova, Kamila Raziyeva, Ayan Nurkesh and Arman Saparov
Department of Medicine, School of Medicine, Nazarbayev University, Nur-Sultan 010000, Kazakhstan

Justin Hui and Sarah Rajani
Department of Biomedical Engineering, Johns Hopkins University, Baltimore, MD 21218, USA

Shivang Sharma
Department of Chemical & Biomolecular Engineering, Johns Hopkins University, Baltimore, MD 21218, USA

Anirudha Singh
Department of Chemical & Biomolecular Engineering, Johns Hopkins University, Baltimore, MD 21218, USA
Department of Urology, The James Buchanan Brady Urological Institute, The Johns Hopkins School of Medicine, Baltimore, MD 21287, USA

Janine Waletzko and Michael Dau
Department of Oral, Maxillofacial and Plastic Surgery, University Medical Center Rostock, 18057 Rostock, Germany

Anika Seyfarth, Rainer Bader and Anika Jonitz-Heincke
Department of Orthopedics, Biomechanics and Implant Technology Research Laboratory, University Medical Center Rostock, 18057 Rostock, Germany

Armin Springer and Marcus Frank
Medical Biology and Electron Microscopy Center, University Medical Center Rostock, 18057 Rostock, Germany
Department Life, Light & Matter, University of Rostock, 18059 Rostock, Germany

Anna Labedz-Maslowska, Zbigniew Madeja and Ewa Zuba-Surma
Department of Cell Biology, Faculty of Biochemistry, Biophysics and Biotechnology, Jagiellonian University, 30-387 Krakow, Poland

Natalia Bryniarska
Department of Cell Biology, Faculty of Biochemistry, Biophysics and Biotechnology, Jagiellonian University, 30-387 Krakow, Poland
Department of Experimental Neuroendocrinology, Maj Institute of Pharmacology, Polish Academy of Sciences, 31-343 Krakow, Poland

Andrzej Kubiak
Department of Cell Biology, Faculty of Biochemistry, Biophysics and Biotechnology, Jagiellonian University, 30-387 Krakow, Poland
Institute of Nuclear Physics, Polish Academy of Sciences, 31-342 Krakow, Poland

Tomasz Kaczmarzyk
Department of Oral Surgery, Faculty of Medicine, Jagiellonian University Medical College, 31-155 Krakow, Poland

Malgorzata Sekula-Stryjewska
Laboratory of Stem Cell Biotechnology, Malopolska Centre of Biotechnology, Jagiellonian University, 30-387 Krakow, Poland

Sylwia Noga
Department of Cell Biology, Faculty of Biochemistry, Biophysics and Biotechnology, Jagiellonian University, 30-387 Krakow, Poland
Laboratory of Stem Cell Biotechnology, Malopolska Centre of Biotechnology, Jagiellonian University, 30-387 Krakow, Poland

Estelle Loing
IFF-Lucas Meyer Cosmetics, Québec, QC G1V 4M6, Canada

Alexandra Damerau and Timo Gaber
Charité—Universitatsmedizin Berlin, corporate member of Freie Universitat Berlin, Humboldt-Universitat zu Berlin, and Berlin Institute of Health, Department of Rheumatology and Clinical Immunology, 10117 Berlin, Germany
German Rheumatism Research Centre (DRFZ) Berlin, a Leibniz Institute, 10117 Berlin, Germany

Mélissa Simard, Sophie Morin, Geneviève Rioux, Rachelle Séguin and Roxane Pouliot
Centre de Recherche en Organogénèse Expérimentale de l'Université Laval/LOEX, Axe Médecine Régénératrice, Centre de Recherche du CHU de Québec—Université Laval, Québec, QC G1J 1Z4, Canada
Faculté de Pharmacie, Université Laval, Québec, QC G1V 0A6, Canada

Teresa Marques-Almeida and Clarisse Ribeiro
CF-UM-UP, Centro de Física das Universidades do Minho e Porto, Campus de Gualtar, Universidade do Minho, 4710-057 Braga, Portugal
CEB, Centro de Engenharia Biológica, Campus de Gualtar, Universidade do Minho, 4710-057 Braga, Portugal

Miguel Gama
CEB, Centro de Engenharia Biológica, Campus de Gualtar, Universidade do Minho, 4710-057 Braga, Portugal

Vanessa F. Cardoso
CF-UM-UP, Centro de Física das Universidades do Minho e Porto, Campus de Gualtar, Universidade do Minho, 4710-057 Braga, Portugal
CMEMS-UMinho, Campus de Azurém, Universidade do Minho, 4800-058 Guimarães, Portugal

Senentxu Lanceros-Mendez
BCMaterials, Basque Center for Materials, Applications and Nanostructures, UPV/EHU Science Park, 48940 Leioa, Spain
IKERBASQUE, Basque Foundation for Science, 48009 Bilbao, Spain

Rafael Garesse
Departamento de Bioquímica, Facultad de Medicina, Universidad Autónoma de Madrid, 28029 Madrid, Spain
Departamento de Modelos Experimentales de Enfermedades Humanas, Instituto de Investigaciones Biomédicas "Alberto Sols" UAM-CSIC, 28029 Madrid, Spain
Centro de Investigación Biomédica en Red (CIBERER), 28029 Madrid, Spain

Teresa Galera-Monge and Francisco Zurita-Díaz
Departamento de Bioquímica, Facultad de Medicina, Universidad Autónoma de Madrid, 28029 Madrid, Spain
Departamento de Modelos Experimentales de Enfermedades Humanas, Instituto de Investigaciones Biomédicas "Alberto Sols" UAM-CSIC, 28029 Madrid, Spain
Centro de Investigación Biomédica en Red (CIBERER), 28029 Madrid, Spain
Instituto de Investigación Sanitaria Hospital 12 de Octubre (i + 12), 28041 Madrid, Spain

Isaac Canals, Marita Gronning Hansen and Henrik Ahlenius
Department of Clinical Sciences, Neurology, Lund Stem Cell Center, Lund University, 221 00 Lund, Sweden

Laura Rufián-Vázquez and Miguel A. Martin
Centro de Investigación Biomédica en Red (CIBERER), 28029 Madrid, Spain
Instituto de Investigación Sanitaria Hospital 12 de Octubre (i + 12), 28041 Madrid, Spain
Laboratorio de enfermedades mitocondriales y Neurometabólicas, Hospital 12 de Octubre, 28041 Madrid, Spain

Johannes K. Ehinger and Eskil Elmér
Mitochondrial Medicine, Department of Clinical Sciences Lund, Faculty of Medicine, Lund University, BMC A13, 221 84 Lund, Sweden

Dariusz Boruczkowski
Polish Stem Cell Bank, 00-867 Warsaw, Poland

M. Esther Gallardo
Departamento de Bioquímica, Facultad de Medicina, Universidad Autónoma de Madrid, 28029 Madrid, Spain
Departamento de Modelos Experimentales de Enfermedades Humanas, Instituto de Investigaciones Biomédicas "Alberto Sols" UAM-CSIC, 28029 Madrid, Spain
Centro de Investigación Biomédica en Red (CIBERER), 28029 Madrid, Spain
Instituto de Investigación Sanitaria Hospital 12 de Octubre (i + 12), 28041 Madrid, Spain
Grupo de Investigación Traslacional con células iPS. Instituto de Investigación Sanitaria Hospital 12 de Octubre (i + 12), 28041 Madrid, Spain

Index

Printed in the USA
CPSIA information can be obtained
at www.ICGtesting.com
JSHW051405091023
49903JS00006B/280